509-521-6002

com

WEEDS
OF THE SOUTH

Edited by Charles T. Bryson

and Michael S. DeFelice

Weeds

OF THE SOUTH

Photographs by Arlyn W. Evans

Illustrations by Michael S. DeFelice

University of Georgia Press · Athens and London

© 2009 by the University of Georgia Press

Athens, Georgia 30602

www.ugapress.org

Designed by Mindy Basinger Hill

Composition by Melissa Bugbee Buchanan
and Mindy Basinger Hill

Set in 9/13.5 pt Adobe Caslon Pro

Printed and bound by Everbest in China
through Four Colour Print Group

The paper in this book meets the guidelines for
permanence and durability of the Committee on
Production Guidelines for Book Longevity of
the Council on Library Resources.

Printed in China

13   12   11   10   09   P   5   4   3   2   1

Library of Congress Cataloging-in-Publication Data
Weeds of the South / edited by Charles T. Bryson and
Michael S. DeFelice ; photographs by Arlyn W. Evans ;
illustrations by Michael S. DeFelice.
        p.   cm.
Includes bibliographical references and index.
ISBN-13: 978-0-8203-3046-4 (pbk. : alk. paper)
ISBN-10: 0-8203-3046-9 (pbk. : alk. paper)
1. Weeds—Southern States—Identification. I. Bryson,
Charles T. II. DeFelice, Michael S.
SB612.S85W44 2009
632'.50975—dc22   2008055517

British Library Cataloging-in-Publication Data available

THIS BOOK HAS BEEN SPONSORED IN PART BY

Southern Weed Science Society

AMVAC Chemical Corporation

BASF Corporation

Bayer CropScience

Dow AgroSciences

Mississippi Weed Science Society

Monsanto Company

Monsanto's Delta & Pineland Business

Pioneer Hi-Bred

Syngenta Crop Protection, Inc.

Valent U.S.A. Corporation

CONTENTS

ACKNOWLEDGMENTS

This book is the much-requested update to the old *Southern Weed Science Society's Weed Identification Guide* three-ring binder. A project that represents more than 30 years of volunteer effort generates many debts. *Weeds of the South* reflects the efforts of dozens of volunteer weed scientists, botanists, agronomists, horticulturalists, foresters, turf and landscape specialists, and land managers. It is impossible to list them all, but they know who they are and they all have our thanks.

This book would not have been created without the pioneering efforts of the first editor of that three-ring binder edition, C. Dennis Elmore, and Chester McWhorter, the past president of the swss, who originally conceived the project. Even more significant are the photographs taken over a 60-year period and graciously donated to this project by Arlyn W. Evans. His untiring work and love of photographing weeds made this book possible. The leadership and many boards of directors of the Southern Weed Science Society have also provided unwavering support for the project over the years. Their willingness to take a chance and invest in the early versions of the three-ring binder and the interactive software version made the realization of this book possible.

The Weed Identification Committee of the Southern Weed Science Society—which has had many members in the 25-plus years of its existence—provided the guidance and energy to keep this project alive. The editors gratefully acknowledge the following committee members for their contributions:

Shawn Askew   David W. Hall   L. Richard Oliver
Mitchell P. Blair   Patricia D. Haragan   E. J. Retzinger Jr.
John W. Boyd   Robert M. Hayes   Walter A. Skroch
David C. Bridges   Peter W. Jordan   Cade Smith
Charles T. Bryson   J. Wayne Keeling   J. F. Stritzke
John Cardina   Clifford H. Koger Jr.   William K. Vencill
Michael S. DeFelice Victor L. Maddox   Theodore M. Webster
C. Dennis Elmore   J. F. Miller   A. F. Wiese
J. D. Green   Edward C. Murdock   R. D. Williams
                     Don S. Murray

Many contributors wrote weed descriptions for the species covered here. Their names and the weed species they described are listed at the back of the book. Stephen J. Darbyshire, Weed Biologist, Agriculture and Agri-Food Canada, graciously provided weed distribution ranges for Canada. Accurate range maps for Canada and the United States would have been impossible without his assistance. Michael S. DeFelice created the weed distribution maps, plant morphology illustrations, and collar illustrations. We are also indebted to Richard Carter, Professor and Director of Herbarium, Department of Biology, Valdosta State University; Victor L. Maddox, postdoctoral associate, Department of Plant and Soil Sciences, Mississippi State University; Robert F. C. Naczi, Curator, Claude E. Phillips Herbarium, Department of Agricultural and Natural Resources, Delaware State University; and Theodore M. Webster, Research Agronomist, USDA-ARS, Crop Protection and Management Research Unit, Coastal Plain Experiment Station, for their expert reviews of the content. Their efforts made this a much better book.

Many support staff were involved in keeping the project organized and moving forward. We especially want to mention Jennifer Paige Goodlett, Technician-Plants, USDA-ARS, Southern Weed Science Research Unit, for helping to keep all the files and formats organized over the many years and iterations of this project.

We are especially grateful to the University of Georgia Press for agreeing to publish this book. The book could not have been completed without the enthusiasm and hard work of Judy Purdy, Acquisitions Editor; Mindy Basinger Hill, Designer; Jane Kobres, Intellectual Property Manager, Jon Davies, Assistant Managing Editor, and the many others at the press working behind the scenes. We also thank our copy editor, Mindy Conner, for her outstanding and meticulous work on the long and detailed manuscript. Last, but certainly not least, we give special thanks to our wives, Nancy B. Bryson and Karen L. DeFelice, for their patience, love, and support during the long hours we spent on this project.

WEEDS
OF THE SOUTH

I immediately went out into the backyard, picked up

the first weed I met with and set about finding out its name.

GEORGE BENTHAM, 1800–1884

INTRODUCTION

Many people define a weed as a plant growing where it is not wanted. More formally, weeds are plants that alter the structure of natural communities; interfere with the function of ecosystems; or have negative effects on people, agriculture, or other societal interests. Any plant can be a weed in one situation and unobtrusive or even desirable in another. In other words, one person's weed is another person's wildflower. *Weeds of the South* was created to help identify the many weedy and invasive plants that interfere with agriculture, industry, and natural ecosystems in the southern United States.

The geographic range covered by this book extends from Virginia along the Ohio River valley to Missouri, across Oklahoma, and south to Texas and Florida (figure 1). This area encompasses climates ranging from the tropics of south Florida to the subtropical climate of much of the southeastern United States, the continental temperate climate of the Ohio River valley and Appalachian Mountains, and the semiarid southern plains of Oklahoma and Texas. An incredible diversity of plant life—and thus a large number of native and introduced weedy species—grows in this region. Some of the weeds described in this book can be found in many habitats all over the world, although the South is also home to many weed species not found elsewhere in the United States or Canada.

Weeds are usually plants that grow spontaneously and prolifically in habitats that have been modified by human activity. The first weed species in the southern United States are thought to have been introduced through trade among the Native American tribes. Pasture hawthorne (also called wild plum), *Crataegus spathulata* Michx., and maypop passionflower, *Passiflora incarnata* L.,

FIGURE 1.
Geographic area
covered by *Weeds
of the South*

are thought to have been introduced into the Southeast by Native Americans as a source of food. The Europeans who arrived in the late 15th century cleared extensive areas and brought—intentionally or accidentally—many foreign plant species from Europe, Asia, and Latin America that became weeds in these new habitats. These plants represent a significant number of the species covered in this book. Clearing for cultivation and urban development has increased ever since and continues to provide habitats for many introduced and native species to become weedy. The recent appearance of tropical soda apple, *Solanum viarum* Dunal., in southeastern pastures and grazing land is an example.

Weed management requires early and accurate identification to ensure the correct control practices. In most cases, it is necessary to identify weeds early in their seedling stages when they are easier to manage. This book emphasizes photographs of seedlings and young plants to allow the reader to identify young weeds using vegetative plant characteristics, although a flower photograph is also usually included for identification of older plants. Seed characteristics are also useful for identifying mature plants, and this book provides photographs of the seeds of most of the weed species treated. Grasses are particularly difficult to identify at vegetative growth stages. The collar region where the leaf meets the stem (culm) is the most useful area for vegetative grass identification, and we have thus included illustrations of the collar region for the grasses.

The common and scientific names used herein conform to the most recent accepted names specified by the Weed Science Society of America. Additional widely used common names and synonyms of scientific names are included in the individual weed descriptions. Weeds are arranged alphabetically by plant family and within family by scientific name. Each species treatment includes a map showing distribution in the continental United States, Canada, and Alaska. The descriptions include plant growth habit and life cycle, important vegetative and reproductive characteristics, special identifying features, and toxic properties of the plant where applicable. The latter information is meant only to warn the reader of possible medical issues, not to diagnose critical health problems. Anyone who suspects plant poisoning should seek professional medical or veterinary help. The book also includes a chapter that illustrates the common morphological characteristics used to describe and identify plants, a glossary of the botanical terminology used, and keys to the plant families.

The editors have made every effort to ensure the accuracy of this book, but errors are almost inevitable in a project of this size. In addition, weeds are frequently "on the move" invading new habitats and adapting to new environments. We welcome reports of errors and new information.

Parts of a dicot plant stem

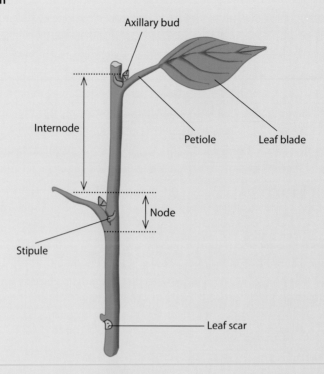

Axillary bud

Internode

Node

Petiole

Leaf blade

Stipule

Leaf scar

Parts of a monocot (grass) collar

The collar region of grasses has the most diagnostic vegetative characteristics.

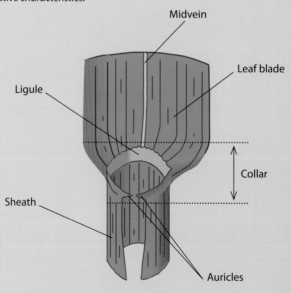

Midvein

Leaf blade

Ligule

Collar

Sheath

Auricles

COMMON TERMINOLOGY

Using the family keys and understanding the species descriptions require a basic familiarity with plant structures and terminology. The most common characteristics and terms used to describe plants are given below. The glossary of botanical terminology at the back of the book includes definitions of terms used in the illustrations.

Structures and growth habits important for identifying weeds and understanding the plant descriptions in this book are illustrated in this section.

Root types

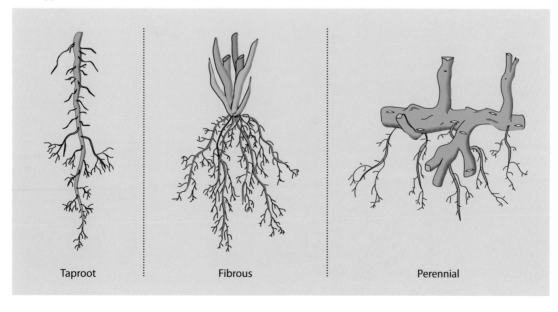

Taproot   Fibrous   Perennial

Perennial stem types

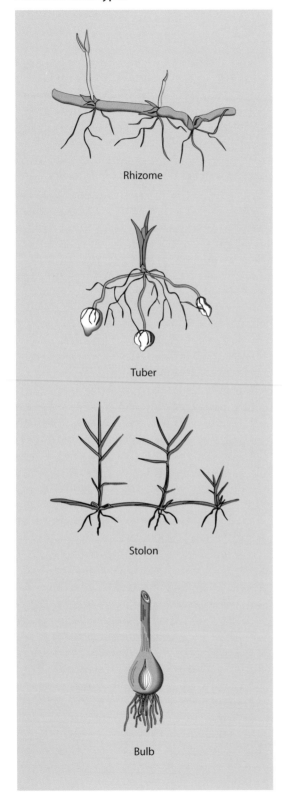

Rhizome

Tuber

Stolon

Bulb

Leaf arrangements

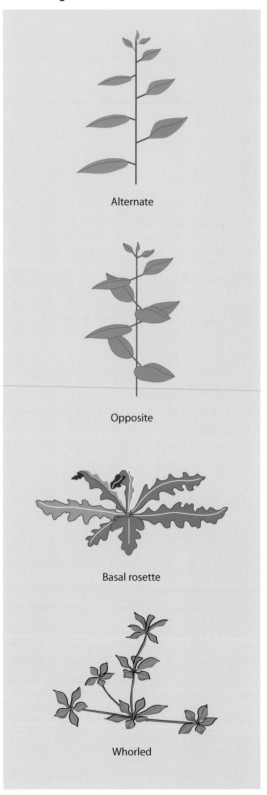

Alternate

Opposite

Basal rosette

Whorled

Leaf shapes

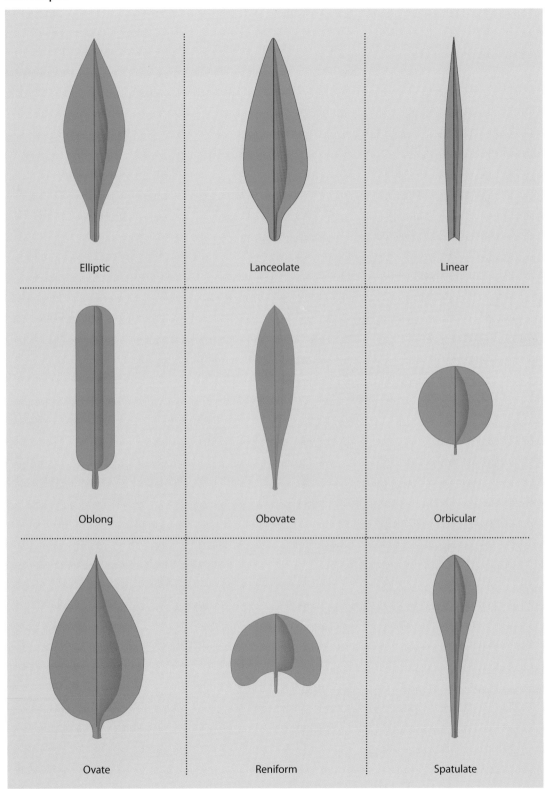

Elliptic

Lanceolate

Linear

Oblong

Obovate

Orbicular

Ovate

Reniform

Spatulate

Leaf margins

Entire   Crenate   Serrate

Palmate   Pinnate

Growth habits

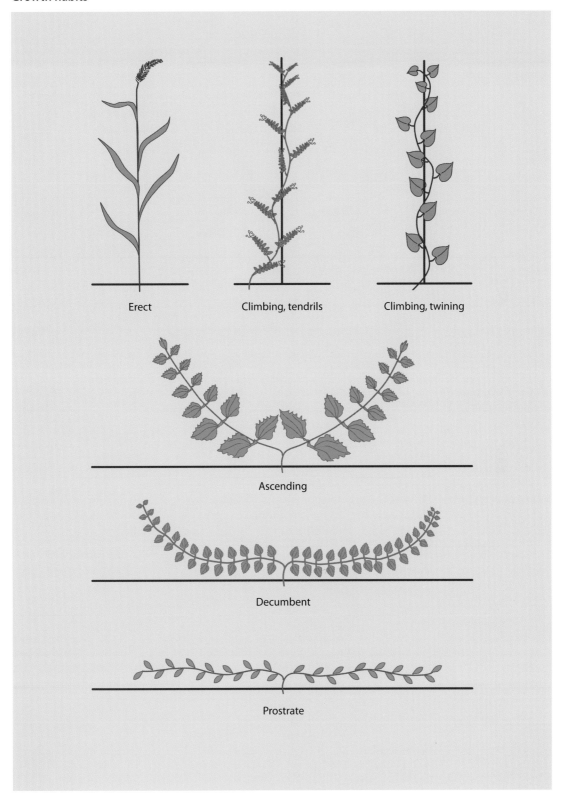

Erect

Climbing, tendrils

Climbing, twining

Ascending

Decumbent

Prostrate

Parts of a dicot flower

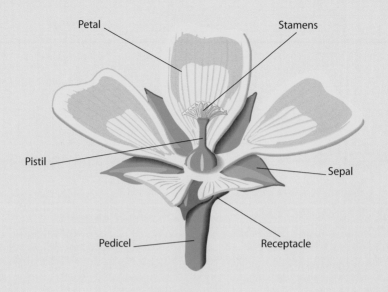

Petal

Stamens

Pistil

Sepal

Pedicel

Receptacle

Parts of a monocot floret

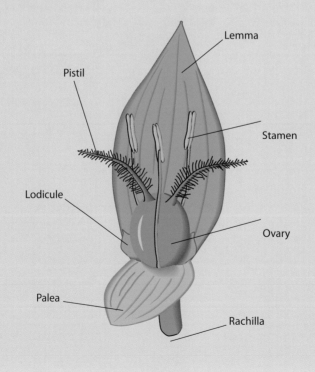

Lemma

Pistil

Stamen

Lodicule

Ovary

Palea

Rachilla

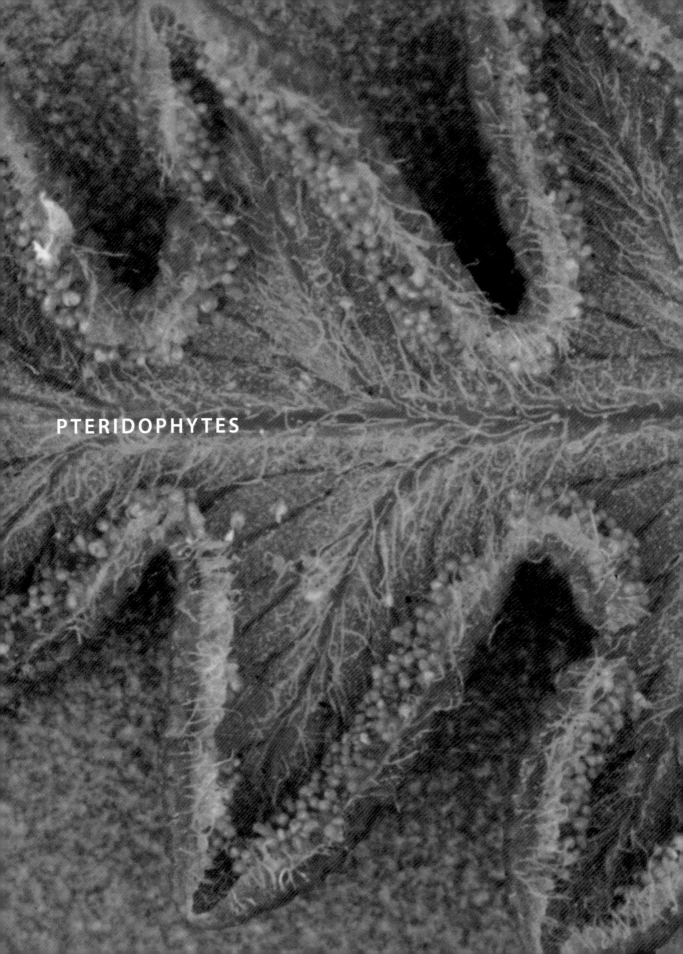

PTERIDOPHYTES

Plant

Field Horsetail

Equisetum arvense L. · Equisetaceae · Horsetail Family

Synonyms

Common horsetail, devil's-guts, horse pipes, horsetail fern, marestail, meadow-pine, pine-grass, foxtail-rush, bottle-brush, snake-grass

Habit, Habitat, and Origin

Erect perennial from rhizomes; plant to 8.0 dm tall; moist woodlands, fields, pastures, meadows, and on banks of streams, rivers, and lakes; native of North America.

Mature Plant Characteristics

ROOTS fibrous, from creeping rhizomes bearing tubers. **STEMS** jointed; 2 forms: sterile stems erect to decumbent, 1.0–8.0 dm tall, branching in whorls, spreading to ascending, central cavity one-fourth stem diameter, vertically ribbed; stomata in 2 broad bands; whorled scalelike sheaths flat or slightly flared, green to blackish; teeth persistent, brown to blackish, deltoid to lanceolate, pointed; fertile stems erect, 1.0–3.2 dm tall, short-lived, unbranched, without chlorophyll, tan to brown, sheaths flaring, pale tan or brownish, teeth dark with narrow, thin margin, lanceolate. **LEAVES** reduced to scales, whorled. **SPORANGIA** conelike structure at tip of stems, 0.5–3.5 cm long, jointed, from peduncle, producing spores. **SPORES** numerous, microscopic.

Special Identifying Features

Perennial from rhizomes; green, sterile, aerial branching stems; pale fertile stem emerging annually, with long peduncle, jointed, spore-bearing cones at tips of stems.

Toxic Properties

Equisetum species contain a neurotoxin leading to muscle weakness, trembling, and collapse but are rarely consumed in quantities sufficient to cause toxicity; they are usually eaten during winter months by cattle, sheep, and horses.

Shoot

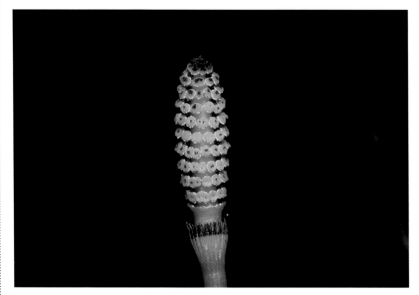
Cone

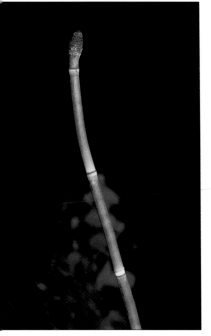
Shoot

Scouringrush

Equisetum hyemale L. · Equisetaceae · Horsetail Family

Synonyms

Common scouringrush, scouringrush horsetail, smooth scouringrush, horsetail; *Equisetum praealtum* Raf.

Habit, Habitat, and Origin

Erect perennial from rhizome with evergreen stems; to 2.0 m tall; open or shaded banks along fields, ditches, rivers, streams, ponds, lakes, roadsides, and railroad beds; native of North America.

Mature Plant Characteristics

ROOTS fibrous from black, round, creeping rhizomes that occasionally bear tubers. STEMS alike, often clustered in dense stands, evergreen, round, rarely branched, erect, 0.2–2.0 m tall, central cavity three-fourths stem diameter, vertically ribbed, stomata in 2 rows, whorled scalelike sheaths flat or flared at top, black or dark brown bands at tip and at or near base, numerous teeth persistent or deciduous, blackish, lanceolate, pointed, with wide hyaline margin. LEAVES reduced to scales. SPORANGIA conelike structure at tip of stem, 0.5–2.4 cm long, jointed, short peduncle to sessile, producing spores. SPORES numerous, microscopic.

Special Identifying Features

Perennial from rhizomes with evergreen aerial stems that rarely branch, jointed spore-bearing cones at tips of fertile stems.

Toxic Properties

See comments under *Equisetum arvense.*

Frond

Brackenfern

Pteridium aquilinum (L.) Kuhn · Polypodiaceae · Fern Family

Synonyms

Bracken, brake, eagle fern, western bracken; *Pteris latiuscula* Desv.

Habit, Habitat, and Origin

Erect terrestrial fern from scaly rhizomes; to 15.0 dm tall; moderately to strongly acid soils in open woods, pastures, abandoned fields, and cutover areas; native of North America.

Mature Plant Characteristics

Young plants rarely observed; small and fernlike, from spores with characteristic fiddleneck; leaves arise annually from rhizome. ROOTS long-creeping subterranean rhizome, with septate trichomes, 5.0 mm diameter. STEMS none. LEAVES fronds, 3.0–15.0 dm long, arising from subterranean rhizome, widely spaced, erect or slightly curved, stiff, blades 2.0–8.0 dm long, widely triangular to oval, tripinnatifid or tripinnate, petioles smooth, reddish toward base and yellowish toward top, spore-producing sori continuous along frond margin. SPORANGIA spore-producing sori along margins of fronds. SPORES numerous per sori, microscopic.

Special Identifying Features

Terrestrial; fronds arising from creeping subterranean rhizomes with scales; sori along frond margins.

Toxic Properties

Plants produce neurotoxins causing neoplasia, bone marrow depression, and retinal degeneration in livestock.

TOP Old frond
BOTTOM LEFT Leaf with sori
BOTTOM RIGHT Emerging frond

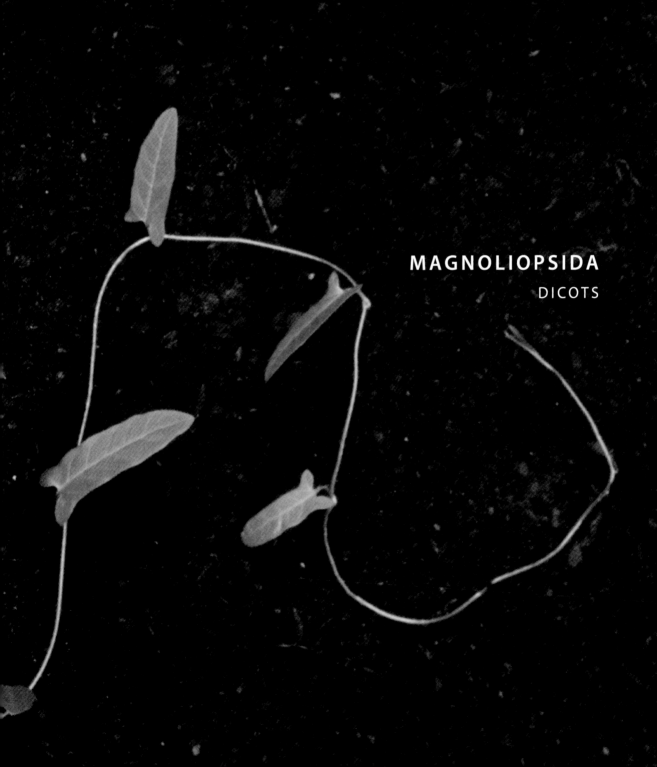

MAGNOLIOPSIDA

DICOTS

Flower

Horse Purslane

Trianthema portulacastrum L. · Aizoaceae · Fig-marigold Family

Synonyms

Sea purslane

Habit, Habitat, and Origin

Fleshy, succulent, prostrate annual herb; to 2.0 m wide; open pastures, fields, and roadsides; native of the tropics.

Seedling Characteristics

Stem pubescent; first leaves glabrous, opposite, unequal.

Mature Plant Characteristics

ROOTS fibrous from taproot. STEMS erect early and prostrate later, to 2.0 m long, weakly and diffusely branched. LEAVES subopposite to opposite, 1.5 cm long, fleshy, bright green, widely elliptic to nearly orbicular, stipules free at apex, in pairs, unequal in size, petiole 1.0–2.0 cm long. INFLORESCENCES solitary, axillary, sessile; sepals pink to rose purple within and green externally, margins hyaline, hoodlike, and bearing a short horn near apex; staminodes absent; stamens 5–10; stigmas 1–2. FRUITS capsule, 2-crested, opening along a circular equatorial line, falling away as a lid when 1–5 seeds are mature. SEEDS squarish, 1.5 mm long, black.

Special Identifying Features

Fleshy, succulent, prostrate annual; leaves opposite and in pairs of unequal size; flowers pink to rose purple.

Toxic Properties

Plants contain oxalic acid, which causes renal failure when consumed in large quantities.

TOP Seeds
MIDDLE Two-leaf seedling
BOTTOM Young seedling

Mature plant

Alligatorweed

Alternanthera philoxeroides (Mart.) Griseb. · Amaranthaceae · Pigweed Family

Synonyms

Alligator grass

Habit, Habitat, and Origin

Decumbent, aquatic or terrestrial perennial forming dense, tangled, floating mats; stems to 1.0 m long; native of South America.

Seedling Characteristics

Reproduction by vegetative means only.

Mature Plant Characteristics

ROOTS fibrous from taproot. **STEMS** decumbent, 0.2–1.0 m long, hollow and slightly flattened at maturity, simple or branched, glabrous except for a few trichomes at leaf base, rooting at nodes. **LEAVES** opposite, 5.0–13.0 cm long, 5.0–20.0 mm wide, entire, linear-elliptic, tips acute, midrib distinctive, glabrous. **INFLORESCENCES** solitary head, axillary or terminal, 13.0 mm diameter; on long peduncle, florets 6–20 per head, white, at least 5.5 cm long and 1.5 mm wide, on stalks, each floret with 5 stamens. **FRUITS** none produced. **SEEDS** none formed.

Special Identifying Features

Aquatic or terrestrial plant with hollow stems; often completely covering waterways or shorelines; leaves opposite; stems rooting from nodes; flowers solitary, white; vegetative reproduction only.

Toxic Properties

None reported.

Young seedling

1mm

TOP Seeds
BOTTOM Two-leaf seedling

Tumble Pigweed

Amaranthus albus L. · Amaranthaceae · Pigweed Family

Synonyms

White pigweed, tumbleweed

Habit, Habitat, and Origin

Erect or decumbent, bushy-branched, summer annual herb; to 1.0 m tall and wide; dry open cultivated areas, fields, pastures, roadsides, and waste sites; native of the Great Plains and prairies, naturalized throughout North America.

Seedling Characteristics

Hypocotyl red to green, glabrous; cotyledons green, glabrous.

Mature Plant Characteristics

ROOTS fibrous from well-developed taproot system; usually not red. STEMS erect or decumbent, to 1.0 m tall, highly branched, whitish to pale green, usually glabrous. LEAVES alternate, 5.0–30.0 mm long, 0.5–1.5 cm wide, simple, pale green, glabrous, petioled. INFLORESCENCES short-dense axillary clusters, not terminal; bracts rigid, subulate, about twice as long as flowers; sepals uneven, commonly 3. FRUITS utricle, 1.3–1.7 mm long, dehiscent at circumference along a seam, finely wrinkled when dry. SEEDS lenticular, 0.7–1.0 mm wide, dark brown, shiny.

Special Identifying Features

Erect or decumbent, bushy-branched summer annual; stems whitish to pale green; inflorescences in axillary clusters.

Toxic Properties

Amaranthus species contain toxins causing myocardial degeneration, renal disease, and nitrate intoxication.

Axillary flowers

TOP Seed and utricle
MIDDLE Flower
BOTTOM Four-leaf seedling

Young plant

Prostrate Pigweed

Amaranthus blitoides S. Wats. · Amaranthaceae · Pigweed Family

Synonyms

Prostrate tumbleweed

Habit, Habitat, and Origin

Prostrate summer annual herb; stems to 0.9 m long; dry open fields, pastures, and roadsides; native of North America.

Seedling Characteristics

Cotyledons long, narrow; lower surface and stem deep red, upper surface shiny, glabrous; first leaf indented at tip.

Mature Plant Characteristics

ROOTS fibrous from taproot. STEMS prostrate with erect tips, 0.3–0.9 m long, fleshy, nearly smooth, reddish, branched at base and less frequently apically. LEAVES alternate, 1.0–4.0 cm long, 0.5–1.5 cm wide, simple, ovate, shiny, broadest at tips. INFLORESCENCES dense axillary clusters, not terminal; bracts oblong to lanceolate, erect, attenuate at apex to short spinose tip, 2.5–3.0 mm long; sepals oblong, acuminate, 1-nerved, green with white margins. FRUITS pyxis with circumscissile utricle. SEEDS nearly circular, 1.5 mm diameter, flattened, shiny, black, notched at narrow end.

Special Identifying Features

Prostrate summer annual; flowers axillary; leaves alternate; utricle circumscissile.

Toxic Properties

See comments under *Amaranthus albus*.

Smooth Pigweed

Amaranthus hybridus L. · Amaranthaceae · Pigweed Family

Synonyms

Carelessweed

Habit, Habitat, and Origin

Erect summer annual herb; to 2.5 m tall; river valleys, cultivated areas, fields, pastures, roadsides, and waste sites; native of eastern North America, Central America, and northernmost South America.

1mm

Seedling Characteristics

Hypocotyls red to green, usually glabrous; cotyledons green, glabrous.

Mature Plant Characteristics

ROOTS fibrous from well-developed taproot; may or may not be red. STEMS erect, 0.5–2.5 m tall, usually branched, stout; glabrous or sparsely pubescent apically. LEAVES alternate, 3.0–15.0 cm long, 1.0–6.0 cm wide, simple, ovate, glabrous, long-petioled. INFLORESCENCES terminal panicles of numerous slender, cylindrical spikes about 1.0 cm thick and up to 15.0 cm long; bracts 2.0–4.0 mm long, equal to or longer than sepals, not stiff; female sepals 5, 1.0–2.5 mm long, straight, with acute tips. FRUITS utricle, wrinkled and rough when dry, 1.5–2.0 mm long, coat dehiscent along circumference. SEEDS lenticular, 1.0 mm diameter, dark brown, numerous per spike.

Special Identifying Features

Erect summer annual; inflorescences in spikes, 1.0 cm or less thick; bracts short; flowering in fall; plants monoecious.

Toxic Properties

See comments under *Amaranthus albus*.

TOP Utricle
BOTTOM Two-leaf seedling

TOP Young seedling
BOTTOM Inflorescence

Palmer Amaranth

Amaranthus palmeri S. Wats. · Amaranthaceae · Pigweed Family

Synonyms

Carelessweed, palmer pigweed

Habit, Habitat, and Origin

Erect, branched summer annual herb; to 2.0 m tall; open river valleys, cultivated areas, fields, pastures, roadsides, and waste sites; native of southern Great Plains of North America.

Seedling Characteristics

Hypocotyl red to green, glabrous to softly pubescent; cotyledons green, glabrous.

Mature Plant Characteristics

ROOTS fibrous from well-developed taproot; may or may not be red. STEMS erect, branched, 0.5–2.0 m tall, may be red, glabrous or pubescent. LEAVES alternate, 3.0–10.0 cm long, egg-shaped, veins prominent beneath, long-petioled. INFLORESCENCES terminal spike to 0.5 m long, lateral spikes few to none, much shorter if present; male and female flowers on separate plants; female bracts subulate, 3.0–6.0 mm long, thick, stiff; male bracts thin, 3.0–4.0 mm long, triangular; female sepals 5, 3.0–4.0 mm long, minutely pointed on tip, obovate, spatulate; male sepals 5, 2.0–3.0 mm long, ovate-lanceolate. FRUITS subglobose, 1.5 mm long, splitting along a seam. SEEDS dark reddish brown, lenticular, 1.0–1.3 mm diameter, numerous per spike.

Special Identifying Features

Erect, branched summer annual; male and female flowers on different plants; unbranched terminal inflorescence to 0.5 m long; plants monoecious.

Toxic Properties

See comments under *Amaranthus albus*.

Four-leaf seedling

Redroot Pigweed

Amaranthus retroflexus L. · Amaranthaceae · Pigweed Family

Synonyms

Carelessweed, common amaranth

Habit, Habitat, and Origin

Erect, stout summer annual herb; to 2.0 m tall; open river valleys, cultivated areas, fields, pastures, roadsides, and waste sites; native of central and eastern North America.

Seedling Characteristics

Hypocotyl red to green, glabrous to softly pubescent; cotyledons green, pubescent.

Mature Plant Characteristics

ROOTS fibrous from well-developed taproot, may or may not be red. STEMS erect, to 2.0 m tall, stout, usually branched, with fine short trichomes. LEAVES alternate, 2.0–15.0 cm long, 1.0–7.0 cm wide, simple, veins with trichomes on underside, long-petioled.

INFLORESCENCES terminal panicle of several densely crowded, stout spikes to 20.0 cm long and 1.0–5.0 cm wide; bracts 4.0–8.0 mm long, two to three times as long as sepals, very stiff; female sepals 5, 3.0–4.0 mm long recurved, obtuse, rounded, truncate or slightly notched at apex. FRUITS utricle, 1.5–2.0 mm long, wrinkled and rough when dry, coat dehiscent at circumference. SEEDS lenticular, 1.0–1.2 mm diameter, dark red-brown, numerous per spike.

Special Identifying Features

Erect, stout summer annual; spikes in inflorescence thick, more than 1.0 cm wide; stiff bracts, two to three times as long as sepals; leaves, young stems pubescent; commonly flowers in summer.

Toxic Properties

See comments under *Amaranthus albus*.

TOP Seeds with utricle
MIDDLE One-leaf seedling
BOTTOM Inflorescence

Four-leaf seedling

TOP Seeds with utricle
BOTTOM Spines and flowers

Spiny Amaranth

Amaranthus spinosus L. · Amaranthaceae · Pigweed Family

Synonyms

Hogweed, spiny pigweed, stickerweed

Habit, Habitat, and Origin

Erect, branched summer annual herb; to 1.5 m tall; open pastures, cultivated areas, fields, gardens, stockyards, roadsides, and waste sites; probably native of tropical America.

Seedling Characteristics

Hypocotyl red to green, glabrous; cotyledons green, glabrous, spoon- or dish-shaped.

Mature Plant Characteristics

ROOTS fibrous from well-developed tap-root, may or may not be red. STEMS erect, 0.4–1.5 m tall, branched, glabrous, pair of sharp spines at nodes 5.0–10.0 mm long. LEAVES alternate, 3.0–6.0 (to 15.0) cm long, 1.5–6.0 cm wide, simple, petiole 3.0–6.0 cm long, glabrous. INFLORES-CENCES terminal with numerous axillary clusters, flowers in terminal spike chiefly male, basal flowers chiefly female; bracts lanceolate or subulate, shorter than sepals; sepals of female flowers 5, oblong, 1.0–1.5 mm long. FRUITS utricle, 1.5–2.0 mm long, coat indehiscent or bursting

irregularly. SEEDS nearly circular, 0.7–1.0 mm diameter, dark brown, numerous per spike.

Special Identifying Features

Erect, branched summer annual; stems with 2 very sharp, stiff spines 5.0–10.0 mm long at most nodes.

Toxic Properties

See comments under *Amaranthus albus*.

Inflorescence

Tall Waterhemp

Amaranthus tuberculatus (Moq.) Sauer · Amaranthaceae · Pigweed Family

Synonyms

Common waterhemp; *Amaranthus rudis* Sauer, *Acnida altissima* (Riddell) Moq. ex Standl.

Habit, Habitat, and Origin

Erect summer annual herb; to 3.0 m tall; cultivated areas, stream banks, lakeshores, floodplains, fields, and waste sites; native of North America.

Seedling Characteristics

Hypocotyl green to red, glabrous to lightly pubescent; cotyledons green to reddish, glabrous.

Mature Plant Characteristics

ROOTS fibrous from well-developed taproot. STEMS erect, occasionally ascending; 0.7–3.0 m tall, slender, green or red (females usually red), glabrous or sparsely pubescent. LEAVES alternate, 1.0–15.0 cm long, 0.5–3.0 cm wide, simple, narrowly ovate to lanceolate, glabrous, long-petioled. INFLORESCENCES staminate and carpellate flowers on separate plants; terminal spike, simple to highly branched; bracts 1.0–1.5 mm long, slender midrib extending beyond lamina; carpellate sepals completely lacking or rudimentary; staminate sepals 5, nearly equal, 2.5–3.0 mm long, greenish. FRUITS utricle, 1.5–2.0 mm long, dehiscing irregularly, smooth or tuberculate. SEEDS lenticular, 0.8–1.0 mm diameter, lustrous, dark reddish to black.

Special Identifying Features

Erect summer annual; dioecious; utricles dehiscing irregularly.

Toxic Properties

See comments under *Amaranthus albus*.

TOP **Seed with utricle**
BOTTOM **Seedling**

Male inflorescence

Female inflorescence

1mm

TOP Utricles
MIDDLE Two-leaf seedling
BOTTOM Four-leaf seedling

Inflorescence

Slender Amaranth

Amaranthus viridis L. · Amaranthaceae · Pigweed Family

Synonyms

Green amaranth; *Amaranthus gracilis* Desf.

Habit, Habitat, and Origin

Erect to spreading, many-branched, summer annual herb; to 1.0 m tall; cultivated areas, open pastures, fields, roadsides, and waste sites; native of tropical America, adventive in southeastern United States.

Seedling Characteristics

Hypocotyl red to green, glabrous; cotyledons green, glabrous.

Mature Plant Characteristics

ROOTS fibrous from well-developed taproot, may or may not be red. STEMS erect, 0.3–1.0 m tall; lower branches spreading to 0.5 m across, glabrous, prostrate; upper branches ascending.

LEAVES alternate, simple, 3.0–7.0 cm long, 2.5–6.0 cm wide, ovate to oblong-ovate or elliptic-ovate, margins entire or slightly undulate, petiole 3.0–6.0 cm long. INFLORESCENCES terminal and axillary spikes few or several, forming panicle 1.0–2.0 dm long; bracts much shorter than flowers; female sepals 3, oblanceolate, shorter than fruit, acute. FRUITS utricle, 1.5–2.0 mm long, rough and wrinkled when dry, indehiscent, splits unevenly on seam. SEEDS circular, 0.8–1.0 mm diameter, dark brown to almost black, numerous per spike.

Special Identifying Features

Erect to spreading, numerous-branched summer annual; bracts not noticeable; utricle indehiscent.

Toxic Properties

See comments under *Amaranthus albus*.

Flowers

Eastern Poison-ivy

Toxicodendron radicans (L.) Kuntze · Anacardiaceae · Cashew Family

Synonyms

Rhus radicans L.

Habit, Habitat, and Origin

Woody perennial; low shrub or trailing or climbing vine; to several meters long; often high into trees; reproducing by seed and root stocks; native of North America.

Seedling Characteristics

Cotyledon and first leaves smooth, difficult to distinguish from seedlings of Virginia-creeper.

Mature Plant Characteristics

ROOTS fibrous from taproot and long subterranean rhizomes; aerial roots on vines. **STEMS** woody, to several meters long, smooth, light brown or grayish. **LEAVES** alternate, compound, 3 large shiny leaflets; leaflets 5.0–10.0 cm long, pointed at tip, leaflet edges either smooth or irregularly toothed, ovate to subrotund, varying to rhombic or el-

liptic; smooth above, pubescent on veins beneath. **INFLORESCENCES** panicles in leaf axis, 2.0–6.0, 2.5–7.5 cm long, individual florets small; petals 5, yellowish green; stamens 5, yellow to yellow-orange; sepals 5, ovary stout. **FRUITS** drupe, 5.0–6.0 mm diameter, glabrous or short trichomes, green turning grayish white or tan at maturity. **SEEDS** irregularly shaped, 1.0–1.2 mm wide, gray to grayish brown, 1 per drupe.

Special Identifying Features

Woody, trailing or climbing perennial vine; variable growth habit; trifoliate compound leaf, pubescent on lower veins; vines climbing by shaggy aerial roots; foliage turns bright red or reddish yellow in fall; fruit a drupe.

Toxic Properties

All parts of live and dead plants contain urushiol oil, which may cause acute contact dermatitis in primates and hamsters.

TOP **Seeds**
MIDDLE **Seedling**
BOTTOM **Green fruit**

Hoary Bowlesia

Bowlesia incana Ruiz & Pavón · Apiaceae · Parsley Family

Synonyms

American bowlesia; *Bowlesia septentri-onalis* J. M. Coult. & Rose

Habit, Habitat, and Origin

Prostrate to suberect, branching or trailing annual herb; to 6.0 dm tall; shaded areas; native of nontropical regions of North and South America.

Seedling Characteristics

Cotyledons ovate to oblong; first true leaves opposite, palmately lobed, covered with stellate trichomes.

Mature Plant Characteristics

ROOTS fibrous from thickened taproot. STEMS weak, trailing, dichotomously branched; glabrous or with slightly stellate pubescence. LEAVES opposite, 0.2–5.0 cm long, 0.5–4.5 cm wide, thin, kidney- to heart-shaped in outline, broader than long; palmately 5–9-lobed, lobes halfway to base; widely lanceolate to round, entire or toothed, pubescent; trichomes stellate, more numerous on ventral surface; petiole slender, 2.0–12.0 cm long. INFLORESCENCES simple umbels on axillary pedicels or sessile; 2–5-flowered, white, lanceolate bracts; calyx lobes minute, less than 0.16 cm diameter, prominently toothed; 5 petals oblong to ovate, pale yellowish green; self-compatible pollination. FRUITS ovate to round, 2.0 mm long, 1.5 mm wide, inflated, flattened, slightly heart-shaped at base, with inconspicuous ridges, cov-ered by dense stellate pubescence. SEEDS broadly oblong, ovate to obovate, 1.0–2.0 mm long, quadrangular, seed face flattened to convex, tan to brown.

Special Identifying Features

Prostrate to suberect, branching or trailing annual; leaves atypical of Apiaceae, opposite, lobed, cordate; flowers inconspicuous.

Toxic Properties

None reported.

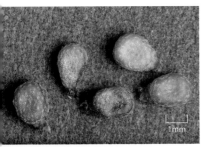

TOP Seeds
BOTTOM Three-leaf seedling

TOP Flowers
BOTTOM Flowering plant

Flowering plant

Poison-hemlock

Conium maculatum L. · Apiaceae · Parsley Family

Synonyms

California fern, Nebraska fern, deadly hemlock, poison parsley, poison stink-weed, snake-weed, spotted hemlock, wode thistle

Habit, Habitat, and Origin

Erect, stoutly branching biennial herb; to 3.0 m tall; moist soils in fields, roadsides, stream banks, and waste sites; native of Eurasia.

Seedling Characteristics

Cotyledons oblong-lanceolate; first leaf pinnately compound, reddish, glabrous.

Mature Plant Characteristics

ROOTS fibrous from white, turniplike taproot. **STEMS** erect, 1.0–3.0 m tall, branching, stout, glabrous, purple-spotted, ridged, hollow except at nodes. **LEAVES** basal rosette, alternate upward, 2.0–4.0 dm long, petioled, broadly triangular-ovate in outline, three to four times pinnately compound, leaflets lanceolate to ovate-oblong, dentate or finely cut, 4.0–10.0 mm long. **INFLORES-CENCES** small and white in large, open compound umbels 4.0–6.0 cm wide, terminal inflorescence blooming first but soon overtopped by others. **FRUITS** schizocarp with 2 mericarps, grayish brown, conspicuous, wavy, somewhat knotted ridges, lacking oil tubes. **SEEDS** encapsulated in fruit.

Special Identifying Features

Erect, stoutly branching biennial; plant resembles wild carrot, parsnip, or hog-weed; purple spots on stems; higly poisonous; herbage has a mouselike odor.

Toxic Properties

Plants produce neurotoxins and are teratogenic; all parts are dangerously toxic to livestock and humans.

TOP Seeds
BOTTOM LEFT Three-leaf seedling
BOTTOM RIGHT Spotted stem

Wild Carrot

Daucus carota L. · Apiaceae · Parsley Family

Synonyms

Bird's nest, Queen Anne's Lace, devil's-plague

Habit, Habitat, and Origin

Erect biennial herb; to 15.0 dm tall; fields, pastures, meadows, and roadsides; native of Asia and the Mediterranean region.

Seedling Characteristics

Cotyledons linear, smooth; first leaf pinnately decompound.

Mature Plant Characteristics

ROOTS fibrous from slightly thickened taproot. STEMS erect, to 15.0 dm tall, freely branching, glabrous or scabrous. LEAVES basal rosette and alternate upward, to 15.0 cm long and 7.0 cm wide, pinnately decompound; oblong in general outline with ultimate divisions linear, lanceolate, or oblong; glabrous or hispid on midribs. INFLORESCENCES in erect terminal umbel, 7.0–15.0 cm wide, on long peduncle; small, white except for a central purple flower. FRUITS umbel rays curve inward after anthesis, producing a congested "birdnest" cluster of fruits. SEEDS ovoid, 3.0–4.0 mm long, about 2.0 mm wide, broadest at the middle, with rows of barblike trichomes.

Special Identifying Features

Erect biennial herb; umbel terminal large; flowers white except central purple flower.

Toxic Properties

Plants produce neurotoxins.

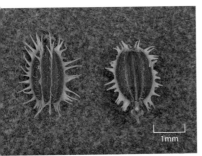

TOP Seeds
MIDDLE Seedling
BOTTOM Inflorescence

Flowers

Flowers

Hemp Dogbane

Apocynum cannabinum L. · Apocynaceae · Dogbane Family

Synonyms

Indian hemp

Habit, Habitat, and Origin

Erect perennial herb, woody at base; to 1.5 m tall; forming colonies in fields, cultivated areas, roadsides, and waste sites; native of North America.

Seedling Characteristics

Opposite, erect, entire leaves; rarely seen.

Mature Plant Characteristics

ROOTS initially fibrous from tap-root; colonizes from long horizontal rootstock. **STEMS** erect, 0.8–1.5 m tall, glabrous, milky sap, much-branched but with well-developed main axis. **LEAVES** opposite, 5.0–12.0 cm long, 1.0–4.0 cm wide, simple, margins entire, ovate, glabrous or sparingly pubescent beneath, milky sap, petiole short. **INFLORESCENCES** in dense terminal cymes, white, perfect, regular, small. **FRUITS** paired follicles, slender, 12.0–20.0 cm long. **SEEDS** thin, flat, 4.0–6.0 mm long, brown to tannish, tuft of soft silky trichomes at one end.

Special Identifying Features

Erect woody perennial; shoots in colonies; leaves opposite, margins entire; leaves and stems with milky sap; frequently branching, unlike most milkweed species.

Toxic Properties

Plants produce cardiotoxins that cause digestive disturbances, diarrhea, impaction, and weakness.

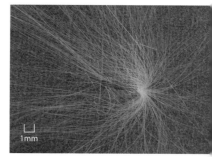

TOP Seed
MIDDLE Sprout
BOTTOM Fruit

Flowers

Common Milkweed

Asclepias syriaca L. · Asclepiadaceae · Milkweed Family

Synonyms

Common silkweed, showy milkweed, silk weed, swallow-wort

Habit, Habitat, and Origin

Erect, branched perennial herb growing from a deep rhizome; 6.0–20.0 dm tall; banks of waterways, floodplains, forest edges, roadsides, and waste sites; native of North America.

Seedling Characteristics

Cotyledons flat; sprout emerging with opposite leaves and prominent midveins.

Mature Plant Characteristics

ROOTS fibrous from taproot with deep rhizome. **STEMS** erect, 5.0–20.0 dm tall, stout, usually single or sparingly branched, covered with fine trichomes, containing milky sap. **LEAVES** opposite, 10.0–30.0 cm long, 5.0–11.0 cm wide, simple, margins entire; oblong, oval, or ovate; finely pubescent beneath, glabrous above, milky sap, petioled. **INFLORESCENCES** in regular head, 20–130 per head, perfect, pedicel 1.5–4.5 cm long; corolla 5-lobed, 11.0–17.0 mm tall, greenish purple to greenish white. **FRUITS** follicles, 2 to several, many-seeded. **SEEDS** oblong, 6.0–8.0 mm long, thin, flattened, brown, with tuft of silky trichomes, 3.0–4.0 cm long.

Special Identifying Features

Erect perennial herb; stems stout; rhizomes deep in soil; milky sap throughout; tufted seeds, wind-dispersed.

Toxic Properties

Asclepias species produce cardiotoxins.

Eastern Whorled Milkweed

Asclepias verticillata L. · Asclepiadaceae · Milkweed Family

Synonyms

Horsetail milkweed, whorled milkweed

Habit, Habitat, and Origin

Erect, slender perennial herb; 2.0–9.0 dm tall; dry or rocky soil in woodlands and roadsides; native of North America.

Seedling Characteristics

Cotyledons flat, glabrous; leaves opposite with prominent midribs.

Mature Plant Characteristics

ROOTS fibrous from shallow taproot with a deep rhizome. **STEMS** erect, 2.0–9.0 dm tall, slender, sparingly branched, containing milky sap, lines of minute trichomes extending down from each leaf node. **LEAVES** whorled, 2.0–6.0 cm long, 1.0–2.0 cm wide, 3–6 per node, simple, blades narrowly linear, apex pointed, rolled downward, margins entire, sessile, milky sap. **INFLORESCENCES** in umbels, borne in clusters at top nodes of stem or in axils of upper leaves; petals 5, greenish white to purple-tinged, elliptic, 4.0–5.0 mm long, hoods roundish to oval. **FRUITS** slender follicle, 7.0–10.0 cm long, 5.0–8.0 mm wide; pedicel erect. **SEEDS** elliptic, 5.0 mm long, with tuft of white trichomes, 2.5 cm long.

Special Identifying Features

Perennial; narrow linear leaves, rolled downward, whorled on slender stem containing milky juice; fruit erect, spindle-shaped; seeds tufted and wind-dispersed.

Toxic Properties

See comments under *Asclepias syriaca*.

TOP Flowers
BOTTOM Flowering plant

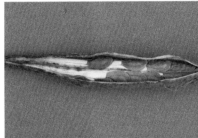

TOP Seeds
BOTTOM Four-leaf seedling

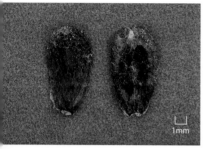

Honeyvine Swallowwort

Cynanchum laeve (Michx.) Pers. · Asclepiadaceae · Milkweed Family

Synonyms

Bluevine, honeyvine milkweed, sandvine; *Ampelamus albidus* (Nutt.) Britt.

Habit, Habitat, and Origin

Deciduous, twining or trailing perennial vine; to 3.0 m long; sandy, clay, rocky, or calcareous soils in cultivated areas, forest edges, thickets, floodplains, and disturbed areas; native of North America.

Seedling Characteristics

Emerging sprout with opposite, entire leaves; glabrous.

Mature Plant Characteristics

ROOTS clustered, fibrous from taproot and rhizomes. STEMS slender, glabrous, twining or trailing vine, to 3.0 m long. LEAVES opposite, simple, ovate, cordate, 5.0–15.0 cm long, 3.0–7.0 cm wide, acuminate tip, petiolate, pedicels tomentose. INFLORESCENCES corymb, 5–40-flowered, pedunculate, axillary; corolla white to light cream, 2.0 mm long; lobes 4.0–7.0 mm long, narrowly oblong to lanceolate, glabrous. FRUITS follicle, 10.0–15.0 cm long, 2.0–3.0 cm wide, smooth, angled, teardrop-shaped. SEEDS ovate, 7.0–9.0 mm long, with tuft of soft trichomes, compressed in follicle.

Special Identifying Features

Perennial, deciduous, twining or trailing vine; leaves opposite, simple, margins entire, no tendrils; follicle smooth, somewhat angled, teardrop-shaped; seeds with tufted trichomes, wind-dispersed.

Toxic Properties

Reported to contain a possible neuro-toxin.

TOP Seeds
BOTTOM Young plant

TOP Fruit
BOTTOM Flower

Flowering plant

Bristly Starbur

Acanthospermum hispidum DC. · Asteraceae · Sunflower Family

Synonyms

Goat-head, goat-spur, Texas sandspur

Habit, Habitat, and Origin

Erect, dichotomously branched summer annual herb; to 8.0 dm tall; sandy disturbed soils, roadsides, and alluvial flats; native of North America.

Seedling Characteristics

Cotyledons oval, pale green.

Mature Plant Characteristics

ROOTS fibrous from slender taproot. **STEMS** erect, 1.0–8.0 dm tall, diffusely branched, pubescent. **LEAVES** opposite, 2.0–12.5 cm long, 1.0–7.0 cm wide, elliptic to oval, margins shallowly toothed, light green, sessile or nearly so, reduced in size near inflorescence, pubescent, both surfaces pilose. **INFLORESCENCES** in heads 4.0–5.0 mm wide, ray florets 5–10, fertile, pale; disk florets 5–15, sterile, yellow; involucral bract elliptic, 4.0–5.0 mm long, inner bracts longer than outer bracts. **FRUITS** burs, 4.0–7.0 mm long, 2-horned, spiny, borne in star-shaped clusters.

Special Identifying Features

Erect, dichotomously branched summer annual; fruit a 2-horned, spiny bur, in star-shaped clusters; achene enclosed in bur.

Toxic Properties

None reported.

TOP **Fruits**
MIDDLE **Four-leaf seedling**
BOTTOM **Flower**

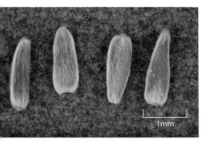

Flowering plant

Common Yarrow

Achillea millefolium L. · Asteraceae · Sunflower Family

Synonyms

Milfoil, western yarrow, yarrow

Habit, Habitat, and Origin

Erect, aromatic, sparsely branched perennial herb; to 0.7 m tall; fields, grasslands, open woods, mildly disturbed areas; native of Europe.

Seedling Characteristics

Leaves villous to woolly villous, pinnately dissected, forming a rosette.

Mature Plant Characteristics

ROOTS fibrous from taproot; weakly spreading rhizomes. STEMS erect, to 0.7 m tall, arising singly or in clusters from rhizomes, sparsely branched. LEAVES alternate, 3.0–15.0 cm long, 0.5–2.5 cm wide, cauline; villous to woolly villous, pinnately dissected, lower leaves with petiole. INFLORESCENCES numerous compact terminal corymbs per plant, rays of flowers pistillate, white to pink, 1.0–3.0 mm long; disk florets 2.0–3.0 mm long, perfect, white or rarely pink. FRUITS achenes compressed parallel to bracts, callous-margined, glabrous, no pappus.

Special Identifying Features

Erect, aromatic, sparsely branched perennial; young plants fernlike; leaves pinnately dissected; inflorescence and seedhead white, flat or round, top terminal.

Toxic Properties

Plant reported to cause contact dermatitis in humans and diarrhea, colic, and tainted milk in cattle.

White Snakeroot

Ageratina altissima (L.) King & H. E. Robins. · Asteraceae · Sunflower Family

Synonyms

Deer-wort-boneset, Indian sanicle, richweed, squaw-weed, white sanicle, whitetop; *Eupatorium rugosum* Houtt.

Habit, Habitat, and Origin

Erect, warm-season perennial; to 1.5 m tall; pastures, abandoned fields, roadsides, and waste areas; native of North America.

Seedling Characteristics

Cotyledons broadly ovate; broadest at base; true leaves opposite; ovate with pointed tip and serrate margins; plant glabrous or covered with short, spreading trichomes.

Mature Plant Characteristics

ROOTS fibrous from taproot and short rhizomes. STEMS erect, 3.0–1.5 m tall, woody base, glabrous or with short, spreading trichomes solitary or a few from a common base. LEAVES opposite, 6.0–12.0 cm long, 1.0–4.0 cm wide, broadly ovate to lanceolate, tip acute, margins variably serrated, glabrous or moderately pubescent; trichomes short, curly, mostly on prominent main veins; petiole long. INFLORESCENCES heads of 9–25 disk florets in flat-topped to dome-shaped panicles; involucre cup-shaped, 8–14 pointed to blunt bracts, glabrous or short-pubescent; corollas bright white, 3.0–4.0 mm long. FRUITS achene, pappus attached, numerous bristles, 1.7–3.0 mm long, glabrous, pentangular, brown to dark brown.

Special Identifying Features

Erect perennial; leaves opposite, margins variably serrated, 3 prominent main veins; heads in flat-topped to dome-shaped panicles; flowers white.

Toxic Properties

Contains toxic oil that causes "trembles" in livestock and "milk-sickness" in humans who consume tainted milk; reported to contain cardiotoxins and hepatotoxins that cause metabolic derangement.

TOP Seeds
BOTTOM Four-leaf seedling

TOP Flowers
BOTTOM Flowering plant

Common Ragweed

Ambrosia artemisiifolia L. · Asteraceae · Sunflower Family

Synonyms

Annual ragweed, bitterweed, hogweed, Roman wormwood, short ragweed, small ragweed

Habit, Habitat, and Origin

Erect to branching summer annual herb; to 2.5 m tall; open disturbed areas and waste sites; native of North America.

Seedling Characteristics

Hypocotyl green usually splotched with purple; cotyledons roundish to oblong with grooved petiole, thick, deep purple underneath, dense pubescence over entire surface.

Mature Plant Characteristics

ROOTS fibrous from shallow taproot. STEMS erect, 0.2–2.5 m tall, sparingly or freely ascending branches, pubescent early. LEAVES opposite near base and alternate apically, 4.0–10.0 cm long, simple or pinnately to bipinnately lobed, pubescent on upper surface and margins; trichomes dense, appressed on lower surface. INFLORESCENCES male and female florets separated; male heads drooping, 2.0–4.0 mm wide, arranged in slender inverted racemes at ends of branches; female heads in axils of upper leaves and bases of leaves, 2–3 florets subtended by small, leaflike bracts. FRUITS woody achene, 3.0–4.0 mm long, 1.8–2.5 mm wide, yellowish brown to reddish brown or bluish near tip, central protuberance 1.0–2.0 mm long surrounded by 4–7 shorter projections, resembling a crown.

Special Identifying Features

Erect to branching summer annual; cotyledons roundish to oblong with underside deep purple; leaves pinnately lobed; achene woody, resembling a queen's crown; strong odor when crushed.

Toxic Properties

Ambrosia species are reported to cause digestive disorders and photosensitization; the pollen is a major contributor to hay fever.

TOP Two-leaf seedling
BOTTOM Four-leaf seedling

Flowering plant

Lanceleaf Ragweed

Ambrosia bidentata Michx. · Asteraceae · Sunflower Family

Synonyms

Southern ragweed

Habit, Habitat, and Origin

Erect, branching, warm-season annual; to 1.2 m tall; open waste sites; native of North America.

Seedling Characteristics

Young leaves sessile, lanceolate, with a single prominent tooth on each margin; lowest leaves sometimes opposite, then alternate above; entire plant rough-pubescent.

Mature Plant Characteristics

ROOTS fibrous from taproot. STEMS erect and much-branched, 0.3–1.2 m tall, rough-pubescent but sometimes smooth near base. LEAVES sometimes opposite lower, otherwise alternate, 1.0–7.0 cm long, simple, sessile, lanceolate, unlobed or usually with single triangular lobe or tooth on each side near base, pubescent,

trichomes sparse to moderate. INFLO-RESCENCES unisexual, heads sessile; male heads in long spikes at ends of branches; female heads in axillary clusters below; florets without a corolla, green; involucre slanted-ovoid, glandular, pubescent, enclosing a floret and a quadrangular conical beak, 5.0–8.0 mm long. FRUITS achene, enclosed in a woody, persistent, 4-sided, pointed involucre as an intact bur, 3.0–4.5 mm long, pappus absent.

Special Identifying Features

Erect, much-branched annual; stems and leaves with rough pubescence; leaves sessile, lanceolate, with 1 triangular lobe on each margin near base; inflorescences with male heads in spikes at ends of branches and female heads in axillary clusters.

Toxic Properties

May be a skin irritant when touched; see comments under *Ambrosia artemisiifolia*.

TOP Seeds
BOTTOM Inflorescence

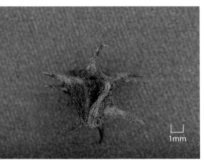

TOP **Seed**
MIDDLE **Three-leaf seedling**
BOTTOM **Five-leaf seedling**

Flowers

Woollyleaf Bursage

Ambrosia grayi (A. Nels.) Shinners · Asteraceae · Sunflower Family

Synonyms

Bur ragweed, woollyleaf ragweed

Habit, Habitat, and Origin

Tends to grow in patches; to 6.0 dm tall; open waste sites and fields; native of North America.

Seedling Characteristics

In colonies; leaves alternate, narrow-lobed, silvery gray.

Mature Plant Characteristics

ROOTS seedling fibrous from taproot, perennial, arising from extensive runnerlike underground rhizomes. **STEMS** erect, 3.0–6.0 dm tall, with silver gray pubescence, in large colonies. **LEAVES** alternate, 6.0–10.0 cm long, 4.0–8.0 cm wide, with deep lobes, ovate to lanceolate, silver gray pubescence, petiole wide. **INFLORESCENCES** male and female flowers separate; male inflorescence spike-like, stalked, 5.0 mm across, 5–9-lobed; female inflorescence in leaf axis of upper leaves, clustered or single, 7.0 mm long, 4.0 mm wide, with 2 flowers. **FRUITS** achene with up to 15 spines, scattered, narrowed to a slender, hooked tip, 7.0 by 4.0 mm.

Special Identifying Features

Erect perennial herb; silver gray pubescence; pistillate heads 2-flowered; spines of pistillate inflorescence scattered over fruiting body.

Toxic Properties

See comments under *Ambrosia artemisiifolia*.

Giant Ragweed

Ambrosia trifida L. · Asteraceae · Sunflower Family

Synonyms

Buffalo-weed, great ragweed, horseweed, kinghead, tall ragweed

Habit, Habitat, and Origin

Erect, summer annual herb; to 5.0 m tall; moist disturbed sites, streamsides, floodplains, roadsides, and ditches; native of Europe.

Seedling Characteristics

Hypocotyl purple; cotyledons roundish to oblong, with grooved petiole 5.0–8.0 mm

long; cotyledons green underneath; first leaves oblanceolate, margins entire with 3 rudimentary lobes, dense pubescence.

Mature Plant Characteristics

ROOTS fibrous from taproot. STEMS erect, 1.0–5.0 m tall, freely branched, rough-pubescent, angled, striate. LEAVES opposite, 0.7–3.0 dm long, 0.1–1.0 dm wide, simple; early ones oblanceolate, later ones ovate or elliptic; palmately divided into 3 or occasionally 5 ovate-lanceolate, serrated lobes; more or less scabrous on both sides; petiole 5.0–15.0 cm long. INFLORESCENCES male and female flowers separated; male flowers abundant in slender racemes in upper terminals, staminate involucres 3-ribbed on one side; female flowers in leaf axils of upper leaves and petals, saucer-shaped, involucre 6.0–13.0 mm long. FRUITS woody achene, obovoid, black, 5.0–10.0 mm long, 2.0–3.0 mm wide, central protuberance surrounded by 4–8 shorter projections, resembling a crown.

Special Identifying Features

Erect summer annual; cotyledons large, roundish to oblong, and three to four times larger than those of common ragweed; leaves opposite, palmately 3–5-lobed, ovate; petiole 5.0–8.0 mm, grooved; woody achene resembling a king's crown.

Toxic Properties

See comments under *Ambrosia artemisiifolia*.

TOP Young seedling
BOTTOM Flowering plant

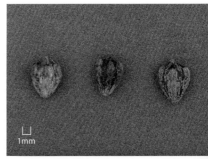

1mm

TOP Seeds
BOTTOM Two-leaf seedling

Flowers

Common Broomweed

Amphiachyris dracunculoides (DC.) Nutt. · Asteraceae · Sunflower Family

Synonyms

Gutierrezia dracunculoides (DC.) Blake

Habit, Habitat, and Origin

Erect, warm-season annual herb; to 1.0 m tall; dry open areas, pastures, and rangeland; native of North America.

Seedling Characteristics

Cotyledons ovate, smooth, dull green; leaves elliptical to spatulate, punctuate, midvein and edges purple toward base.

Mature Plant Characteristics

ROOTS fibrous from strong taproot; to 0.6 m deep; feeder roots in upper 0.3 m of soil. STEMS erect, 0.4–1.0 m tall, tough, woody, branched near top of plant, forming a rather round uni-form crown. LEAVES alternate, simple, 2.0–5.0 cm long, 0.2–0.4 cm wide, entire, linear-lanceolate, sticky resin on smooth surface. INFLORESCENCES head 0.4–1.5 cm across; flowers numerous, tiny, bright yellow, at tips of small branches; ray flowers 5–7, fertile; disk flowers 10–35, not fertile. FRUITS achene, shaped like an inverted pyramid and covered with upright, stiff white trichomes.

Special Identifying Features

At bloom, stalk and branches are bare except for flowers.

Toxic Properties

A toxic compound in the plant causes pinkeye-like inflammation of the eye in humans and livestock.

Flowers

Mayweed Chamomile

Anthemis cotula L. · Asteraceae · Sunflower Family

1mm

Synonyms

Dogfennel, mayweed, stinking chamomile, stinking mayweed

Habit, Habitat, and Origin

Erect or decumbent winter annual herb; to 6.0 dm tall; cultivated areas, fields, pastures, roadsides, and waste sites; native of North America.

Seedling Characteristics

Cotyledon leaves opposite without obvious petiole; first true leaves finely dissected.

Mature Plant Characteristics

ROOTS fibrous from taproot. STEMS erect to decumbent, 2.0–6.0 dm tall, coarse, highly branched. LEAVES alternate, 1.5–6.0 cm long, 0.5–3.0 cm wide, finely bi- or tripinnately dissected. INFLORESCENCES in numerous heads at terminals of small branches; ray florets white and 3-toothed; disk florets yellow. FRUITS achene, obovoid, 1.5–2.5 mm long, truncate at summit, slightly quadrangular, distinctly ribbed, glabrous, pappus absent, aggregated in terminal heads.

Special Identifying Features

Erect or decumbent winter annual; leaves finely dissected; flowers white with yellow center; distinctive odor when crushed.

Toxic Properties

Plants reported to cause contact irritation and to taint milk.

TOP Seeds
MIDDLE Four-leaf seedling
BOTTOM Five-leaf seedling

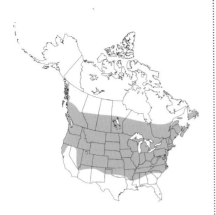

Flower

Common Burdock

Arctium minus Bernh. · Asteraceae · Sunflower Family

Synonyms

Clotbur, cuckoo-button, lesser burdock, smaller burdock, wild burdock

Habit, Habitat, and Origin

Erect biennial herb with numerous basal leaves; to 1.5 m tall; disturbed sites, stream valleys, pastures, fields, and feedlots; native of Europe.

Seedling Characteristics

Cotyledons spatulate; first leaves long, broadly ovate, pubescent, with hollow petiole.

Mature Plant Characteristics

ROOTS fibrous from large, fleshy taproot. STEMS erect, 0.5–1.5 m tall, branched, ridged, pubescent, trichomes small. LEAVES alternate, 15.0–45.0 cm long, 10.0–35.0 cm wide, broadly ovate, margins serrate to dentate, loosely pubescent on underside, lower petioles hollow. INFLORESCENCES head 1.5–3.0 cm diameter, arranged in axillary and terminal racemes, perfect, discoid, purple to lavender or occasionally white; involucral bracts overlapping, with hooked, barbed tips. FRUITS achene, mottled dark gray to black, tapered from base to apex, flattened, 3–5-sided, pappus bristles barbed.

Special Identifying Features

Erect biennial; leaves large, broadly ovate, basal; lower petioles hollow; stems with bitter sap; involucral bracts with hooked, barbed tips; flowers purple to lavender.

Toxic Properties

None reported.

Spanishneedles

Bidens bipinnata L. · Asteraceae · Sunflower Family

Synonyms

Spanish beggarticks

Habit, Habitat, and Origin

Erect, branching, warm-season annual herb; to 1.7 m tall; moist to dry, fertile, semishaded soils in disturbed and waste sites, fencerows, fields, pastures, and gardens; native of North America.

Seedling Characteristics

Cotyledons linear to narrowly oblanceolate.

Mature Plant Characteristics

ROOTS fibrous from taproot. STEMS erect, 0.3–1.7 m tall, branching, quadrangular, slender, glabrous or minutely pubescent. LEAVES opposite, 4.0–20.0 cm long including petiole 2.0–5.0 cm long, simple, pinnately dissected two or three times; primary leaf thin, broadly deltoid to ovate, cuneate overall, segments deltoid to deltoid-lanceolate; segment lobes rounded, margins toothed, glabrous to minutely pubescent. INFLORESCENCES narrow heads, solitary, at ends of branches; ray florets few, inconspicuous, yellow, shorter than disk florets; disk florets 5-lobed; outer involucre bracts 7–10, linear; inner bracts linear-lanceolate, longer than outer bracts. FRUITS achene, linear, black to dark brown, quadrangular, sparsely pubescent, 3–4 barbed awns, yellow, 10.0–13.0 mm long.

Special Identifying Features

Erect, branching, warm-season annual; leaves opposite, bi- or tripinnately dissected, glabrous to minutely pubescent; disk florets 4.0–6.0 mm wide, 5-lobed; ray florets few, inconspicuous, yellow; seed awns adhere to fur and clothing.

Toxic Properties

None reported.

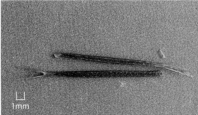

TOP Seeds
BOTTOM Four-leaf seedling

TOP Flower
BOTTOM Inflorescence

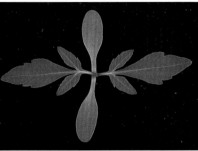

TOP Two-leaf seedling
MIDDLE Four-leaf seedling
BOTTOM Flowering plant

Devils Beggarticks

Bidens frondosa L. · Asteraceae · Sunflower Family

Synonyms

Devil's pitchfork, common beggarticks, sticktight

Habit, Habitat, and Origin

Erect annual herb; to 1.5 m tall; moist wooded areas, stream banks, roadsides, and waste sites; native of North America.

Seedling Characteristics

Hypocotyl glabrous and reddish; cotyledons large, spatulate, and fleshy.

Mature Plant Characteristics

ROOTS many-branched, fibrous from shallow taproot. **STEMS** erect, 0.4–1.5 m tall, branching near top, appearing 4-sided, glabrous or nearly so. **LEAVES** opposite, to 10.0 cm long, 3.0 cm wide, 4-ranked on petioles; deeply pinnately divided into 3–7 lanceolate segments, 1.0–6.0 cm long; acuminate tips, margins serrate, occasionally pubescent; trichomes sparse, short, on lower surface. **INFLORESCENCES** head saucer-shaped, 7.0–15.0 mm wide, surrounded by leaf bracts; pubescent orange-yellow ray florets around outer edge sometimes absent; brownish yellow, 2.5–4.0 mm disk florets in center. **FRUITS** achene, flat, wedge-shaped, 5.0–10.0 mm long, 3.0–4.0 mm wide; strongly 1-nerved on each side; dark brown or blackish; nearly glabrous or appressed-pubescent; with 2 slightly divergent to erect barbed spines, 2.0–5.0 mm long, resembling cow horns.

Special Identifying Features

Erect annual; stem appears square in cross section; leaves deeply pinnately divided with serrate leaflets; fruits with pappus of 2 slightly divergent to erect barbed spines; seeds adhere to fur and clothing.

Toxic Properties

None reported.

Fruit

LEFT Flower
RIGHT Flowering plant

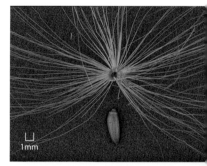

1mm

TOP Seed
BOTTOM Three-leaf seedling

Musk Thistle

Carduus nutans L. · Asteraceae · Sunflower Family

Synonyms

Musk-thistle, nodding thistle

Habit, Habitat, and Origin

Erect, coarse, biennial or rarely annual
herb; to 2.0 m tall; open areas, pastures, and roadsides; native of North
America.

Seedling Characteristics

Basal rosette; leaves waxy, pale green,
prominent veins, coarsely lobed, 3–5
points per lobe.

Mature Plant Characteristics

ROOTS fibrous from large, shallow, fleshy,
corky, hollow taproot. STEMS erect,
0.5–2.0 m tall; branched; spiny, lobed
leaves extending down stem, winged
appearance, spine at tip of each lobe.
LEAVES alternate, 15.0–40.0 cm long,
5.0–12.0 cm wide, dark green with light
green midribs and white margins, deeply
lobed with 3–5 points per lobe; spines

prominent on each lobe, white or yellow, 2.0–5.0 mm long. INFLORESCENCES
mostly solitary heads nodding at ends
of branches, 30.0–50.0 mm wide; disk
florets deep rose to violet or purple;
involucral bracts often tinged purple,
conspicuously broad, 2.0–8.0 mm wide,
long flat or reflexed, spine-pointed tips.
FRUITS achene, straw-colored, oblong,
glabrous, 5–10-nerved or nerveless,
protrusion at attachment of plumeless,
whitish pappus.

Special Identifying Features

Erect, coarse biennial or rarely annual;
stem winged, leaflike, appearing spined;
leaves coarsely lobed, spines on each
lobe; flower heads nodding, solitary,
conspicuously broad with long flat or
reflexed, spine-pointed tips.

Toxic Properties

Plants associated with nitrate accumulation; may cause mechanical injury.

Flower

TOP Seeds
BOTTOM Six-leaf seedling

Spotted Knapweed

Centaurea biebersteinii DC. · Asteraceae · Sunflower Family

Synonyms

Jersey knapweed; *Centaurea maculosa* auct. non Lam., *Centaurea stoebe* L. subsp. *micranthos* (S. G. Gmel. ex Gugler) Hayek

Habit, Habitat, and Origin

Erect or ascending biennial or short-lived perennial herb; to 1.2 m tall; open areas, grasslands, pastures, roadsides, and waste sites; native of Europe.

Seedling Characteristics

Cotyledons spatulate to obovate.

Mature Plant Characteristics

ROOTS fibrous from taproot. **STEMS** erect, spreading, or ascending; 0.3–1.2 m tall, freely branching, branches slender and wiry, rough-pubescent. **LEAVES** alternate, 10.0–15.0 cm long, pinnatifid with long, narrow divisions, gray-green to gray, tomentose, upper often linear and with entire margins, rough-pubescent. **INFLORESCENCES** numerous terminal and axillary heads, lavender to pinkish, perfect; receptacle bristly, flat; involucre 8.0–12.0 mm long, 6.0–10.0 mm wide; bracts striate; middle and outer bracts with short, dark, pectinate tips. **FRUITS** achene, obovoid, olive green to blackish brown with 4 yellow longitudinal lines, 2.5–3.0 mm long, 1.1–1.5 mm wide, pubescent, prominent scar notch immediately above one side of base; pappus 1.0–2.0 mm long.

Special Identifying Features

Erect or ascending biennial or short-lived perennial; all leaves pinnately lobed; bracts not spiny; achene pappus much reduced.

Toxic Properties

None reported, but may cause mechanical injury.

Cornflower

Centaurea cyanus L. · Asteraceae · Sunflower Family

Synonyms

Batchelor's button, blue bottle, cornflower, raggedy sailors

Habit, Habitat, and Origin

Erect annual herb; to 1.2 m tall; fields, pastures, roadsides, and waste sites; native of Europe.

Seedling Characteristics

Hypocotyl stout, glabrous; cotyledons glabrous; first true leaves grayish, pubescent, with a cottony envelope.

Mature Plant Characteristics

ROOTS fibrous from taproot. STEMS erect, 0.2–1.2 m tall, freely branching, pubescent, slender, angled in cross section. LEAVES alternate, 5.0–15.0 cm long, less than 1.0 cm wide, narrowly lance-shaped or linear, margins entire or slightly toothed, pubescent. INFLORESCENCES numerous solitary heads, 2.5–5.0 cm diameter, on long peduncle; ray florets blue, occasionally pink, purple, or white, radiate, sterile, 1.5–3.0 cm long; inner florets smaller, perfect, with involucre, ovoid, 0.8–2.5 cm long. FRUITS achene, 3.0–3.8 mm long, 1.5–2.0 mm wide, lustrous, purplish gray in middle, yellowish basally; pappus 2.0–3.0 mm long, cinnamon brown.

Special Identifying Features

Erect annual herb; plant covered with loose white trichomes; outer ray florets blue, 1.5–3.0 cm long, deeply toothed at apex.

Toxic Properties

See comments under *Centaurea biebersteinii.*

TOP Seeds
BOTTOM Two-leaf seedling

TOP Purple flower
BOTTOM White flower

Yellow Starthistle

Centaurea solstitialis L. · Asteraceae · Sunflower Family

Synonyms

Barnaby's thistle, yellow star-thistle

Habit, Habitat, and Origin

Erect or spreading annual or biennial herb; to 8.0 dm tall; open fields, pastures, roadsides, and waste sites; native of Eurasia.

Seedling Characteristics

Cotyledons elliptic, with long petioles; leaves lobed to deeply lobed; pubescent.

Mature Plant Characteristics

ROOTS fibrous from taproot. **STEMS** erect, 4.0–8.0 dm tall, freely branched, winged from leaf base down, tomentose. **LEAVES** alternate, most less than 20.0 cm long and 5.0 cm wide; basal leaves lyrate pinnatifid, tomentose. **INFLORESCENCES** solitary heads on long, leafy peduncles, yellow, 6.0–15.0 mm tall and wide; large, rigid yellow spines, 1.0–3.0 cm long, flanked by pair of small upright spines. **FRUITS** achene, grayish brown, obovoid, 2.5 mm long, 0.7 mm wide, turgid, glabrous; white pappus in central florets 3.0–5.0 mm long.

Special Identifying Features

Erect or spreading annual or biennial; stems winged; foliage with persistent trichomes and spines; flower heads solitary, pale yellow.

Toxic Properties

See comments under *Centaurea biebersteinii*.

TOP Seeds
BOTTOM Five-leaf seedling

TOP Flower
BOTTOM Flowering plant

Flower

Chicory

Cichorium intybus L. · Asteraceae · Sunflower Family

Synonyms

Blue daisy, blue sailors, coffee-weed, succory, wild chicory, wild succory

Habit, Habitat, and Origin

Erect, branched perennial from deep taproot; to 1.5 m tall; open areas, roadsides, railroad beds, and moist disturbed areas; native of Mediterranean region of Europe.

Seedling Characteristics

Rarely seen; forms a rosette with rough, oblong to lance-shaped and toothed leaves, 5.0–15.0 cm long.

Mature Plant Characteristics

ROOTS fibrous from deep, fleshy taproot; simple or branched; sometimes sub-rhizomatous. STEMS erect, 0.5–1.5 m tall, branched, smooth, with milky sap. LEAVES basal leaves form a rosette; upper leaves alternate, 3.0–35.0 cm long, 1.0–12.0 cm wide, sessile or short-petioled, slightly lobed or entire. INFLORESCENCES heads single or in raceme at ends of branches, and in axils of leaves; rays bright blue, rarely white, 3.0–4.0 cm wide, with square, lobed ends.

FRUITS achene, weakly wedge-shaped, smooth, dark brown, slightly ribbed; pappus of 2 or 3 minute scales.

Special Identifying Features

Erect, branched perennial from deep, fleshy taproot; flowers bright blue with square, lobed rays; milky sap.

Toxic Properties

None reported.

Flowering branch

TOP Seeds
BOTTOM Five-leaf seedling

TOP Seeds
MIDDLE Seedling
BOTTOM Flowering plant

Flower

Canada Thistle

Cirsium arvense (L.) Scop. · Asteraceae · Sunflower Family

Synonyms

Creeping thistle; *Carduus arvensis* (L.) Robson

Habit, Habitat, and Origin

Erect perennial; 0.6–1.5 m tall; open areas, pastures, roadsides, bottomlands, and waste sites; native of Eurasia.

Seedling Characteristics

Cotyledons club-shaped, 16.0 mm long, 8.0 mm wide; first leaves also club-shaped, coarse, margin serrate with spines on tips of serrations.

Mature Plant Characteristics

ROOTS fibrous from taproot; to 1.0 m deep; forming extensive creeping horizontal growth. STEMS erect, 0.6–1.5 m tall, branching toward top, grooved, glabrous early but pubescent with maturity. LEAVES alternate, 3.0–30.0 cm long, 1.0–6.0 cm wide, simple, sessile, slightly clasping, oblong to lanceolate, somewhat lobed, margins crinkled and spiny. INFLORESCENCES male and female florets on separate plants; heads numerous, 2.0–2.5 cm diameter, compact in terminal clusters with lavender, rose-purple, or white disk florets only; involucre bracts numerous, spineless. FRUITS achene, brown, smooth-coated, slightly tapered, 4.5 mm long, with ridge around blossom end, attached to a tannish pappus, easily detached.

Special Identifying Features

Erect perennial thistle; flowers rose-purple, 2.5 cm diameter; plants forming patches from horizontal roots.

Toxic Properties

Plants associated with nitrate accumulation; may cause mechanical injury.

Flowers

Field Thistle

Cirsium discolor (Muhl. ex Willd.) Spreng. · Asteraceae · Sunflower Family

Synonyms

Carduus discolor (Muhl. ex Willd.) Nutt.

Habit, Habitat, and Origin

Erect, robust biennial or perennial; to 3.0 m tall; meadows, pastures, and open areas; flowering late summer to fall; native of North America.

Seedling Characteristics

Fibrous-rooted; cotyledons spatulate; young leaves with some trichomes and spiny tips on leaf margins.

Mature Plant Characteristics

ROOTS fibrous network of secondary roots from taproot. **STEMS** erect, 2.0–3.0 m tall, thickened, robust, villous, leafy to near summit. **LEAVES** alternate, 1.0–2.0 dm long, 4.5 cm wide, pinnately dissected, elliptic, villous above, white tomentose beneath, margins revolute, cauline leaves deeply pinnatifid with linear-lanceolate lobes. **INFLORESCENCES** corymbose, lavender to dark purple; involucres 2.0–3.5 cm long, 2.5–3.5 cm wide, weakly arachnose; inner involucral bracts with attenuate tips. **FRUITS** achene, 3.0–4.5 mm long, tawny, with plumose white pappus.

Special Identifying Features

Erect, robust biennial or perennial; flower heads few to many per plant, corymbose, corolla lavender purple.

Toxic Properties

None reported, but may cause mechanical injury.

TOP **Seeds**
MIDDLE **Four-leaf seedling**
BOTTOM **Leaves**

Flowers

Yellow Thistle

Cirsium horridulum Michx. · Asteraceae · Sunflower Family

Synonyms

Bristly thistle, bull thistle; *Carduus spinosissimus* Walter

Habit, Habitat, and Origin

Erect, stout biennial or perennial herb; to 1.2 m tall; open areas, fields, pastures, and roadsides; native of North America.

Seedling Characteristics

Forming rosette early; leaves elliptic, dissected, divisions spine-tipped, to 30.0 cm long, pubescent.

Mature Plant Characteristics

ROOTS fibrous from taproot; some roots fleshy and thickened. **STEMS** erect, 0.4–1.2 m tall, stout, pubescent with long delicate trichomes, leafy to near summit, occasionally branched. **LEAVES** rosette at base, alternate upward, 10.0–30.0 cm long, 3.0–10.0 cm wide, elliptic, dissected, each division spine-tipped, villous above, grayish arachnoid below. **INFLORESCENCES** racemose, up to 10 per plant, sessile in a whorl of leaflike bracts, 3.0–4.5 cm long, 3.0–5.5 cm broad; corollas buff yellow, reddish purple or rarely white, yellow form usually restricted to coastal plain. **FRUITS** achene with plumose pappus, 3.0–4.0 cm long, grayish black.

Special Identifying Features

Erect biennial or perennial; outer whorl of leaflike bracts spiny, more or less enclosing head; corolla buff yellow, reddish purple, or rarely white.

Toxic Properties

None reported, but may cause mechanical injury.

TOP Seed
MIDDLE Two-leaf seedling
BOTTOM Rosette

Bull Thistle

Cirsium vulgare (Savi) Ten. · Asteraceae · Sunflower Family

Synonyms

Bull-thistle, common thistle, Scotch thistle; *Carduus lanceolatus* L.

Habit, Habitat, and Origin

Erect biennial herb; to 2.0 m tall; open areas, pastures, roadsides, and waste sites; native of Europe.

Seedling Characteristics

Leaves forming rosette; first leaves oblong, fringed with spines.

Mature Plant Characteristics

ROOTS fibrous from fleshy taproot. **STEMS** erect, 0.5–2.0 m tall, often-branched, conspicuously spiny-winged by decurrent leaf bases, pubescent, stalked in second season. **LEAVES** rosette first year, alternate from stem second year, 0.8–3.0 dm long, 4.0–10.0 mm wide, elliptic to oblanceolate, dissected, pinnatifid; larger leaves with lobes again toothed or lobed, scabrous-hispid above, scattered grayish or white-tomentulose or hirsute below. **INFLORESCENCES** heads corymbose, 3 to many, 2.5–4.0 cm high, 1.8–3.5 cm wide, rose to reddish purple; disk florets surrounded by spiny-tipped bracts. **FRUITS** achene with plumose pappus, deciduous in a ring, 3.0–4.0 mm long, light straw-colored, nerveless but with brown stripes.

Special Identifying Features

Erect biennial; stem conspicuously spiny-winged; upper leaf surface scabrous-hispid.

Toxic Properties

None reported, but may cause mechanical injury.

TOP Seeds
BOTTOM Three-leaf seedling

TOP Flower
BOTTOM Flowering plant

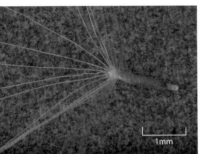

1mm

TOP Seed
MIDDLE Three-leaf seedling
BOTTOM Flower

Flowering plant

Hairy Fleabane

Conyza bonariensis (L.) Cronq. · Asteraceae · Sunflower Family

Synonyms

Asthmaweed, flax-leaved fleabane; *Erigeron bonariensis* L.

Habit, Habitat, and Origin

Erect warm-season annual or biennial herb; to 1.0 m tall; fields, pastures, roadsides, and waste sites; native of tropical America.

Seedling Characteristics

Rosette with narrow, pubescent, wrinkled leaves; branching early.

Mature Plant Characteristics

ROOTS fibrous from shallow taproot. STEMS erect, 0.5–1.0 m tall, glabrate below, becoming villous-ciliate apically. LEAVES alternate, 3.0–12.0 cm long, 1.0–25.0 mm wide, simple, narrow, ascending, slightly grayish, lower leaves prominently lobed or toothed, mostly entire near inflorescence. INFLORESCENCES head-bearing region of plant cylindrical, round-topped; heads to 7.0 mm diameter, numerous, borne at terminal ends of stems and each composed of 10–20 disk florets; corollas slightly shorter than pappus. FRUITS achene, 1.0–1.3 mm long, yellow to tan, tapered and slightly flattened, with attached pappus 2.0–2.8 mm long, pappus white to tan with minute barbs directed forward and upward.

Special Identifying Features

Erect warm-season annual or biennial; often confused with horseweed (*Conyza canadensis*) but shorter and with larger, more distinct flowers; more-branched habit starting from early vegetative growth; narrower leaves, often wrinkled or distorted, usually pubescent; and darker green foliage.

Toxic Properties

None reported.

68 · MAGNOLIOPSIDA · DICOTS

Young plant

Flowering plant

1mm

TOP Seed
BOTTOM Young seedling

Horseweed

Conyza canadensis (L.) Cronq. · Asteraceae · Sunflower Family

Synonyms

Butterweed, Canadian fleabane, fireweed, hogweed, mare's tail; *Erigeron canadensis* L.

Habit, Habitat, and Origin

Erect, coarse annual herb; to 2.0 m tall; fields, cultivated areas, pastures, roadsides, and waste sites; native of North America.

Seedling Characteristics

Cotyledons green, lacking evident veins, smooth; first leaves in rosette, entire; subsequent leaves toothed with a short apical projection, blades green on upper surface and light green on smooth lower surface.

Mature Plant Characteristics

ROOTS fibrous from taproot. **STEMS** erect; 1.0–2.0 m tall, simple, solid, trichomes bristly. **LEAVES** basal rosette, alternate upward, 2.5–10.0 cm long, 1.0–15.0 mm wide, simple, linear to oblanceolate, sessile, crowded on stems, margins entire or toothed, more or less pubescent or at least coarsely ciliate near base. **INFLORESCENCES** numerous small heads arranged on elongated panicle; involucres 2.0–3.0 mm long, 2.0–5.0 mm wide; ray florets 1.0–2.0 mm long, white to lavender; disk florets yellow, receptacle 1.0–2.0 mm wide; bracts acuminate. **FRUITS** achene, 1.0 mm long, tapered from base, yellowish, pubescent with ribbed margins; pappus 2.0–2.3 mm long, white at apex, with numerous slender white to tan bristles, antrorsely barbed.

Special Identifying Features

Erect herb initially forming rosette; stem with frequent alternate and sessile leaves; highly variable with several named varieties.

Toxic Properties

Plants rarely cause intoxication in sheep and cattle.

TOP **Seed**
MIDDLE **Four-leaf seedling**
BOTTOM **Inflorescence**

Flowering plant

Dwarf Fleabane

Conyza ramosissima Cronq. · Asteraceae · Sunflower Family

Synonyms

Low horseweed, purple horseweed, spreading fleabane; *Erigeron divaricatus* Michx.

Habit, Habitat, and Origin

Erect, spreading or ascending, many-branched annual herb; to 3.0 dm tall; in dry fields, prairies, and waste sites; native of North America.

Seedling Characteristics

Cotyledons sessile, spatulate, lightly pilose, fleshy; first true leaves spatulate, lightly pilose, with petioles.

Mature Plant Characteristics

ROOTS fibrous from taproot. **STEMS** erect, highly branched with no central axis, 1.0–3.0 dm tall, slender, grayish, with antrorsely appressed trichomes. **LEAVES** alternate, to 4.0 cm long, 2.0 mm wide, narrowly linear, uppermost reduced to bracts, grayish pubescence. **INFLORESCENCES** numerous terminal heads; involucre 3.0–4.0 mm and strongly imbricate; outer bracts pubescent; inner bracts glabrous, small, purplish rays in a single series equal to or slightly longer than dull white pappus. **FRUITS** achene, 1.0 mm long, yellowish tan, linear, pappus attached; pappus nearly twice as long as achene.

Special Identifying Features

Erect, spreading or ascending annual; many-branched, low and bushy with no main axis; stems with grayish pubescence; leaves reduced to small bracts in uppermost branches; ray florets in single row.

Toxic Properties

None reported.

Plains Coreopsis

Coreopsis tinctoria Nutt. · Asteraceae · Sunflower Family

Synonyms

Common tickseed, eye-flower, tickseed

Habit, Habitat, and Origin

Erect, freely branched annual herb; to 1.2 m tall; fields, prairies, roadsides, moist ditches, and waste sites; native of North and South America.

Seedling Characteristics

Cotyledons spatulate; first leaves forming rosette, opposite, glabrous.

Mature Plant Characteristics

ROOTS fibrous from taproot. STEMS erect, freely branched, 0.4–1.2 m tall, smooth, slightly angled. LEAVES opposite, sub-sessile or short-petiolate; basal and lower leaves pinnately to bipinnately divided, forming a rosette; uppermost leaves 6.0–10.0 cm long, undivided, blades narrowly linear or linear-lanceolate. INFLORESCENCES diffuse corymb of many heads; peduncles 1.0–6.0 cm long; ray florets yellow or occasionally orange or deep red, reddish brown spot at base, conspicuously toothed; disk florets dark red. FRUITS achene, linear-oblong, incurved, wingless, awnless, 1.2–4.0 m long, black with white scar.

Special Identifying Features

Erect annual; stems freely branched, glabrous; leaves long, narrow; ray florets yellow or occasionally orange or deep red, reddish brown spot at base.

Toxic Properties

None reported.

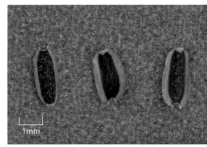

TOP **Seeds**
BOTTOM **Four-leaf seedling**

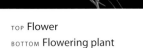

TOP **Flower**
BOTTOM **Flowering plant**

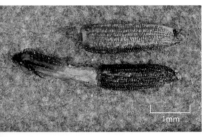

Coneflower

Dracopis amplexicaulis (Vahl) Cass. · Asteraceae · Sunflower Family

Synonyms

Clasping coneflower; *Rudbeckia amplexi-caulis* Vahl.

Habit, Habitat, and Origin

Erect, freely branching annual herb; 30.0–80.0 cm tall; low, moist, disturbed prairies and open areas; native of the south-central United States.

Seedling Characteristics

Cotyledons ovate; first leaves ovate to spatulate, margins and petiole with scattered stiff trichomes.

Mature Plant Characteristics

ROOTS fibrous from taproot; minimal secondary root growth. STEMS erect, 30.0–80.0 cm tall, simple or loosely many-branched, branches ascending, glaucous, glabrous. LEAVES alternate, 10.0 cm long, 4.0 cm wide, margins entire or occasionally toothed, 1-ribbed, ovate, clasping stem, glabrous. INFLORESCENCES few or solitary dark heads, 1.0–1.5 cm wide; ray florets 6–10, 1.0–3.0 cm long, spreading to reflexed, yellow, usually with orange or brown base; disk region cylindrical. FRUITS disk achene, elongate, elliptic, thin, circular to quadrangular in cross section, glabrous, wrinkled.

Special Identifying Features

Erect, many-branched annual; distinguished from *Rudbeckia* species by tall, nearly cylindrical disk region of flower head; upper leaves cordate, clasping, and extending proximal ends toward stem.

Toxic Properties

None reported.

Flower

Flowering plant

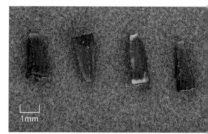

TOP Seeds
MIDDLE Seedling
BOTTOM Fruit and flower

Eclipta

Eclipta prostrata (L.) L. · Asteraceae · Sunflower Family

Synonyms

Eclipta alba (L.) Hassk.

Habit, Habitat, and Origin

Erect, spreading, or prostrate summer annual herb; to 1.5 m tall; damp, sandy, or muddy soils, cultivated areas, fields, and roadsides; native of Asia.

Seedling Characteristics

Hypocotyl light green to light purple to pink; cotyledons small, green, spatulate, glabrous, slightly thickened, midvein evident on lower surface.

Mature Plant Characteristics

ROOTS fibrous from shallow taproot. STEMS erect, spreading, or prostrate; 0.3–1.5 m tall, freely branched, rooting at nodes, green becoming reddish brown or purplish, transversely ridged at nodes; trichomes stiff, antrorse-appressed. LEAVES opposite, 3.0–13.0 cm long, 0.5–3.0 cm wide, simple, elliptic to lanceolate, petiole lacking, slightly thick-ened, midvein evident, margins serrate, lower surface pubescent throughout, appressed pubescence at base of first 3–4 leaves, remainder pubescent. INFLORES-CENCES 2 solitary heads, terminal head matures before elongated axillary head, perfect, rounded; female ray florets many, white, 1.0–2.0 mm long, with trichomes at junction of blade and tube; disk florets numerous, 4.0–6.0 mm wide; involucral bracts 10–12, 3.0–7.0 mm long, pubescent. FRUITS achene, 1.8–2.5 mm long, brown, widest at apex and tapering to base; pappus attached at apex, inconspicuous tuft of very short trichomes.

Special Identifying Features

Erect to spreading annual; cotyledons spatulate; leaves opposite, margins remotely toothed; flowers round, green early and then white with short rays; leaves and stems turn black when broken from plant.

Toxic Properties

None reported.

Philadelphia Fleabane

Erigeron philadelphicus L. · Asteraceae · Sunflower Family

Synonyms

Common fleabane

Habit, Habitat, and Origin

Erect biennial or short-lived perennial herb; to 1.0 m tall; in cultivated and hay fields, pastures, fencerows, and waste areas; native of North America.

Seedling Characteristics

First leaves lanceolate, pubescent, with petioles.

Mature Plant Characteristics

ROOTS fibrous from taproot. STEMS erect; 0.1–1.0 m tall; soft, flatten under light pressure. LEAVES lower ones in rosette, 10.0–15.0 cm long, 1.0–3.0 cm wide, oblong or narrowly obovate, crenate or dentate, pubescent and toothed; upper leaves cauline, smaller, alternate, oblong, sessile, round-based, clasping, decreasing with size apically. INFLORESCENCES heads few to many, nodding in bud, 1.5–2.5 cm wide; ray flowers pink to rosy or whitish; disk corollas 2.5–3.5 mm long, yellow. FRUITS achene, 0.6–1.0 mm long, tan or light brown, 2-nerved, pappus attached.

Special Identifying Features

Erect biennial or short-lived perennial; leaves pubescent, upper ones clasping stem; flowers white with yellow center; flowering in spring, not odorous.

Toxic Properties

None reported.

Flowering plant

Goldenweed

Grindelia papposa G. L. Nesom & Y. B. Suh · Asteraceae · Sunflower Family

Synonyms

Grindelia ciliata (Nutt.) Spreng.

Habit, Habitat, and Origin

Erect annual or biennial herb or shrub; to 1.5 m tall; open areas, roadsides, railroad beds, pastures, and waste sites; native of the Americas.

Seedling Characteristics

Cotyledons oval, lustrous; first leaves ovate, serrate, covered with fine pubescence.

Mature Plant Characteristics

ROOTS fibrous from taproot. **STEMS** erect and often shrubby, 0.4–1.5 m tall, glabrous, leafy throughout. **LEAVES** alternate, 3.0–8.0 cm long, 1.0–4.0 cm wide, simple, sessile, clasping, oblong to ovate or elliptic-obovate, rounded or obtuse at tip, spinose-dentate. **INFLORESCENCES** heads, few, large; disk 1.5–3.0 cm wide; involucre bracts slightly or moderately imbricate; outer bracts with long, loose, or spreading green tip, larger bracts generally well over 1.0 mm wide; ray florets yellow, 25–50, 1.0–2.0 cm wide. **FRUITS** achene, 2.0–4.0 mm long, 4–5-angled, ellipsoid to oblong, glabrous, with pappus of numerous rigid bristles, central ones sterile.

Special Identifying Features

Erect, bushy annual or biennial; leaves clasping, sessile; inflorescences few, large.

Toxic Properties

Plants associated with secondary accumulation of selenium.

TOP Seeds
BOTTOM Four-leaf seedling

Flowers

TOP Seeds
BOTTOM Seedling

Curlycup Gumweed

Grindelia squarrosa (Pursh) Dunal · Asteraceae · Sunflower Family

Synonyms

None

Habit, Habitat, and Origin

Erect biennial or weakly perennial herb; to 1.0 m tall; open areas, roadsides, railroad beds, pastures, and waste sites; native of the Americas.

Seedling Characteristics

Cotyledons oblong, glabrous; first leaves initially with entire margins, then toothed, pubescent.

Mature Plant Characteristics

ROOTS fibrous from taproot. STEMS erect, 0.1–1.0 m tall, single or branching from base, glabrous, sticky. LEAVES alternate, 1.5–7.0 cm long, 4.0–20.0 mm wide, oblong or ovate, linear-oblong to oblanceolate, acute tips, sessile, somewhat clasping, firm, margins serrulate to coarsely toothed. INFLORESCENCES heads, several to numerous, 0.7–30.0 cm wide, bright yellow, ray florets 12–38, 7.0–15.0 mm long, 0.7–2.5 mm wide; disk florets 5-toothed, perfect, central ones sterile; bracts in 48 series, linear, pointed, upper portion spreading or strongly reflexed, secreting sticky resin. FRUITS achene, 2.3–3.0 mm long, oblong, smooth or with longitudinal furrows, tan to brown, usually quadrangular, deep ridges, striate, with linear 2-awned pappus, 3.0–5.0 mm long.

Special Identifying Features

Erect, branching biennial or weak perennial; involucre surrounding flower with curved bracts; plants covered with sticky resin.

Toxic Properties

Plants associated with secondary accumulation of selenium.

TOP Flowering plant
BOTTOM Flower

Flowers

Broom Snakeweed

Gutierrezia sarothrae (Pursh) Britt. & Rusby · Asteraceae · Sunflower Family

Synonyms

Common broomweed, matchbrush, matchweed, kindlingweed, perennial snakeweed, slinkweed, turpentine weed, yellow top

Habit, Habitat, and Origin

Erect or sprawling, short-branched perennial shrub; to 1.0 m tall; native of North America.

Seedling Characteristics

Cotyledons elliptic; first leaves alternate, entire, lanceolate to oblanceolate, in threadlike clusters.

Mature Plant Characteristics

ROOTS fibrous from taproot, from woody caudex. **STEMS** erect to ascending, 0.6–1.0 m tall, branching from woody caudex at base, striate, scabrous, often resinous. **LEAVES** alternate, 5.0–70.0 mm long, simple, margins entire and inrolled, sessile, first linear-lanceolate then linear-filiform in threadlike clusters, glabrous or lightly scabrous, often sticky from resin secreted by glands. **INFLORESCENCES** cluster of 3–10 small heads at ends of branches, rounded, loose to dense corymb; involucre 3.0–6.0 mm tall, obconic; bracts narrow; ray florets 3–8, fertile, 2.0–5.4 mm long, yellow; disk florets 2–9, fertile, 2.0–3.5 mm long. **FRUITS** achene, 0.9–2.0 mm long, narrowly obovoid to obconic, tan to brown, pubescent, with a pappus of 8–10 whitish scales.

Special Identifying Features

Short, branching perennial shrub; usually sticky; leaves in threadlike clusters; flowers a cluster of small yellow heads at ends of branches.

Toxic Properties

Plants reported to cause abortion, weak offspring, and impaired reproduction in livestock.

TOP **Seeds**
MIDDLE **Young seedling**
BOTTOM **Flowering plant**

TOP Seeds
MIDDLE Young seedling
BOTTOM Flowering plant

Flowers

Bitter Sneezeweed

Helenium amarum (Raf.) H. Rock · Asteraceae · Sunflower Family

Synonyms

Bitterweed; *Helenium tenuifolium* Nutt.

Habit, Habitat, and Origin

Erect, usually highly branched, annual herb; to 8.0 dm tall; fields, prairies, waste areas, and openings in woods; native of North America.

Seedling Characteristics

Cotyledons small, ovate, glabrous; first true leaves entire, pubescent; later basal leaves linear with lobed margin, finely pubescent.

Mature Plant Characteristics

ROOTS fibrous from taproot. STEMS erect, usually much-branched, 2.0–8.0 dm tall, glabrous, green. LEAVES alternate, 2.0–7.0 cm long, 1.0–4.0 mm wide, glabrous; basal leaves usually not present on flowering plants; upper leaves linear, filiform, entire. INFLORESCENCES several or numerous on short, naked peduncles; ray florets pistillate, yellow, 5.0–10.0 mm long; disk florets perfect, yellow. FRUITS achene, 1.0–1.5 mm long excluding pappus, 4–5-angled, brown, pubescent, with attached pappus, scales hyaline, awned.

Special Identifying Features

Erect, usually highly branched annual; flowers yellow, in a mass; leaves linear, filiform.

Toxic Properties

Helenium species contain toxins causing metabolic derangement, digestive disturbance, loss of appetite, neurological and cardiological effects, and tainted milk.

Common Sneezeweed

Helenium autumnale L. · Asteraceae · Sunflower Family

Synonyms

Autumn sneezeweed, oxeye, sneezeweed, sneezewort, staggerwort, staggerweed, swamp sunflower, yellow star, yellow oxeye

Habit, Habitat, and Origin

Erect, simple perennial herb; to 1.3 m tall; moist open areas, fields, pastures, and roadsides; native of North America.

Seedling Characteristics

Cotyledons broadly oblong to obovate, glabrous; first true leaves lanceolate to elliptic, slightly lobed margins, pubescent.

Mature Plant Characteristics

ROOTS fibrous from taproot and sub-rhizomatous rootstocks. **STEMS** erect, unbranched to narrowly branching near top of plant, 0.3–1.3 m tall, winged, glabrous to finely pubescent. **LEAVES** alternate, 3.0–10.0 cm long, 1.0–4.0 cm wide, simple, lanceolate to elliptic, tip pointed, glabrous to moderately or sparsely pubescent. **INFLORESCENCES** heads in clusters at branch terminals, 2.5 cm diameter; peduncles long; involucre bracts in 2 series, approximately 18 bracts, pointed to a tapering to slender tip; florets daisylike, domed; disk flowers yellow-green, 1.0 mm long; ray florets wedge-shaped, bright yellow, 3-lobed at tips. **FRUITS** achenes, 1.0 mm long, 4–5-angled, pubescence on conspicuous ribs, tapered from base to apex; pappus 1.5 mm long with 5 brown scales, tapering to awnlike tip.

Special Identifying Features

Erect perennial; stems distinctly winged, branched near top of plant.

Toxic Properties

See comments under *Helenium amarum*.

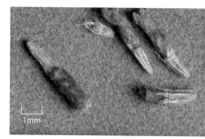

TOP Seeds
MIDDLE Four-leaf seedling
BOTTOM Flower

Flowering plant

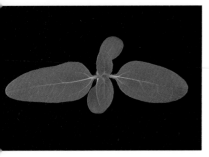

TOP Seeds
MIDDLE Two-leaf seedling
BOTTOM Flowering plant

Common Sunflower

Helianthus annuus L. · Asteraceae · Sunflower Family

Synonyms

Sunflower, wild sunflower

Habit, Habitat, and Origin

Erect annual herb; to 3.5 m tall; open sites, fields, roadsides, and waste areas; native of North America.

Seedling Characteristics

Hypocotyl stout, green, with stiff white trichomes; cotyledons glabrous, oblong to oval, prominent midvein; lower true leaves opposite.

Mature Plant Characteristics

ROOTS fibrous from taproot. STEMS erect, 1.0–3.5 m tall, green, stout, coarse, pubescent, usually branching above. LEAVES mostly alternate, some opposite at base, 10.0–30.0 cm long, simple, ovate to obtuse, base truncate to cordate, 3 main veins, pubescent, rough on both surfaces, margins serrate. INFLORESCENCES showy heads, 21–35, yellow to orange-yellow; ray florets 3.0–6.0 cm long, numerous, red to dark reddish brown; disk florets perfect, receptacle nearly flat, chaffy, 4.0 cm or more diameter; involucral bracts overlapping, oblong to ovate, densely ciliate with pubescent backs and tapering tips. FRUITS achene, 4.0–8.0 mm long, 2.2–2.6 mm wide; white, grayish, or brown with black mottling; ovate to wedge-shaped, flattened, enclosed in nutlet 5.0–15.0 mm long, quadrangular to flattened, glabrous except for a few short trichomes at apex; pappus of 2 thin scales, chaffy, deciduous.

Special Identifying Features

Erect annual; cotyledons glabrous, oblong to oval, prominent midvein; stem pubescent; leaves pubescent, cordate, margins serrate, 3 prominent veins; flowers showy, rays yellow, disks reddish brown.

Toxic Properties

Young plants may be toxic to livestock if eaten in large quantities, possibly from nitrate accumulation and bioactive sesquiterpene lactones.

Flower

LEFT AND RIGHT Flower

1mm

TOP Seeds
BOTTOM Sprouts

Texas Blueweed

Helianthus ciliaris DC. · Asteraceae · Sunflower Family

Synonyms

Blueweed sunflower

Habit, Habitat, and Origin

Erect, summer perennial herb; to 7.0 dm tall; dry or damp, sandy, open, disturbed sites; native of North America.

Seedling Characteristics

Stem glabrous and glaucous with blue-green cast; opposite leaves; arising from rhizomes as well as seeds.

Mature Plant Characteristics

ROOTS fibrous from taproot, then from creeping rootstocks and slender rhizomes. STEMS erect, 2.0–7.0 dm tall, 1 to several in clusters, glabrous, glaucous. LEAVES mostly opposite, 3.0–7.0 cm long, 0.5–2.0 cm wide, narrow to broadly lanceolate, ciliate margin, hispid. INFLO-RESCENCES in head, 1.5–2.5 cm across; ray florets absent or usually 10–18, 1.0 cm long, yellow; disk florets reddish, 5.0–6.0 mm long. FRUITS achene, 3.0 mm long, black or grayish, nutlet enclosing achene, chaffy bracts entire or 3-toothed, pappus of 2 broadly ovate-acuminate scales.

Special Identifying Features

Erect perennial; proliferating from slender rhizomes and tending to grow in patches; stems glaucous; leaves blue-green with ciliate margins; usually with yellow ray florets and reddish disk florets.

Toxic Properties

Many species in the genus *Helianthus* are grown for food. Young plants may be toxic to livestock if eaten in large quantities, possibly as a result of nitrate accumulation.

TOP Seeds
MIDDLE Seedling
BOTTOM Flower

Flowering plant

Woodland Sunflower

Helianthus divaricatus L. · Asteraceae · Sunflower Family

Synonyms

None

Habit, Habitat, and Origin

Erect, warm-season perennial herb; to 1.5 m tall; dry woods and roadsides; native of North America.

Seedling Characteristics

Cotyledons elliptic to lanceolate; pubescent.

Mature Plant Characteristics

ROOTS fibrous from taproot, then long rhizomes. STEMS erect, 0.5–1.5 m tall, glabrous below inflorescence. LEAVES opposite, 5.0–18.0 cm long, 1.0–8.0 cm wide, cauline, lanceolate to lance-ovate, margins serrate to entire, pubescent below on veins, scabrous above, rounded base, sessile or rarely with short petiole up to 5 mm long. INFLORESCENCES 1 to several heads terminating on stiff, cymose branches; involucres 10.0–15.0 mm long, 10.0–20.0 mm wide; ray florets yellow, 3.0–4.0 cm long; disk florets yellow, 1.5–3.0 cm long; bracts lanceolate. FRUITS achene, 4.0–4.5 mm long, dark brown to black, nutlet enclosing achene, rounded at apex, glabrous, pappus 1.5–3.3 mm long.

Special Identifying Features

Erect perennial; leaves opposite; ray florets bright yellow and slightly darker than disk florets.

Toxic Properties

See comments under *Helianthus ciliaris*.

Sawtooth Sunflower

Helianthus grosseserratus Martens · Asteraceae · Sunflower Family

Synonyms

Helianthus instabilis E. Wats.

Habit, Habitat, and Origin

Erect perennial herb; to 5.0 m tall; prefers bottomlands, damp prairies, and other moist sites; native of North America.

Seedling Characteristics

Cotyledons linear-lanceolate, tapering to stem, entire; first true leaves similar but more lanceolate than cotyledons.

Mature Plant Characteristics

ROOTS fibrous from taproot then branching, woody, and rhizomatous. STEMS erect, 1.0–5.0 m tall, stout, smooth, glabrous, often glaucous. LEAVES alternate, 10.0–30.0 cm long, 1.5–10.0 cm wide, simple, lanceolate to oblong-ovate with 1 main pinnately branched nerve, lower leaves often larger than middle and upper ones, sessile, margins deeply and saliently serrated especially on lower leaves, pubescent, often tapering to a winged petiole 1.5–9.0 cm long. INFLORESCENCES numerous showy, yellow heads, 1.0–2.5 cm wide, in loose paniculate clusters, with appressed pubescence; ray florets 10–20, 2.0–4.5 cm long, yellow; disk florets yellow; involucral bracts loose, longer than disk florets. FRUITS achene, 3.0–5.0 mm long, glabrous, lanceolate, nutlet enclosing achene, flattened, pale brown to dark gray, with pappus of 2 lanceolate awns.

Special Identifying Features

Erect perennial; rhizomes well-developed; main stem glabrous; leaves lanceolate, covered with stiff short pubescence, scarcely scabrous, margins coarsely serrated; flowers in showy yellow heads.

Toxic Properties

See comments under *Helianthus ciliaris.*

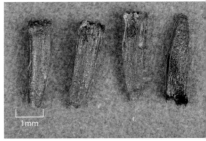

TOP **Seeds**
BOTTOM **Seedling**

TOP **Flowering plant**
BOTTOM **Flower**

TOP Seeds
MIDDLE Two-leaf seedling
BOTTOM Flowering plant

Flower

Prairie Sunflower

Helianthus petiolaris Nutt. · Asteraceae · Sunflower Family

Synonyms

Kansas sunflower

Habit, Habitat, and Origin

Erect annual herb; to 1.2 m tall; fields, roadsides, railroad beds, and waste areas; native of North America.

Seedling Characteristics

Cotyledons lanceolate; first true leaves lanceolate, margins shallowly serrated.

Mature Plant Characteristics

ROOTS fibrous from taproot; light brown. STEMS erect, 0.5–1.2 m tall, simple or sometimes branched near base, strigose or occasionally glabrate. LEAVES alternate, 4.0–15.0 cm long, 1.0–8.0 cm wide, simple, deltoid-ovate to lanceolate, margins entire to shallowly serrate, strigose on both surfaces, older leaves glaucous, peti-ole 1.0–12.0 cm long. INFLORESCENCES head with disk and ray florets, 1.0–2.5 cm across; ray florets 15–30, 2.0 cm long; disk florets reddish purple or rarely yellow; bracts chaffy, 3-forked, center ones densely white-pubescent at apex; involucral bracts lanceolate to ovate-lanceolate, 3.0–4.0 mm wide. FRUITS achene, 3.5–5.0 mm long, mottled dark and light brown, nutlet enclosing achene, pappus of pubescent tufts forming 2 lateral terminal "ears," awns 3.5–5.0 mm long.

Special Identifying Features

Erect annual; involucral bracts lanceolate to subulate, tapering, less than 4.0 mm wide; central chaffy bracts white-pubescent at tips.

Toxic Properties

See comments under *Helianthus ciliaris*.

Jerusalem Artichoke

Helianthus tuberosus L. · Asteraceae · Sunflower Family

Synonyms

None

Habit, Habitat, and Origin

Erect perennial herb; to 3.0 m tall; native of North America.

TOP **Flower**
BOTTOM **Flowering plant**

Seedling Characteristics

Arising from rootstock or seed; young leaves opposite and glabrous.

Mature Plant Characteristics

ROOTS fibrous from taproot then producing ovoid, fleshy, tuber-bearing rhizomes. **STEMS** erect, 1.0–3.0 m tall, branched apically, pubescent. **LEAVES** lower ones opposite, upper ones alternate, 10.0–25.0 cm long, 4.0–12.0 cm wide, broadly lanceolate to broadly ovate, 3 prominent veins, tapering to well-developed more or less winged petiole 2.0–8.0 cm long, margins serrate, scabrous above, short-pubescent beneath. **INFLORESCENCES** numerous heads in a corymb; disk florets yellow, 1.5–2.5 cm wide; ray florets 10–20, 2.0–4.0 cm long; involucral bracts generally dark. **FRUITS** achene, 5.0–6.0 mm long, tan to light brown; nutlets enclosing achene, mottled light and dark brown, rounded to truncate at apex, awns 3.0–4.0 mm long.

Special Identifying Features

Erect perennial; rootstock with tubers; leaf with 3 prominent veins, petioles winged.

Toxic Properties

Jerusalem artichoke is occasionally grown as a crop. Rootstocks are used for human food, and the roots and foliage are fed to livestock. See additional comments under *Helianthus ciliaris*.

1mm

TOP **Seeds**
BOTTOM **Four-leaf seedling**

Flower

Oxeye

Heliopsis helianthoides (L.) Sweet · Asteraceae · Sunflower Family

Synonyms

False sunflower, hardy zinnia, rough heliopsis, smooth oxeye

Habit, Habitat, and Origin

Erect perennial; to 3.0 m tall; tallgrass prairies, savannahs, moist to mesic areas, open woods, roadsides, and waste sites; native of North America.

Seedling Characteristics

Cotyledons ovate with slight indentation at tip; first true leaves simple, 3 main veins, margins somewhat serrated becoming more so during maturity.

Mature Plant Characteristics

ROOTS fibrous from thick taproot. STEMS erect, 2.0–3.0 m tall, unbranched or branched apically, rough. LEAVES opposite, sometimes whorled, 5.0–15.0 cm long, 3.0–8.0 cm wide, simple, ovate to lanceolate, margins serrate, scabrous on both sides, tip pointed with subtruncated base attached to petiole. INFLORESCENCES borne on a stout peduncle in summer, resembling a small sunflower; ray florets to 16, dark yellow, 4.0–5.0 cm long; disk florets numerous, 4.0–5.0 mm long, yellow to orange, fertile. FRUITS achene, 3.5–4.5 mm long, brown; nutlet enclosing achene 4.0–6.0 mm long, four-sided, glabrous to slightly pubescent, pappus absent or a crown of a few scales.

Special Identifying Features

Erect perennial with several varieties; leaves opposite or sometimes whorled, shaped like arrowheads, occasionally with small paired leaves at base; flower resembles a small sunflower.

Toxic Properties

Plants accumulate toxic levels of nitrates but are seldom a danger to animals.

Florida Wild Lettuce

Lactuca floridana (L.) Gaertn. · Asteraceae · Sunflower Family

Synonyms

Woodland lettuce

Habit, Habitat, and Origin

Erect annual or biennial herb; to 2.0 m tall; fields, roadsides, and forest edges; native of North America.

Seedling Characteristics

Cotyledons oval, with pubescent margins, milky sap; first leaves thin, with petiole.

Mature Plant Characteristics

ROOTS fibrous from taproot. STEMS erect, 0.5–2.0 m tall, glabrous, usually reddish, leafy, milky sap. LEAVES alternate, 8.0–30.0 cm long, 3.0–20.0 cm wide, elliptic to triangular-ovate, margins toothed or deeply lobed, often pubescent on veins beneath, mostly petiolate. INFLORESCENCES diffuse panicle to 8.0 dm long and about as wide, branches long and spreading; heads 1.5–2.0 cm wide, 11–17-flowered; involucre 8.0–14.0 mm long, 4.0–6.0 mm wide; ray florets blue to violet. FRUITS achene, 3.0–5.5 mm long, brown; nutlet enclosing achene elliptic to oblanceolate, 3–5-ribbed, 4.0–6.0 mm long, tapering to a short, stout beak; white pappus.

Special Identifying Features

Erect annual or biennial; stems with milky sap; leaf margins toothed or deeply lobed; flowers 11–17, rays blue or violet; fruit with white pappus.

Toxic Properties

None reported.

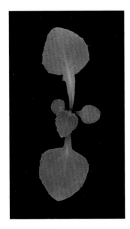

TOP **Seed**
BOTTOM **Three-leaf seedling**

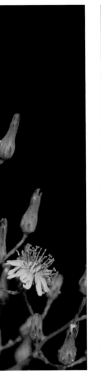

Flowers

Flowering plant

Prickly Lettuce

Lactuca serriola L. · Asteraceae · Sunflower Family

Synonyms

Compass plant, lobed prickly lettuce, wild lettuce

Habit, Habitat, and Origin

Erect winter annual or biennial herb; to 2.0 m tall; disturbed sites and waste areas; native of Europe.

Seedling Characteristics

Leaves forming basal rosette; clasping petioles; spiny edges; pale green; prominent midvein.

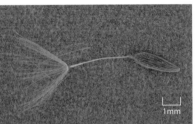

Mature Plant Characteristics

ROOTS fibrous from taproot. STEMS erect, 1.0–2.0 m tall, pale green or straw-colored, hollow, lower stems often prickly, milky juice. LEAVES alternate, oblong-lanceolate, glaucous, clasping stem with base extended into pair of arrowhead-shaped projections, lobed or wavy; margins toothed and prickly, midveins prickly underneath.

INFLORESCENCES panicle, 2.0–8.0 dm long, branches widely spreading, 16–24 heads; involucres 9.0–15.0 mm long, 3.0–6.0 mm wide; ray florets yellow to pale yellow, often blue after drying, 8.0–10.0 mm diameter. FRUITS achene, flat, oblanceolate, grayish yellow to yellowish brown, 3.0–4.0 mm long, 1.1–1.5 mm wide, 5–7 lengthwise ribs on each side, ribs pubescent toward apex, beak 3.0–5.0 mm long; white pappus, 4.0–5.0 mm long, easily detached, almost always absent, leaving a short remnant.

Special Identifying Features

Erect winter annual or biennial; stem hollow with milky sap; leaves glaucous, clasping, margins and underside midvein prickly; ray florets yellow.

Toxic Properties

None reported.

TOP Seed
MIDDLE Three-leaf seedling
BOTTOM Flowering plant

Flowers

Flower

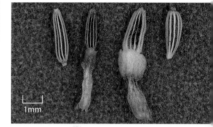

Oxeye Daisy

Leucanthemum vulgare Lam. · Asteraceae · Sunflower Family

Synonyms

Field daisy, moon daisy, white daisy; *Chrysanthemum leucanthemum* L.

Habit, Habitat, and Origin

Erect, warm-season perennial herb; to 8.0 dm tall; fields, roadsides, pastures, and waste sites; native of Eurasia.

Seedling Characteristics

Erect, not highly branched.

Mature Plant Characteristics

ROOTS fibrous from taproot and rhizomes. STEMS erect, becoming slightly branched, 2.0–8.0 dm tall, glabrous or inconspicuously pubescent, forming clumps from rhizomes. LEAVES alternate, 4.0–15.0 cm long, simple, oblanceolate or spatulate basally, pinnately lobed, margins cleft or blunt-toothed, petioled or sessile, pungent odor. INFLORESCENCES arranged in 1 to a few solitary heads; ray florets 15–30, 10.0–25.0 mm long, pistillate, white to yellow; disk florets perfect, fertile, yellow, midrib sometimes greenish. FRUITS achenes, 1.5 mm long, terete, black with 8–10 gray ribs; nutlet enclosing achene blackish, oblanceolate, resin-dotted; pappus forming very short crown.

Special Identifying Features

Erect perennial; leaves alternate, primarily basal, reduced in size on upper main stem, pinnately lobed, cleft, and/or toothed; flowers daisylike, white with yellow center; pungent odor when crushed.

Toxic Properties

None reported.

TOP **Seeds**
MIDDLE **Four-leaf seedling**
BOTTOM **Flowering plant**

Pineapple-weed

Matricaria discoidea DC. · Asteraceae · Sunflower Family

Synonyms

Rayless chamomile, wild marigold; *Matricaria matricarioides* (Less.) Porter

Habit, Habitat, and Origin

Erect, freely branched annual herb; to 4.5 dm tall; disturbed fields, cultivated areas, roadsides, gardens, lawns, and waste sites; dispersed eastward from the Pacific states, native of the western United States.

Seedling Characteristics

Cotyledons small, narrow, spatulate, and fused together.

Mature Plant Characteristics

ROOTS fibrous from taproot. STEMS erect, much-branched, 1.5–4.5 dm tall, mostly round, smooth to sparsely pubescent. LEAVES alternate, 1.0–6.0 cm long, 0.5–2.0 cm wide, very finely pinnately divided into linear segments, bright green. INFLORESCENCES numerous, on ends of branches, without ray florets, greenish yellow cone-shaped head about 6.0 mm diameter; bracts subtending a complete floret cluster, oblong with broad transparent margins. FRUITS achene, 1.0–1.2 mm long, 3–5-ribbed with 2 marginal and 1 rather weak ventral nerve; nutlet enclosing achene, 1.5 mm long, slightly curved, greenish brown with reddish brown line on either side.

Special Identifying Features

Erect, freely branched annual; leaves finely pinnately divided; inflorescence and seedheads emit distinct pineapple odor when crushed.

Toxic Properties

None reported.

TOP Seeds
MIDDLE Four-leaf seedling
BOTTOM Flowers

Flowering plant

Flowering plant

Cressleaf Groundsel

Packera glabella (Poir.) C. Jeffrey · Asteraceae · Sunflower Family

Synonyms

Butterweed, ragwort, squawweed; *Senecio glabellus* Poir.

Habit, Habitat, and Origin

Erect, succulent annual herb; to 1.0 m tall; pastures, fencerows, moist and shaded waste areas; native of North America.

Seedling Characteristics

Cotyledons round; first true leaves roundish, glabrous, with petioles.

Mature Plant Characteristics

ROOTS fibrous from taproot. **STEMS** erect, 0.2–1.0 m tall, glabrous, soft, succulent, hollow, sometimes cobwebby in leaf axils. **LEAVES** basal rosette, alternate upward, 5.0–20.0 cm long, 2.0–7.0 cm wide, rapidly diminishing in size up the stem; basal and lower leaves with petioles, upper leaves sessile; lower leaves lyrate-pinnatifid with flabelliform, bluntly notched lateral divisions; terminal lobe much larger and more rounded than lateral lobes. **INFLORESCENCES** radiate and discoid, yellow; ray florets 1.0–1.3 cm long, without pappus; disk florets perfect, lobes 0.5–0.7 mm long, tube 4.0–5.0 mm long, with pappus. **FRUITS** achene, brown, narrowly ellipsoid, 1.4–1.5 mm long, 0.4–0.5 mm wide, glabrous or pubescent; nutlet enclosing achene with white pappus, 3.0–4.0 mm long, deciduous.

Special Identifying Features

Erect annual; stem hollow, glabrous; leaves pinnately dissected; flowers showy, yellow.

Toxic Properties

Senecio species contain toxins causing liver disease.

TOP Seed
MIDDLE Seedling
BOTTOM Inflorescence

Ragweed Parthenium

Parthenium hysterophorus L. · Asteraceae · Sunflower Family

Synonyms

False ragweed, Santa Maria feverfew

Habit, Habitat, and Origin

Erect, much-branched annual herb; to
1.0 m tall; croplands, disturbed fields,
roadsides, and other open areas; native
of tropical America.

Seedling Characteristics

Leaves in basal rosette; pinnatifid to bipinnatifid; pubescence similar on leaves
and stems.

Mature Plant Characteristics

ROOTS fibrous from taproot. **STEMS** erect,
0.3–1.0 m tall, hirsute, with paniclelike
branching, longitudinally striate. **LEAVES**
alternate, 3.0–18.0 cm long, 1.0–9.0
cm wide, whitish green, pinnatifid to
usually bipinnatifid, lower leaves forming basal rosette, upper leaves entire
to slightly lobed, short pubescence
on both sides, dotted with glands.
INFLORESCENCES head small, slightly
convex on top, numerous in an often
leafy inflorescence, densely pubescent;
disk florets 3.0–5.0 mm wide, whitish;
ray florets minute. **FRUITS** achene, 2.0
mm or less long, obovate, black; nutlet
enclosing achene with pappus as 2
lateral scales.

Special Identifying Features

Erect annual; plants with a basal rosette
and stem leaves; leaves whitish green,
pinnately to usually bipinnately dissected; inflorescence white.

Toxic Properties

Plants cause acute dermatitis in humans
and digestive and metabolic derangement in livestock producing diarrhea,
muscular twitching, and increased
excitability.

TOP **Seeds**
BOTTOM **Four-leaf seedling**

TOP **Flowering plant**
BOTTOM **Flower**

Carolina Falsedandelion

Pyrrhopappus carolinianus (Walt.) DC. · Asteraceae · Sunflower Family

Synonyms

False dandelion

Habit, Habitat, and Origin

Erect, branched, winter annual or biennial herb; to 6.0 dm tall; fallow fields, lawns, and open disturbed areas; native of Mexico and southern United States.

Seedling Characteristics

Cotyledons roundish; first true leaves entire, lanceolate, pubescent.

Mature Plant Characteristics

ROOTS fibrous from thickened taproot. **STEMS** erect, 2.0–6.0 dm tall, freely branched, smooth below, minutely pubescent above. **LEAVES** alternate, 8.0–25.0 cm long, 1.0–5.0 cm wide, oblanceolate to lanceolate, acute to acuminate tips, margins entire to deeply dissected, pubescent, base cuneate, basal leaves with petiole. **INFLORESCENCES** pedunculate head, involucre round, 1.5–2.5 cm long, 1.0–2.0 cm wide; inner bracts in 1 series, bilobed or winged, pubescent; outer bracts shorter, in several series; ray florets 2.0–2.5 cm long, yellow. **FRUITS** achene, 4.0–4.5 mm long, 1.0–1.1 mm wide, cylindrical, tapered at both ends, pubescent, 5-grooved, with beak 9.0–11.0 mm long; nutlet enclosing achene with pappus, 8.0–10.0 mm long, bristles whitish tan.

Special Identifying Features

Erect annual; stems green, mostly smooth; basal leaves petioled; upper leaves sessile; flowers yellow; plant with milky sap.

Toxic Properties

None reported.

TOP Flower
BOTTOM Flowering stem

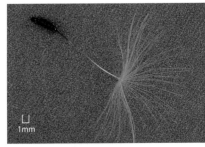

1mm

TOP Seed
BOTTOM Seedling

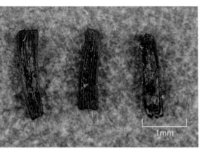

TOP Seeds
MIDDLE Four-leaf seedling
BOTTOM Flowering plant

Flower

Blackeyed-Susan

Rudbeckia hirta L. · Asteraceae · Sunflower Family

Synonyms

Yellow daisy

Habit, Habitat, and Origin

Erect, warm-season, biennial or short-lived perennial, rarely annual; to 1.0 m tall; disturbed roadsides, prairies, pastures, and waste sites; native of North America.

Seedling Characteristics

Erect; 1 to several plants per crown.

Mature Plant Characteristics

ROOTS fibrous from taproot. STEMS erect, 0.3–1.0 m tall, simple or much-branched, terete or slightly angled, dense, spreading, with bristly pubescence. LEAVES basal leaves whorled, alternating upward, 4.0–18.0 cm long, 1.0–8.0 cm wide, simple, slightly serrate to entire, basal leaves ovate with petioles, middle leaves ovate becoming linear and sessile, margins coarsely toothed or entire. INFLORESCENCES arranged in 1 to a few solitary heads; ray florets 2.0–4.0 cm long, orange to orange-yellow, sometimes darker at base; disk florets 1.2–2.0 cm wide, fertile, ovoid, dark purple or brown. FRUITS achene, 2.5–2.7 mm long, black, quadrangular, tapering from base to apex; nutlets enclosing achene with pappus, short crown, almost obsolete.

Special Identifying Features

Erect biennial or short-lived perennial, rarely annual; leaves numerous basally and alternate above, pubescent; leaves elliptic, lanceolate, or ovate; flowers orange to orange-yellow with dark disk florets.

Toxic Properties

None reported.

Common Groundsel

Senecio vulgaris L. · Asteraceae · Sunflower Family

Synonyms

Common ragwort, garden groundsel, grand mouron, groundsel, old-man-in-the-spring

Habit, Habitat, and Origin

Erect, cool-season annual herb; to 6.0 dm tall; moist soils, cultivated areas, waste sites, gardens, and roadsides; native of Eurasia.

TOP Flowering plant
BOTTOM Flowers

Seedling Characteristics

Cotyledons elliptic to ovate; first true leaves obovate, margins dentate.

Mature Plant Characteristics

ROOTS fibrous from taproot. STEMS erect, 10.0–60.0 cm tall, freely branching, covered with sparse matted trichomes, then glabrous when mature. LEAVES basal rosette, alternate upward, 1.5–10.0 cm long, 0.5–4.5 cm wide, soft, fleshy, obovate to oblanceolate, margins dentate to coarsely and irregularly pinnately lobed; pubescent, tomentose, or subglabrous; petiole short toward plant base, sessile and clasping apically. INFLORESCENCES loose cluster of 8–20 discoid heads at end of each main branch; no ray florets; disk florets golden yellow, each head 2.5–10.0 mm wide; bracts 4.0–9.0 mm long, 4.0–6.0 mm wide, approximately 21, involucral bracts black-tipped. FRUITS achene, 1.5–2.5 mm long, 0.4–0.5 mm wide, sparsely pubescent along angles, brown; nutlet enclosing achene, pappus, 5.0–6.0 mm long, copious, white.

Special Identifying Features

Erect annual; stem single or branched near base in corymbose fashion; leaves pinnatifid, coarse irregular lobes; disk florets golden yellow; no ray florets; overall ratty appearance.

Toxic Properties

Alkaloids found in many *Senecio* species can cause liver disease in livestock. Humans are susceptible to exposure in herbal teas and dietary supplements. The concentration of alkaloids is greatest in flowers but significant throughout the plant.

TOP Seed
BOTTOM Four-leaf seedling

Flower

Blessed Milkthistle

Silybum marianum (L.) Gaertn. · Asteraceae · Sunflower Family

Synonyms

Lady's thistle, milk thistle

Habit, Habitat, and Origin

Erect winter annual or biennial herb; to 2.0 m tall; native of the Mediterranean region.

Seedling Characteristics

Cotyledons obovate; leaves glabrous, white-mottled, with spiny margins.

Mature Plant Characteristics

ROOTS fibrous from shallow taproot. STEMS erect, 0.5–2.0 m tall, simple or sparingly branched, glabrous, not winged. LEAVES alternate, 1.5–8.0 dm long, 0.2–3.0 dm wide, mottled white especially along veins, bases clasping; margins coarsely lobed, wavy; spines numerous, yellow. INFLORESCENCES in discoid heads terminating branches, 3.0–6.0 cm wide, purple; individual florets 2.5–3.5 cm long, perfect, tubes 1.0–2.5 cm long, throats campanulate, 2.0–3.0 mm long, lobes 4.0–10.0 mm long. FRUITS achene, 6.0–7.0 mm long, glabrous, somewhat compressed, brown, black-speckled; nutlet enclosing achene with pappus of numerous subpaleaceous bristles, 15.0–20.0 mm long, slender, unequal, with deciduous ring.

Special Identifying Features

Erect annual or biennial; stem not spiny-winged; leaves white-mottled, wavy; flowers in purple heads; fruits with pappus bristles not plumose.

Toxic Properties

Plants associated with nitrate accumulation in livestock; a minor vegetable or medicinal herb for human ailments.

Flowers

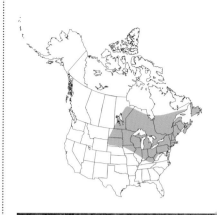

Canada Goldenrod

Solidago canadensis L. · Asteraceae · Sunflower Family

Synonyms

Common goldenrod

Habit, Habitat, and Origin

Erect perennial herb; to 2.5 m tall; pastures, fallow fields, meadows, and roadsides; native of North America.

Seedling Characteristics

Cotyledons small, elliptic, entire; first true leaves thin, serrate, with short pubescence on midrib.

Mature Plant Characteristics

ROOTS fibrous from taproot; then forming extensive, creeping, slender rhizomes. **STEMS** erect, 0.2–2.5 m tall, sparsely to densely pubescent, lower half of stem generally smooth, growing singly or in clusters from creeping rhizomes. **LEAVES** alternate, 3.0–15.0 cm long, 1.0–3.0 cm wide, mainly on stem, sessile, narrowly lanceolate to elliptic, 3-nerved, thinnish, margins serrate to subentire; pubescence short, soft on midrib above and below; smooth to scabrous above. **INFLORESCENCES** panicle-like with strongly recurved branches; head small; involucre 2.0–3.0 mm tall; disk florets 3–7, yellow; ray florets 10–18, 1.0–1.5 mm long; bracts thin. **FRUITS** achene, 1.0–1.5 mm long, pubescent; nutlet enclosing achene, short-pubescent pappus, barbed.

Special Identifying Features

Erect perennial from rhizomes; lower half of stem smooth; leaves thin, margins serrate to subentire; inflorescence terminal, panicle-like; flowers yellow.

Toxic Properties

See comments under *Solidago canadensis* var. *scabra*.

TOP **Seeds**
MIDDLE **Seedling**
BOTTOM **Flowering plant**

Tall Goldenrod

Solidago canadensis L. var. *scabra* Torr. & Gray · Asteraceae · Sunflower Family

Synonyms

Solidago altissima L.

Habit, Habitat, and Origin

Erect perennial herb; to 2.0 m tall; pastures, fallow fields, meadows, and roadsides; native of North America.

Seedling Characteristics

Cotyledons small; first true leaves ovate, entire, pubescent.

Mature Plant Characteristics

ROOTS fibrous from taproot; then forming extensive, creeping, slender rhizomes. **STEMS** erect, 1.2–2.0 m tall, coarse, terete, densely pubescent, from a creeping rhizome. **LEAVES** alternate, 6.0–15.0 cm long, 1.0–4.0 cm wide, narrow, serrate, chiefly on stem, numerous, base cuneate or attenuate, tip acute or acuminate, pubescent below. **INFLORESCENCES** terminal, panicle-like, from strongly recurved branches; involucre 2.5–4.0 mm long; ray florets 10.0–17.0 mm long, yellow; disk florets 2.0–8.0 mm long; bracts thin, acute or obtuse, appressed, glabrous. **FRUITS** achene, 1.0–1.5 mm long, pubescent; nutlet enclosing achene with short pappus, 2.0–4.0 mm long, pubescent.

Special Identifying Features

Erect perennial from rhizomes; stems pubescent; leaves pubescent below; inflorescence panicle-like, terminal.

Toxic Properties

Solidago species contain neurotoxins with variable expressions.

TOP Seed
MIDDLE Five-leaf seedling
BOTTOM Young plant

Flowering plant

Lawn Burweed

Soliva sessilis Ruiz & Pavón · Asteraceae · Sunflower Family

Synonyms

Burweed, spurweed; *Soliva pterosperma* (Juss.) Less.

Habit, Habitat, and Origin

Low-growing, branched, winter annual herb; rarely to 15.0 cm tall; lawns, turf, and shaded open disturbed areas; native of South America.

Seedling Characteristics

Cotyledon broadly elliptic or ovate; first true leaves spatulate, entire, pubescent, and opposite.

Mature Plant Characteristics

ROOTS fibrous from thickened taproot. **STEMS** low-growing, decumbent or spreading, 4.0–15.0 cm tall, freely branching, internodes 1.0–4.5 cm long, glabrous, usually not rooting at nodes. **LEAVES** opposite, 1.0–3.5 cm long, 7.0–1.5 cm wide, bipinnately dissected, glabrous. **INFLORESCENCES** small, inconspicuous solitary head, 3.0–7.0 mm wide; disk florets 3.0–6.0 mm wide, green to yellowish green; bracts lance-ovate, pubescent. **FRUITS** achene, brown at maturity; nutlet enclosing achene, 1.0–4.5 cm long, obovoid, marginal wings 2-lobed, glabrous or sparsely pubescent at apex; lateral and terminal spines conspicuous, spines 1.5–2.0 mm long, vexatious.

Special Identifying Features

Low-growing winter annual; fruit with sharp spines that may cause injury and readily attach to fur, hair, or clothing.

Toxic Properties

None reported, but fruit can cause mechanical injury.

1mm

TOP **Seeds**
BOTTOM **Flower**

TOP **Four-leaf seedling**
BOTTOM **Three-leaf seedling**

1mm

TOP Seed
MIDDLE Seedling
BOTTOM Mature plant

Flowers

Perennial Sowthistle

Sonchus arvensis L. · Asteraceae · Sunflower Family

Synonyms

Corn sow-thistle, field sowthistle

Habit, Habitat, and Origin

Erect perennial herb; to 1.5 m tall; cultivated areas, fields, roadsides, and disturbed areas; native of Europe and western Africa.

Seedling Characteristics

Hypocotyl glabrous; cotyledons spatulate, pale green; first true leaves with glaucous underside and spine-tipped edges.

Mature Plant Characteristics

ROOTS fibrous from long, deep, vertical taproot; then forming spreading horizontal rhizomes; producing lateral roots from buds; to 6.0 dm deep. STEMS erect, 0.5–1.5 m tall, stout, large, hollow, milky sap, glaucous, gland-tipped trichomes on upper stem and peduncles. LEAVES alternate, 10.0–20.0 cm long, 3.0–8.0 cm wide, shiny green, sinuate pinnatifid with 2–5 lobes, spiny edges, bases sessile and clasping. INFLORESCENCES corymblike terminal clusters, yellow to golden, 14.0–22.0 mm tall, 3.0–5.0 cm wide. FRUITS achene, dark reddish brown, 2.0–3.5 mm long, 0.6–0.7 mm wide, oblong, ribbed; nutlet enclosing achene, easily dislodged pappus 10.0–14.0 mm long.

Special Identifying Features

Erect perennial; deep taproot; stem hollow and containing milky sap, horizontal branches producing new plants; leaf base clasping; flowers in yellow heads, 3.0–5.0 cm wide.

Toxic Properties

None reported, but can cause mechanical injury.

Spiny Sowthistle

Sonchus asper (L.) Hill · Asteraceae · Sunflower Family

Synonyms

Annual sowthistle, prickly sowthistle, spinyleaf sowthistle

Habit, Habitat, and Origin

Erect, cool-season annual herb; to 15.0 dm tall; cultivated areas, gardens, and roadsides; native of Europe.

Seedling Characteristics

Cotyledons roundish, grayish green; first true leaves form a rosette, blades oblong to ovate, delicate prickles around margins, pale green on underside.

Mature Plant Characteristics

ROOTS fibrous from short taproot. **STEMS** erect, 1.0–15.0 dm tall, coarse, hollow, milky sap, upper half of main stem often branched. **LEAVES** alternate, 0.6–3.0 dm long, 2.0–10.0 cm wide, lanceolate to oblanceolate, lower leaves deeply lobed, gradually reduced apically with irregularly toothed margins and sharp-pointed prickles, rounded earlike projections clasp stem at leaf base. **INFLORESCENCES** numerous small heads, 9.0–13.0 mm tall, 1.2–2.5 cm wide; ray florets bright yellow; bracts numerous, 9.0–13.0 mm long. **FRUITS** achene, 2.5–3.0 mm long, reddish brown, flat, 3–5-ribbed on each side, nutlet with achene and pappus crowned with tufts of long white trichomes.

Special Identifying Features

Erect annual; stems hollow, with milky sap; leaves alternate, deeply lobed with prickly spines along leaf margins; flowers bright yellow, borne in small heads.

Toxic Properties

None reported, but can cause mechanical injury.

Flowers

TOP **Seeds**
MIDDLE **Seedling**
BOTTOM **Flowering plant**

Annual Sowthistle

Sonchus oleraceus L. · Asteraceae · Sunflower Family

Synonyms

Colewort, common sowthistle, hare's lettuce, hare's thistle

Habit, Habitat, and Origin

Erect annual herb; to 2.0 m tall; cultivated and disturbed areas, roadsides; native of Europe.

Seedling Characteristics

Cotyledons smooth, oblong to ovate; first true leaves form a rosette, slightly lobed, bluish gray, paler on underside.

Mature Plant Characteristics

ROOTS fibrous from short taproot. STEMS erect, 0.1–2.0 m tall, glabrous, hollow, milky sap, fleshy, usually somewhat branched. LEAVES alternate, 6.0–30.0 cm long, pinnately lobed, occasionally toothed, margins with scarce to numerous prickles; all but lowermost leaves with prominent rounded, earlike lobes at base that become sharply pointed with age, leaves progressively smaller on upper portion of stem. INFLORESCENCES numerous small heads, 9.0–13.0 mm tall, 1.5–2.5 cm wide; ray florets pale yellow; bracts numerous, 9.0–13.0 mm long. FRUITS achene, 2.5–3.0 mm long, flat, 3–5-ribbed on each side, rugose-terminated, nutlet including achene and pappus with tufts of long white trichomes.

Special Identifying Features

Erect annual; stems glabrous, with milky sap; leaves pinnately lobed or toothed, with earlike lobes at base, margins weakly prickly; flowers borne in small heads, yellow.

Toxic Properties

None reported, but can cause mechanical injury.

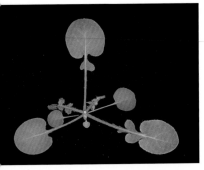

TOP Seed
MIDDLE Four-leaf seedling
BOTTOM Flowering plant

Flowers

Flower

Flowering plant

Western Salsify

Tragopogon dubius Scop. · Asteraceae · Sunflower Family

Synonyms

Goat's-beard, yellow goat's-beard, yellow salsify

Habit, Habitat, and Origin

Erect or ascending biennial herb; to 8.0 dm tall; fields, railroad beds, and roadsides; native of Europe.

Seedling Characteristics

Grasslike when young; producing a rosette; leaves pubescent.

Mature Plant Characteristics

ROOTS fibrous from deep, turnip-shaped taproot. **STEMS** erect or ascending, 3.0–8.0 dm tall, basally 3.0–9.0 mm thick, 1 or a few originating from near stem base, sparingly branched above, coarse and very tough; upper branches 5.0–15.0 cm long, leafless, containing milky sap. **LEAVES** alternate, to 30.0 cm long, lance-linear, grasslike, gradually tapering from base to a fine point at apex. **INFLORESCENCES** head, 8.0–12.0 cm diameter; ray florets perfect, fertile, 5-toothed, terminal; disk florets absent; bracts usually 13 per head, in 2 series, subequal, long, thin, tapering to a fine point, the longer ones always surpassing the rays; receptacle flat, rough, not chaffy. **FRUITS** achene, 25.0–36.0 mm long including beak, narrowly spindle-shaped, rough 5-ribbed basal body, outer pale brown, inner straw-colored; nutlet including achenes and persistent pappus of white, plumose, dorsiventrally flattened bristles.

Special Identifying Features

Erect or ascending biennial; branch ends in a long, hollow, cone-shaped peduncle capped by numerous lemon yellow flowers.

Toxic Properties

None reported.

TOP Seed
BOTTOM Seedling

Flowers

Crownbeard

Verbesina encelioides (Cav.) Benth. & Hook. f. ex Gray · Asteraceae · Sunflower Family

Synonyms

Golden crownbeard, skunk daisy, wild sunflower, yellowtop

Habit, Habitat, and Origin

Erect annual herb; to 1.7 m tall; open areas, disturbed and waste sites; native of the Americas.

Seedling Characteristics

Cotyledons obovate, glabrous; first true leaves opposite, ovate to deltoid-ovate, toothed margins, covered with trichomes.

Mature Plant Characteristics

ROOTS fibrous from taproot. STEMS erect, 0.3–1.7 m tall, densely gray pubescent. LEAVES lower opposite, upper alternate, 3.0–12.0 cm long, 3.0–10.0 cm wide, ovate or deltoid-ovate, sometimes narrowly so; margins coarsely serrate, especially near base; veins prominent with dense white pubescence on underside, fine gray pubescence on upper side; petioles as long as leaves. INFLORESCENCES globose head on pedicels; ray florets yellow, 1.5–2.5 cm long, 1.0–1.5 cm wide, 3–5-toothed or lobed at apex; disk florets yellow; bracts, 1.0–2.0 cm long, linear to lanceolate, pubescent. FRUITS achene, 4.0–6.0 mm long, 1.0–1.7 mm wide, black or dark brown, pubescent, awn 1.0–2.0 mm long; nutlet enclosing achene, subglobose to hemispheric, on long peduncle.

Special Identifying Features

Erect annual; stems and lower leaves covered with fine white pubescence. Can be differentiated from garden sunflower by the smaller flowers, typically shorter growth habit, and opposite leaves on lower plant parts.

Toxic Properties

Plants contain toxins causing edema related to cardiac insufficiency; consumption can kill livestock.

Western Ironweed

Vernonia baldwinii Torr. · Asteraceae · Sunflower Family

Synonyms

Baldwin's ironweed

Habit, Habitat, and Origin

Erect perennial herb; to 2.0 m tall; prairies and fields; native of North America.

Seedling Characteristics

Cotyledons spatulate; first true leaves alternate, resin dots apparent.

Mature Plant Characteristics

ROOTS fibrous from taproot. STEMS erect, 1.0–2.0 m tall, simple, 1 or more from base, branched at top. LEAVES alternate, 8.0–20.0 cm long, 2.0–6.0 cm wide, pinnately veined, elliptic to lanceolate, margins serrate, narrowed at base; glabrous to tomentose with crooked, slender trichomes, more below; dotted with sessile or nearly sessile resin globules. INFLORESCENCES heads in terminal flat-topped aggregates, usually 5.0–14.0 mm diameter, 18–34 aggregates per head; ray florets absent; disk florets 18–34, perfect, fertile; bracts closely imbricate, often purplish green, darker along margins and prominent narrow midvein, with a broad patch of resinous globules, spreading to recurved, margins pubescent. FRUITS achene, 3.0 mm long, 6–10-ribbed, with resin globules in furrows between ribs; nutlet with achene and pappus; pappus grayish to rusty white, persistent, double, outer scales less than 1.0 mm long, inner pappus of numerous coarse bristles to 6.0 mm long.

Special Identifying Features

Erect perennial; leaves, bracts, and achenes covered with resin dots; flowers purple.

Toxic Properties

None reported.

1mm

TOP Seed
BOTTOM Seedling

TOP Flowering plant
BOTTOM Flowers

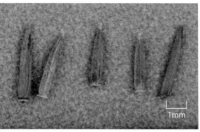

Flowers

Tall Ironweed

Vernonia gigantea (Walt.) Trel. · Asteraceae · Sunflower Family

Synonyms

Giant ironweed; *Vernonia altisima* Nutt.

Habit, Habitat, and Origin

Erect, warm-season perennial herb; to 3.0 m tall; damp soils in fields, pastures, roadsides, and disturbed areas; native of North America.

Seedling Characteristics

Cotyledons oblong; first leaves oblanceolate to obovate.

Mature Plant Characteristics

ROOTS from fibrous taproot and rhizomes. STEMS erect, 0.8–3.0 m tall, branched at top, trichomes minute, older stems dark red. LEAVES alternate, 6.0–30.0 cm long, 1.0–4.0 cm wide, lanceolate to oblanceolate, tapered at both ends, margins finely serrated, white midrib below, downy trichomes near midrib, glabrous above. INFLORESCENCES spreading, flat-topped head with 13–30 florets, at tips of branches; involucre short-cylindrial to bell-shaped, 3.0–7.0 mm long; disk florets reddish, purple-tipped, with 5 lobes; bracts 2.0–5.5 mm long, ovate to oblong-lanceolate, blunt-pointed, pressed close together, purplish-tinged to dark purple. FRUITS achene, 2.8–3.5 mm long, oblong to triangular, acute tip, grayish brown to brown; nutlet surrounding achene 3.0–4.0 mm long, cylindrical, with longitudinal grooves; pappus purple-white, tan, to purplish brown.

Special Identifying Features

Erect warm-season perennial; reproduces by seeds primarily; difficult to uproot, thus the name "ironweed"; leaves with trichomes on lower surface and prominent midrib; flowers red-purple.

Toxic Properties

No toxicity confirmed in North America, but some *Vernonia* species in South America and Africa produce toxic sequiterpene lactones that irritate skin.

Spiny Cocklebur

Xanthium spinosum L. · Asteraceae · Sunflower Family

Synonyms

Bathurst bur, daggerweed, Spanish thistle, spiny burweed, spiny clotbur, thorny burweed

Habit, Habitat, and Origin

Erect annual herb; to 1.2 m tall; damp, fertile, disturbed soils and waste areas; native of South America.

Seedling Characteristics

Cotyledons linear-lanceolate.

Mature Plant Characteristics

ROOTS fibrous from taproot. STEMS erect, 0.3–1.2 m tall, branching above, slender, pubescent, with stout yellow spines 1.0–2.0 cm long, 3 in each leaf axil. LEAVES alternate, 2.5–6.0 cm long, 5.0–25.0 mm wide, lance-shaped, irregularly toothed or lobed, downy pubescent below. INFLORESCENCES borne in separate male and female heads, both inconspicuous; male heads in axils of uppermost leaves; female heads in axils of lower leaves; ray florets absent. FRUITS achene within 2-chambered oblong bur covered with hooked, thornlike prickles and terminal beaks; prickles to 1.0 cm long, inconspicuous or apparent; achenes not usually separable.

1mm

Special Identifying Features

Erect annual; stems armed with stout yellow spines 1.0–2.0 cm long, 3 in each leaf axil; fruit a bur covered with thorny prickles.

Toxic Properties

Seedpods, seeds, and new sprouts contain a toxin fatal to swine and calves; spines on seedpods and stems can cause mechanical injury.

TOP **Fruit**
BOTTOM **Seedling**

TOP **Spines and fruit**
BOTTOM **Mature plant**

TOP **Fruit**
MIDDLE **Seedling**
BOTTOM **Flowers**

Mature plant

Common Cocklebur

Xanthium strumarium L. · Asteraceae · Sunflower Family

Synonyms

Heartleaf cocklebur, rough cocklebur

Habit, Habitat, and Origin

Erect, coarse, summer annual herb; to 2.0 m tall; open areas, cultivated fields, waste sites, streambeds, and floodplains; native of North America.

Seedling Characteristics

Hypocotyl stout, purple at base, green above; cotyledons lanceolate, waxy, smooth, thick, dark green upper surface, 1.5–4.2 cm long, 0.3–1.0 cm wide; first true leaves opposite.

Mature Plant Characteristics

ROOTS fibrous from freely branched taproot. STEMS erect, 0.2–2.0 m tall, green with maroon to black lesions or flecks; stout, roughened, dense, short, stiff, ascending pubescence and minute bumps. LEAVES alternate, 5.0–15.0 cm long and wide, simple, triangular-ovate, margins irregularly 3–5-lobed, pubescent, veins evident on upper surface, petiole 5.0–15.0 cm long. INFLORESCENCES heads terminal on short axillary branches; male heads rounded, at ends of axillary branches; female heads below male florets; involucre completely enclosing 2 florets. FRUITS achene, usually 2 per bur, 1.0–1.5 cm long, brown to black, oval to elliptic, upper larger than lower, surfaces with 5 longitudinal ridges; bur 1.0–3.5 cm long, oval to oblong, 2-chambered, covered with stout, hooked prickles, terminated by 2 incurved beaks, pubescent or glabrous.

Special Identifying Features

Erect, stout annual; cotyledons waxy, lanceolate; stems with maroon to black lesions; fruit a bur, conspicuously prickly, that clings readily to fur, hair, and clothing.

Toxic Properties

See comments under *Xanthium spinosum*.

Trumpetcreeper

Campsis radicans (L.) Seem. ex Bureau · Bignoniaceae · Bignonia Family

Synonyms

Cow-itch vine, trumpetvine

Habit, Habitat, and Origin

Trailing or climbing, woody, deciduous perennial vine; to 12.0 m long; roadsides, cultivated fields, stream banks, thickets, fencerows, and open woods; native of North America.

Seedling Characteristics

Leaves opposite; first pair of leaves simple, entire or coarsely toothed, smooth; second pair of leaves pinnately compound.

Mature Plant Characteristics

ROOTS fibrous from taproot; then forming woody, vigorously running, extensive rhizomes. STEMS trailing or climbing woody vine, to 12.0 m long, bark on younger stems smooth, light brown. LEAVES opposite, 3.0–15.0 cm long, 1.0–3.5 cm wide, pinnately compound, leaflet margins coarsely toothed. INFLORESCENCES in terminal clusters from stout pedicel; calyx 5-parted, 1.5–2.0 cm long, firm, tubular-campanulate, teeth lanceolate, 4.0–5.0 mm long; corolla 6.0–8.0 cm long, narrowly funnelform, red or red-orange, 5-lobed, much shorter than tube; fertile stamens 4. FRUITS capsule, 1.0–1.8 dm long, 2.0–3.0 cm diameter, fusiform. SEEDS broadly winged, about 15.0 mm long, several rows per pod.

Special Identifying Features

Perennial vine; leaves opposite, pinnately compound; flowers showy, red-orange, trumpet-shaped.

Toxic Properties

Plants may cause transient dermatitis.

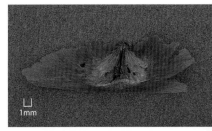

1mm

Flowers and fruit

TOP Seed
MIDDLE Seedling
BOTTOM Young plant

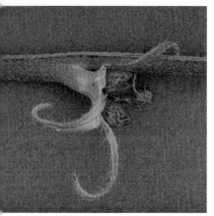

Flower

Corn Gromwell

Buglossoides arvense (L.) I. M. Johnston · Boraginaceae · Borage Family

Synonyms

Bastard alkanet, gromwell; *Lithospermum arvense* L.

Habit, Habitat, and Origin

Erect, winter annual or biennial herb, branching at base; 2.0–7.0 dm tall; fields, waste areas, and roadsides; native of Asia and Europe.

Seedling Characteristics

Cotyledons large, 20.0 mm long, 5.0–7.0 mm wide, stalked.

Mature Plant Characteristics

ROOTS fibrous or little-branched from taproot. STEMS erect, 2.0–7.0 dm tall, slender, simple or branching at base, minutely roughened, pubescent, leafy at top. LEAVES alternate, 1.0–3.0 cm long, 10.0–15.0 mm wide, simple, entire, sessile, lanceolate to linear, pubescent on both sides. INFLORESCENCES corolla white to cream, funnel-shaped, with 5 rounded spreading lobes, 5.0–7.0 mm long, finely pubescent; sepals distinct, 5.0–7.0 mm long, remaining intact and surrounding fruit. FRUITS formed as 4 small nutlets, 3.0–6.0 mm long, wrinkled, surrounded by intact sepals. SEEDS nutlet, 2.7–3.0 mm long, conical, rough, brown or grayish tan.

Special Identifying Features

Erect winter annual or biennial; troublesome winter weed, especially in wheat; plants bright green and pubescent; flowers small, white; fruit 4 small, erect nutlets in terminals of stems.

Toxic Properties

None reported.

Houndstongue

Cynoglossum officinale L. · Boraginaceae · Borage Family

Synonyms

Common hound's tongue

Habit, Habitat, and Origin

Erect to inclined biennial herb; to 1.2 m tall; fields and rocky waste areas; native of Europe.

Seedling Characteristics

Cotyledons linear to oblanceolate; leaves opposite, forming rosette, pubescent.

Mature Plant Characteristics

ROOTS fibrous from thick, woody taproot. **STEMS** erect to inclined, 0.3–1.2 m tall, herbaceous, pubescent, single-stemmed, leafy throughout. **LEAVES** alternate, 15.0–30.0 cm long, 2.0–5.0 cm wide, pubescent, margins entire; lower leaves in rosette, elliptic to oblanceolate base tapering into petiole; upper leaves lanceolate, 2.0–7.0 cm long, acute or acuminate tips, sessile and clasping.

INFLORESCENCES corolla dull reddish purple, 6.0–10.0 mm wide, subsessile and crowded at anthesis; calyx 4.0–6.0 mm long; sepals elliptic to ovate, rounded to blunt tips, hirsute. **FRUITS** mericarp, 5.0–8.0 mm long, flat on upper surface, splitting away at maturity but remaining attached to beaklike style, covered by prickles. **SEEDS** nutlet, 1–4 per fruit, 5.0–7.0 mm long, ovoid, flattened above with scar extending to near middle of ventral surface.

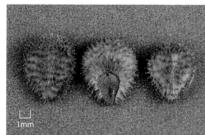

Special Identifying Features

Erect to inclined biennial; plants 2.0–4.0 dm tall; stem single; leafy and pubescent throughout; flowers reddish purple with musty odor; prickly fruits attach readily to fur, hair, and clothing.

Toxic Properties

Plants and seeds contain toxins causing liver disease.

TOP **Seeds**
MIDDLE **Seedling**
BOTTOM **Inflorescence**

Flowers

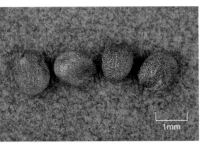

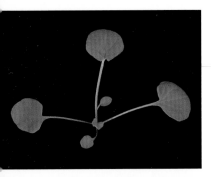

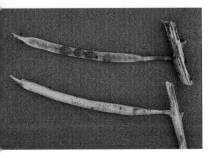

TOP **Seeds**
MIDDLE **Seedling**
BOTTOM **Fruit**

Flowering plant

Yellow Rocket

Barbarea vulgaris Ait. f. · Brassicaceae · Mustard Family

Synonyms

Common winter-cress, winter cress

Habit, Habitat, and Origin

Erect biennial or winter annual herb; to 9.0 dm tall; disturbed fields, roadsides, and railroad beds; native of Europe.

Seedling Characteristics

Cotyledons ovate, slightly notched at tip, on long stalks.

Mature Plant Characteristics

ROOTS fibrous from taproot. STEMS erect, 3.0–9.0 dm tall, green, glabrous, numerous, branched at top. LEAVES alternate, to 15.0 cm long, pinnately dissected, variable; basal leaves with large terminal lobe; lateral leaves with 4–8 lobes, becoming smaller and less lobed apically, petiole glabrous. INFLORESCENCES in elongated raceme, to 3.0 dm long; petals 7.0–10.0 mm long, bright yellow. FRUITS silique, 1.5–3.0 cm long, 1.5–2.0 mm wide, squarish in cross section, beak 1.0–2.0 mm long, borne on pedicel 3.0–6.0 mm long. SEEDS oblong to squarish, 1.0–1.5 mm long, light brown to gray.

Special Identifying Features

Erect biennial or winter annual; flowers bright yellow; fruit a silique, slender and slightly curved.

Toxic Properties

Plants cause digestive tract irritation.

Black Mustard

Brassica nigra (L.) W. J. D. Koch · Brassicaceae · Mustard Family

Synonyms

Shortpod mustard, true mustard

Habit, Habitat, and Origin

Erect, branching, winter annual herb; to 7.0 dm tall; fields and waste sites; native of southern Mediterranean region.

Seedling Characteristics

Cotyledons kidney-shaped, glabrous; first true leaves ovate to obovate, margins crenate, long petioles, glabrous or sparsely covered with trichomes.

Mature Plant Characteristics

ROOTS fibrous from taproot. STEMS erect, 1.0–7.0 dm tall, simple, branching, usually bristly below, glabrate or glabrous above. LEAVES alternate, 6.0–15.0 cm long, 2.0–6.0 cm wide, ovate to obovate, margins on lower leaves lobed, margins on upper leaves dentate, all petioled. INFLORESCENCES on elongate terminal raceme to 6.0 dm tall; petals 4, 8.0–10.0 mm wide, yellow; mature pedicels erect, 3.0–4.0 mm long. FRUITS erect or appressed, valves with stout midnerve, quadrangular, beak slender, 2.5–4.0 mm long. SEEDS round, 1.5–2.0 mm diameter, brown, minutely roughly reticulate.

Special Identifying Features

Erect winter annual; upper leaves not clasping stem; fruit erect or appressed, valves with prominent midnerve and appearing quadrangular.

Toxic Properties

Brassica species can cause nitrate and oxalate poisoning, and contain toxins causing anemia, acute respiratory distress syndrome, blindness, bloat, polioencephalomalacia, reproductive problems, and photosensitization.

1mm

TOP **Seeds**
BOTTOM **Seedling**

TOP **Flowers**
BOTTOM **Fruit**

TOP Seeds
MIDDLE Two-leaf seedling
BOTTOM Four-leaf seedling

Birdsrape Mustard

Brassica rapa L. · Brassicaceae · Mustard Family

Synonyms

Argentine rape, canola, colza, field mustard, Polish rape, rape, rapeseed, turnip; *Brassica napus* (L.) Koch

Habit, Habitat, and Origin

Erect winter or summer annual herb; to 8.0 dm tall; fields, pastures, and waste sites; introduced and cultivated as a commercial crop; native of Europe.

Seedling Characteristics

Cotyledons rounded to heart-shaped; leaves alternate, pubescent.

Mature Plant Characteristics

ROOTS fibrous from taproot. STEMS erect, 3.0–8.0 dm tall, usually branched at tip, sparsely pubescent. LEAVES alternate, 7.0–30.0 cm long, 3.0–10.0 cm wide; basal and lower leaves pinnately dissected, with 5–11 rounded segments; upper leaves entire to coarsely toothed, clasping, auriculate, progressively smaller apically. INFLORESCENCES elongate raceme, 4–20 branches; petals 4, 13.0–20.0 mm wide, obovate, yellow; bud clusters compact, below uppermost florets. FRUITS slender pod, 4.5–7.5 cm long, ascending, quadrangular, beak flattened. SEEDS round, 1.3–2.0 mm diameter, smooth; color variable from yellow to yellowish brown, dark brown, or bluish black.

Special Identifying Features

Erect winter or summer annual; leaves clasping stem; flower yellow, 4–20 floret branches per plant.

Toxic Properties

See comments under *Brassica nigra*.

Flowers

Lesser Swinecress

Coronopus didymus (L.) Sm. · Brassicaceae · Mustard Family

Synonyms

Swine wart cress; *Lepidium didymium* L.

Habit, Habitat, and Origin

Prostrate, procumbent, spreading or ascending, highly branched, annual or biennial herb; to 4.0 dm tall; cultivated areas, fields, roadsides, lawns, flowerbeds, and waste sites; native of Europe.

Seedling Characteristics

Cotyledons smooth, club-shaped; first true leaves simple and entire initially, becoming pinnatifid.

Mature Plant Characteristics

ROOTS fibrous from taproot. **STEMS** prostrate, procumbent, spreading, or ascending; 1.0–4.0 dm tall; much-branched, branches to 4.0 dm long or longer, glabrous or pubescent, trichomes retrorsely spreading. **LEAVES** basal rosette, alternate upward, 1.0–4.0 cm long, 0.5–1.5 mm wide, oblong in general outline, pinnatifid, the segments entire or with a few deep teeth, petiole short. **INFLORESCENCES** raceme, 1.0–3.0 cm long; petals minute, white, filiform; stamens 2; pedicels 2.0 mm long. **FRUITS** silique formed from pair of mericarps, 1.7 mm long, 3.0 mm wide, distended over seed, cordate base, notched at tip, coarsely wrinkled. **SEEDS** snail-like in outline, 1.0–1.4 mm long, flattened, wrinkled.

Special Identifying Features

Prostrate, procumbent, or spreading annual or biennial; leaves and stems produce pungent aromatic odor; seedpods distinctively notched.

Toxic Properties

Reported to taint milk.

TOP **Inflorescence**
BOTTOM **Flowers**

TOP **Fruit**
BOTTOM **Seedling**

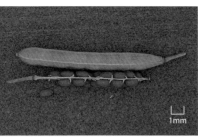

Pinnate Tansymustard

Descurainia pinnata (Walt.) Britt. · Brassicaceae · Mustard Family

Synonyms

Green tansy mustard, short-fruited tansy mustard

Habit, Habitat, and Origin

Erect spring or winter annual herb; to 8.0 dm tall; open dry areas, prairies, sparsely wooded areas, roadsides, fields, and waste sites; native of North America.

Seedling Characteristics

Hypocotyl reddish green to purple; cotyledons stalked, divided leaves with segments narrow to spatulate.

Mature Plant Characteristics

ROOTS fibrous from taproot. STEMS erect, 5.0–8.0 dm tall, simple to much-branched, densely pubescent, trichomes branched or glandular. LEAVES alternate, 1.0–7.0 cm long, 0.5–2.0 cm wide, greenish; lower leaves deeply two or three times divided with narrow to ovate segments, upper ones smaller and less divided. INFLORESCENCES elongate raceme, up to 3.0 dm long; petals 4, yellow, yellow-green, or white, broad or spatulate at tip; sepals 4, smaller or equal to petals; pedicels to 1.0 cm long, divergent. FRUITS slender silique, 5.0–7.0 mm long, 1.0–1.5 mm wide, clublike, with prominent midrib. SEEDS elliptic to oblong, 0.5–0.7 mm long, dark reddish brown, finely reticulate, produced in 2 rows.

Special Identifying Features

Erect spring or winter annual; stems with branched or glandular trichomes; leaves lacking glandular trichomes; fruit clublike with seeds in 2 rows.

Toxic Properties

Flowering plants contain neurotoxins that cause blind staggers, anorexia, weakness, tremors, and tongue paralysis in livestock.

TOP Seeds
BOTTOM Inflorescence

TOP Young plant
BOTTOM Flowering plant

Seedling

Flixweed

Descurainia sophia (L.) Webb. ex Prantl · Brassicaceae · Mustard Family

Synonyms

Herb sophia, tansy mustard

Habit, Habitat, and Origin

Erect winter or summer annual herb; to 8.0 dm tall; fields, pastures, and waste sites; native of Eurasia.

Seedling Characteristics

Hypocotyl with a few star-shaped trichomes; cotyledons pale green, slightly rough to touch, with star-shaped trichomes, oval to club-shaped.

Mature Plant Characteristics

ROOTS fibrous from taproot. STEMS erect, 4.0–8.0 dm tall, branched above, branches to 3.9 dm long, canescent, grayish green throughout, pubescent, trichomes branched. LEAVES alternate, 3.0–10.0 cm long, 1.5–5.0 cm wide; two or three times pinnately divided into narrow segments; divisions lanceolate, oblanceolate, or linear; tips acute, upper leaves usually narrower, grayish, stalked. INFLORESCENCES dense raceme; petals 4, 2.0–3.0 mm long, yellow; sepals 4, longer than petals; pedicels 1.5 cm long. FRUITS narrow linear silique, 15.0–25.0 mm long, 0.5–1.0 mm wide, ascending at right angle from stem, pedicels 6.0–11.0 mm long. SEEDS oblong-ellipsoid, 6.0–7.0 mm long, bright orange, 10–20 per row in each half of fruit.

Special Identifying Features

Erect winter or summer annual; stems and leaves grayish green with pubescence throughout, trichomes without glands, with aromatic odor; fruit pod ascending at right angles, seeds in a row.

Toxic Properties

Consumption can produce neurological effects, but this species is less likely to be consumed than *Descurainia pinnata*.

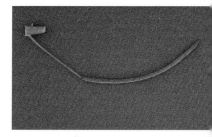

TOP **Seeds**
MIDDLE **Fruit**
BOTTOM **Flowering plant**

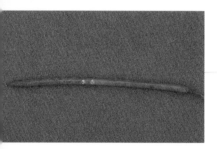

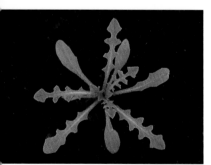

TOP Seeds
MIDDLE Fruit
BOTTOM Seedling

Flowers

Bushy Wallflower

Erysimum repandum L. · Brassicaceae · Mustard Family

Synonyms

Treacle mustard

Habit, Habitat, and Origin

Erect winter annual; to 4.0 dm tall; prairies, fields, roadsides, and waste sites; native of Eurasia.

Seedling Characteristics

Hypocotyl pale green; cotyledons small, spoon-shaped, tapering at base and slightly notched at top.

Mature Plant Characteristics

ROOTS fibrous from a shallow taproot. STEMS erect, 2.0–4.0 dm tall, spreading, rarely simple, branched; zigzagged at nodes, pubescent, trichomes forked. LEAVES alternate, 2.0–12.0 cm long, 2.0–12.0 mm wide, linear to narrowly oblanceolate; margins entire, dentate, or wavy; attenuate, pubescent, trichomes forked. INFLORESCENCES raceme elongating at maturity; corolla 8.0–12.0 mm long, pale yellow, pubescent with numerous star-shaped trichomes, 4.0–5.5 mm long. FRUITS silique, 6.0–12.0 cm long, slightly quadrangular, widely spreading, pubescent, trichomes forked, stalk short and thick. SEEDS oblong to narrowly ovoid, 1.1–1.4 mm long, orangish to reddish brown, grooved, often winged.

Special Identifying Features

Erect winter annual; stem zigzagged at nodes; flowers yellow with 4 sepals; fruit widely spreading.

Toxic Properties

None reported.

Greenflower Pepperweed

Lepidium densiflorum Schrad. · Brassicaceae · Mustard Family

Synonyms

Green pepperweed, miner's cress, prairie pepperweed, wild tongue

Habit, Habitat, and Origin

Erect annual or sometimes biennial herb; to 7.0 dm tall; dry, sandy fields and waste sites; native of Asia.

Seedling Characteristics

Cotyledons incumbent, petiolate, elliptical-oblong; first true leaves entire, petiolate, oblong-lanceolate; later leaves slightly dentate.

Flowering plant

Mature Plant Characteristics

ROOTS fibrous from slender taproot. **STEMS** erect, 2.0–7.0 dm tall, round, usually single from base and branching above; gray-green. **LEAVES** basal rosette, alternate above, 1.0–10.0 cm long, simple, long-petiolate, oblong-lanceolate, margins deeply serrated; middle leaves with short petioles, sometimes nearly clasping stem, oblong-lanceolate, serrated, 2.0–3.0 cm long; upper leaves cauline, sessile, linear-lanceolate, 1.0–2.0 cm long, margins slightly serrated and ciliate basally. **INFLORESCENCES** narrow racemes, ascending, dense, to 1.5 dm tall, florets whitish or greenish; petals 4, rudimentary or none; sepals 4, ovate. **FRUITS** silicle, 2.0–4.0 mm long, oval to obovate, compressed, narrowly winged, notched, 2-seeded, 9–15 silicles per centimeter of raceme. **SEEDS** ovoid, 1.0–2.0 mm long, compressed, yellow-brown.

Special Identifying Features

Erect annual or sometimes biennial; cotyledons incumbent; flower petals rudimentary, racemes dense; fruit silicles round-obcordate to obovate and narrowly winged at apex; plant nearly odorless.

Toxic Properties

None reported.

1mm

TOP Seeds and fruit
MIDDLE Four-leaf seedling
BOTTOM Inflorescence

Flowers and fruit

Virginia Pepperweed

Lepidium virginicum L. · Brassicaceae · Mustard Family

Synonyms

Bird's pepper, pepper-grass, poorman's pepper, tongue-grass

Habit, Habitat, and Origin

Erect annual or winter annual herb; to 9.0 dm tall; cultivated areas, fields, pastures, gardens, lawns, roadsides, and waste sites; native of North America.

Seedling Characteristics

Hypocotyl short, not noticeable after second leaf stage; cotyledons with long-stalked, toothed or lobed, oval blades unequal in size.

Mature Plant Characteristics

ROOTS fibrous from slender taproot. **STEMS** erect, 3.0–9.0 dm tall, green, smooth to scantily pubescent, much-branched. **LEAVES** basal rosette, alternate above, 2.0–10.0 cm long, 5.0–20.0 mm wide; rosette margins deeply double-toothed to -lobed or deeply dissected, withering when plant is mature; leaves on stem linear to lanceolate, margins toothed or entire, glabrous. **INFLORES-CENCES** elongated raceme; corolla 4-petaled, white, small, or absent; stamens 2–4. **FRUITS** silique, 2.0–3.5 mm long, orbicular, flattened, notched at top with a short style protruding at apex. **SEEDS** oval, 1.7–2.0 mm long, orangish yellow to reddish brown, flattened, minutely pitted, 2 per fruit.

Special Identifying Features

Erect annual or winter annual; stems much-branched; flowers small in crowded racemes; fruit a nearly circular silique.

Toxic Properties

Reported to cause digestive tract irritation.

Wild Radish

Raphanus raphanistrum L. · Brassicaceae · Mustard Family

Synonyms

Jointed charlock, jointed radish, jointed wild radish

Habit, Habitat, and Origin

Erect winter annual; to 8.0 dm tall; cultivated areas, fields, pastures, and roadsides; native of Eurasia.

Seedling Characteristics

Cotyledons rounded with a distinct notch at apex; first true leaves irregularly lobed, sparsely hispid.

Mature Plant Characteristics

ROOTS fibrous from stout taproot. STEMS erect, 4.0–8.0 dm tall, coarse, stout, freely branched, pubescent; trichomes short, stiff, retrorse near base. LEAVES basal rosette, alternate above, 1.0–22.0 cm long, 1.0–5.0 cm wide; lower leaves obovate-oblong in outline, lyrate or pinnatifid into 5–15 segments, lateral lobes 1–4 on each side of midvein, oblong or ovate and smaller than terminal lobe, margins dentate, apex acute or obtuse; upper leaves much smaller, margins entire or dentate, undivided or occasionally few-lobed. INFLORESCENCES raceme to 4.0 dm long; petals 5.0–15.0 mm long, yellow becoming white with age; pedicels 8.0–15.0 mm long. FRUITS silique, body 2.0–5.0 cm long, beak 1.0–3.0 cm long, nearly cylindrical when fresh, prominently several-ribbed and constricted between seeds at maturity, 4–10 seeds per pod. SEEDS ovoid, 2.0–3.0 mm long, brown.

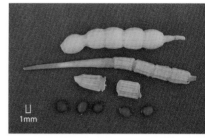

Special Identifying Features

Erect winter annual; stems coarse, stout; flowers yellow becoming white with age; fruit constricted between seeds in seedpod.

Toxic Properties

Seeds reported to contain toxins causing digestive tract irritation, colic, and diarrhea.

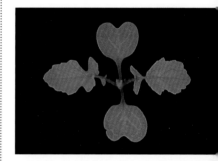

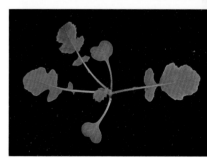

TOP Seeds and fruit
MIDDLE Two-leaf seedling
BOTTOM Four-leaf seedling

Flowers

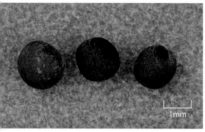

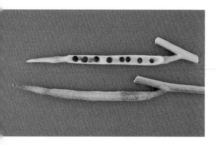

TOP **Seeds**
MIDDLE **Fruit**
BOTTOM **Seedling**

Wild Mustard

Sinapis arvensis L. · Brassicaceae · Mustard Family

Synonyms

Brassica kaber (DC.) L. C. Wheeler, *B. arvensis* Rabenh.

Habit, Habitat, and Origin

Erect winter or summer annual herb; to 8.0 dm tall; disturbed areas, fields, pastures, and waste sites; native of Europe.

Seedling Characteristics

Cotyledons rounded to heart-shaped; true leaves alternate, pubescent.

Mature Plant Characteristics

ROOTS fibrous from taproot. STEMS erect, 4.0–8.0 dm tall, usually branched at tip, sparsely pubescent. LEAVES alternate, 3.0–25.0 cm long, 1.5–7.0 cm wide, obovate, lanceolate, lyrate-pinnatifid or undivided; lower leaves lobed, margins coarsely toothed; upper leaves oblong to ovate, progressively smaller apically, sessile, clasping; petiole 1.0–7.0 cm or lacking apically, hispid. INFLORESCENCES on ascending or suberect stout pedicels; petals 4, 0.8–1.7 cm long, 3.0–7.5 mm wide, bright or pale yellow, obovate; filaments 3.0–6.0 mm long; anthers 1.2–1.5 mm long, oblong; sepals 4.5–7.0 mm long, 1.0–1.8 mm wide, yellow or green, narrowly oblong, spreading or reflexed. FRUITS silique, 2.5–4.5 cm long, slender, ascending, almost upright, beak at tip, flattened, quadrangular. SEEDS oblong, 1.5 mm wide, smooth, blackish blue to dark brown.

Special Identifying Features

Erect, stout winter or summer annual; flowers in conspicuous clusters at branch terminals, yellow; fruit cylindrical on long stalk.

Toxic Properties

Plants contain toxins causing digestive tract irritation.

Flower

Flower

1mm

Tumble Mustard

Sisymbrium altissimum L. · Brassicaceae · Mustard Family

Synonyms

Jim Hill mustard, tall hedge mustard, tall rocket

Habit, Habitat, and Origin

Erect winter annual or biennial herb; to 1.5 m tall; fields, roadsides, and waste sites; native of Europe.

Seedling Characteristics

Rosette leaves deeply lobed, resembling basal leaves of mature plant.

Mature Plant Characteristics

ROOTS fibrous from slender taproot. STEMS erect, 0.5–1.5 m tall, simple below, much-branched above, bushy in appearance, pubescent toward base; trichomes simple, spreading. LEAVES basal rosette, alternate above, 2.0–16.0 cm long, 1.0–4.0 cm wide, lower leaves deeply pinnately lobed, gradually changing upward to linear filiform segments.

INFLORESCENCES loosely racemose; petals 6.0–9.0 mm long, pale yellow, longer than sepals. FRUITS silique, 5.0–10.0 cm long, 1.0–1.5 mm wide, round to quandrangular in cross section, straight, long-linear, extending at same angle as pedicel, glabrous, beak 2.0–5.0 mm long. SEEDS oblong, 1.0 mm long, numerous, yellow to olive brown, finely reticulate.

Special Identifying Features

Erect annual or biennial with bushy appearance; leaves deeply lobed; flowers small, light yellow; at maturity whole plant breaks off at base and tumbles with the wind.

Toxic Properties

Plants contain toxins that cause digestive tract irritation.

TOP **Seeds**
MIDDLE **Seedling**
BOTTOM **Flowering plant**

Four-leaf seedling

London Rocket

Sisymbrium irio L. · Brassicaceae · Mustard Family

Synonyms

London hedgemustard

Habit, Habitat, and Origin

Erect annual or biennial herb; to 6.0 dm tall; fields, roadsides, and waste sites; native of Europe.

Seedling Characteristics

Hypocotyl pale green; cotyledons stalked, blades broadly elliptic to oblong, margins toothed or entire.

Mature Plant Characteristics

ROOTS fibrous from taproot. STEMS erect, 1.0–6.0 dm tall, branched, spreading, glabrous or pubescent; trichomes simple, sparse to dense. LEAVES alternate, 1.5–16.0 cm long, 0.5–9.0 cm wide, lower leaves deeply pinnately divided, upper ones fewer and smaller; lateral lobes 1–8 on each side of midvein, oblong to ovate, smaller than terminal lobe; petiole 0.5–6.0 cm long. INFLORESCENCES in raceme; pedicels divaricate or ascending, slender, much narrower than flower or fruit; petals 3.0–4.0 mm long, yellow, obovate to spatulate, clawlike, slightly exceeding sepals; sepals 2.0–3.0 mm long, oblong, erect. FRUITS silique, 2.5–4.0 mm long, 0.9–1.1 mm wide, slender, rounded to quandrangular in cross section, tipped with persistent style, seeds in 1 row. SEEDS oblong, 0.8–1.0 mm long, 0.5–0.6 mm wide, plump, wingless, smoothish.

Special Identifying Features

Erect annual or biennial; fruits slender, widely spreading, projecting above flowers.

Toxic Properties

See comments under *Sisymbrium altissimum*.

TOP Seeds
MIDDLE Two-leaf seedling
BOTTOM Flowering plant

Field Pennycress

Thlaspi arvense L. · Brassicaceae · Mustard Family

Synonyms

Bastard cress, fan-weed, Frenchweed, field penny-cress, field thlaspi, stinkweed

Habit, Habitat, and Origin

Erect annual or winter annual herb; to 8.0 dm tall; pastures, roadsides, fields, and open disturbed areas; native of Eurasia.

Seedling Characteristics

Hypocotyl long-stalked, pubescent; cotyledons bluish green, stalked, ovate to oblong, with prominent midvein, blades slightly toothed.

Mature Plant Characteristics

ROOTS fibrous from taproot. **STEMS** erect, 4.0–8.0 dm tall, smooth, simple or much-branched. **LEAVES** basal rosette, alternate above, 2.0–10.0 cm long, 0.3–3.0 cm wide; basal leaves lanceolate to oblanceolate, stalked, soon withering; middle and upper leaves smooth, simple, margins toothed or entire, sessile, clasping. **INFLORESCENCES** elongated raceme, at ends of branches; corolla 3.0–4.0 mm long; petals 4, oblong to spatulate, in shape of cross, white; sepals 4, small, ascending. **FRUITS** silique, 8.0–12.0 mm wide, circular or rounded-oblong, short style persisting in notch of broadly winged fruit. **SEEDS** ovoid, 2.0–2.3 mm long, dark reddish brown, with many concentric granular ridges on each side.

Special Identifying Features

Erect annual or winter annual; leaves smooth, clasping stem; fruit roundish-oblong, notched at top; plants emit distinct odor when crushed.

Toxic Properties

Plants contain toxins causing digestive tract irritation, colic, diarrhea, photosensitization, and blood in urine.

TOP **Seeds and fruit**
BOTTOM **Seedling**

TOP **Flowers**
BOTTOM **Inflorescence**

Spreading Pricklypear

Opuntia humifusa Raf. · Cactaceae · Cactus Family

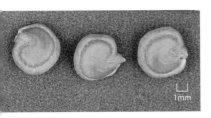

Synonyms

Devil's-tongue, eastern pricklypear, low pricklypear, pricklypear, smooth pricklypear

Habit, Habitat, and Origin

Mat-forming, succulent perennial; to 4.0 dm tall and to 1.0 m wide; associated with dry sandy or rocky soils, or bedrock such as sandstone or chalk; native of North America.

Seedling Characteristics

Cotyledons pointed, very fleshy; first spines thin; first stems cylindrical, becoming broader with age.

Mature Plant Characteristics

ROOTS fibrous from creeping stems; detached stem segments root easily. STEMS mat-forming, 4.0 dm tall, branches to 1.0 m long; segments (sometimes called cladodes) flattened, ellipsoid, ovoid, or oblanceolate, 2.5–20.0 cm long, 2.5–10.0 cm wide, glochidiate. LEAVES small, inconspicuous, conical, succulent, present only during early stem development. INFLORESCENCES solitary, 5.0–7.0 cm diameter; corolla yellow, sometimes with red center, symmetrical; petals obovate to obcordate. FRUITS berry, 1.5–5.0 cm long, 1.5–3.0 cm diameter, green, sessile, reddish brown to purple at maturity. SEEDS dark, 0.2–0.5 cm diameter.

Special Identifying Features

Mat-forming perennial; flattened succulent stems covered with spines, to 1 per node; flowers single and bright yellow, sometimes with red center.

Toxic Properties

Spine punctures can lead to infection.

Flower

Flower

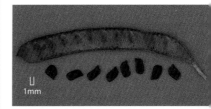

Partridgepea

Chamaecrista fasciculata (Michx.) Greene · Caesalpiniaceae · Caesalpinia Family

Synonyms

Cassia chamaecrista L., *Cassia fasciculata* Michx.

Habit, Habitat, and Origin

Erect annual herb; to 1.2 m tall; fields, woodland edges, pastures, and disturbed sites; native of North America.

Seedling Characteristics

Cotyledons large, rounded; first leaf pinnately compound, glabrous.

Mature Plant Characteristics

ROOTS fibrous from taproot. **STEMS** erect, 0.4–1.2 m tall, slender, few to diffusely branched, glabrous or pubescent; trichomes incurved or villous, to 2.0 mm long. **LEAVES** alternate, pinnately compound, usually 8–36 leaflet pairs; leaflets linear-oblong, 1.0–2.5 cm long, 2.0–6.0 mm wide; petiole with gland below first leaflet pair, stipule persistent, leaves fold when touched. **INFLORESCENCES** solitary on 2–5-flowered raceme; petals 5, 10.0–20.0 mm long, yellow; stamens 10, not equal; pedicels 10.0–25.0 mm long. **FRUITS** legume, 2.5–7.0 cm long, 4.0–6.0 mm wide, erect, linear-oblong, variously glabrous or pubescent. **SEEDS** 4–20 per fruit, 3.5–5.0 mm long, flattened, oval or somewhat triangular in outline, smooth, lustrous.

Special Identifying Features

Erect annual; leaves pinnately compound; flowers large, stamens 10; sensitive to touch.

Toxic Properties

Plants contain toxins causing mild to moderate digestive tract irritation, colic, and diarrhea.

TOP Seeds and pod
MIDDLE Seedling
BOTTOM Flowering plants

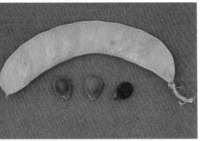

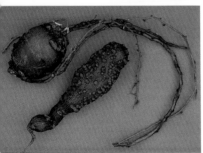

Flower

Hogpotato

Hoffmanseggia glauca (Ortega) Eifert · Caesalpiniaceae · Caesalpinia Family

Synonyms

Indian rushpea, pignut; *Hoffmanseggia densiflora* Benth.

Habit, Habitat, and Origin

Low-growing perennial herb; seldom more than 30.0 cm tall; clay and alkaline soil in open areas, old fields, and floodplains; native of southwestern United States.

Seedling Characteristics

Cotyledons small, green, glabrous, ovate in outline; first true leaves compound; new growth emerging from rootstock often ruby red.

Mature Plant Characteristics

ROOTS fibrous from deep taproot, expanding with tuberous or nutlike propagules in top 30.0 cm of soil. STEMS weakly erect to procumbent, seldom more than 30.0 cm tall, often covered with soft pubescence, herbaceous but may become slightly woody. LEAVES alternate, 7.0–12.0 cm long, bipinnately compound; leaflets 13–20, linear to broadly linear, four times longer than wide; stipules ovate, persistent, papery, and covered with fine trichomes. INFLORESCENCES irregular terminal racemes; petals yellow, distinctly separate, lower portion with many red glands. FRUITS legume, 2.0–4.0 cm long, 4.0–7.0 mm wide, compressed, curved, nerves prominent. SEEDS oval in outline but pointed on one end, 3.0–5.0 mm long, brown, gray, occasionally black, glabrous or sometimes slightly pubescent.

Special Identifying Features

Low-growing perennial herb often found in isolated patches; flowers many, showy, yellow.

Toxic Properties

None reported.

Sicklepod

Senna obtusifolia (L.) H. S. Irwin & Barneby · Caesalpiniaceae · Caesalpinia Family

Synonyms

Coffeebean; *Cassia obtusifolia* L.

Habit, Habitat, and Origin

Erect annual herb; to 2.0 m tall; cultivated areas, fields, pastures, roadsides, and waste sites; native of American tropics.

Seedling Characteristics

Cotyledons rounded, with 3–5 distinct veins; leaflets on young plants may develop wrinkles.

Mature Plant Characteristics

ROOTS fibrous from taproot. STEMS erect, 0.3–2.0 m tall, branched, green, round, glabrous. LEAVES alternate, pinnately compound; leaflets 4–6, most commonly 6, 2.0–7.0 cm long, terminal pair largest, basal pair smallest, wider at apex, gland between or just above petiole of longest pair of leaflets, stipules deciduous. INFLORESCENCES usually 1 or 2 borne in leaf axils; petals 0.8–2.0 cm long, yellow, showy; sepals unequal, 4.0–10.0 mm long, 2.0–5.0 mm wide. FRUITS legume, 8.0–20.0 cm long, 3.0–5.0 mm wide, slender, curved, somewhat rounded, glabrous, tetragonal appearance due to seed shape. SEEDS angular, 4.0–5.0 mm long, brownish, shiny.

Special Identifying Features

Erect annual; leaves pinnately compound, terminal pair of leaflets largest; seed angular; leaves and stems with distinct odor when crushed.

Toxic Properties

Plants contain toxins causing digestive tract irritation and cardiac muscle degeneration.

TOP Flowers
BOTTOM Flowers and fruit

1mm

TOP Seeds
BOTTOM Seedling

Coffee Senna

Senna occidentalis (L.) Link · Caesalpiniaceae · Caesalpinia Family

Synonyms

Bricho; *Cassia occidentalis* L.

Habit, Habitat, and Origin

Erect annual herb; up to 3.0 m tall; cultivated areas, fields, pastures, road-sides, and waste sites; native of the tropics.

Seedling Characteristics

Cotyledon rounded; glabrous above, white trichomes below distinguishing from sicklepod; veins prominent.

Mature Plant Characteristics

ROOTS fibrous from taproot. STEMS erect, 1.0–3.0 m tall, glabrous, furrowed at maturity. LEAVES alternate, 20.0–30.0 cm long, pinnately compound; leaflets 8–12, 3.0–7.0 cm long, 1.5–3.0 cm wide, ovate to ovate-lanceolate, acute to acuminate, terminal pair largest, basal pair smallest, margins entire, petiole 3.0–5.0 cm long; stipules 4.0–6.0 mm long, lance-shaped, entire; glabrous, deciduous, spherical gland near base of petiole in axils. INFLO-RESCENCES solitary in terminal or in axils of upper leaves or borne in few-flowered racemes; petals 1.0–2.0 cm long, yellow, perfect; stamens 6–7, two distinct sizes; sepals 6.0–9.0 mm long. FRUITS legume, 8.0–14.0 cm long, 5.0–9.0 mm wide, linear, straight to slightly curved, com-pressed, glabrous or minutely pubescent. SEEDS elliptical, 4.0–5.0 mm long, dull brown to brown, compressed.

Special Identifying Features

Erect annual herb; leaves pinnately compound, spherical gland near base of petiole in axils.

Toxic Properties

See comments under *Senna obtusifolia*.

Flowers

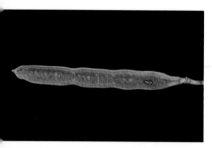

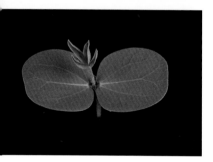

TOP Seeds
MIDDLE Fruit
BOTTOM Seedling

Flower

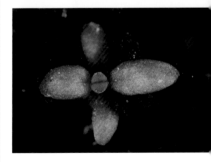

Gooseweed

Sphenoclea zeylanica Gaertn. · Campanulaceae · Bellflower Family

Synonyms

Chicken-spike

Habit, Habitat, and Origin

Erect annual or perennial herb; to 1.0 m tall; open shallow water on edges of ponds and streams and in rice fields; native of Asia.

Seedling Characteristics

Cotyledons spatulate; first true leaves appearing opposite initially, glabrous.

Mature Plant Characteristics

ROOTS fibrous and fleshy with aerenchyma when growing in water. STEMS erect, 0.2–1.0 m tall, branched above base, coarse, hollow, glabrous. LEAVES alternate, 4.0–12.0 cm long, 2.0–5.0 cm wide, variable in size, elliptic, acute, margins entire, base cuneate, petiole 5.0–20.0 mm long. INFLORESCENCES dense cylindrical terminal spike, 1.5–10.0 cm long, 4.0–12.0 mm wide; corolla 5-lobed, sessile, petals yellowish to whitish; stamens 5; calyx 5-lobed; sepals 5.0 mm long, green. FRUITS capsule, 2-locular, dehiscing circumscissilely, containing numerous seeds. SEEDS minute, oblong, 0.4–0.5 mm long, surface minutely roughened, somewhat lustrous.

Special Identifying Features

Erect annual or perennial; stems hollow; flower spike dense, terminal, and cylindrical; plants growing in shallow water.

Toxic Properties

None reported.

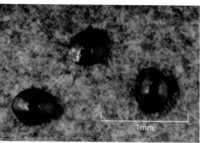

TOP Seeds
MIDDLE Lower flower
BOTTOM Flower

Common Venus's Lookingglass

Triodanis perfoliata (L.) Nieuwl. var. *perfoliata* · Campanulaceae · Bellflower Family

Synonyms

Clasping bellwort, clasping Venus's looking-glass, round-leaved triodanis; *Specularia perfoliata* (L.) DC.

Habit, Habitat, and Origin

Erect to ascending annual; to 1.0 m tall; dry sandy or gravelly soils, prairies, pastures, rangeland, woodlands, stream banks, turf, railroad beds, and roadsides; native of North America.

Seedling Characteristics

Cotyledons broadly ovate to orbicular, apex slightly notched; first leaves opposite, broadly elliptic to ovate, sessile or short-petiolate, apex rounded or bluntly pointed.

Mature Plant Characteristics

ROOTS fibrous. STEMS erect to ascending, 0.1–1.0 m tall, simple or branched from base, pentangular, glabrous or short trichomes along angles. LEAVES opposite initially, broadly elliptic to ovate, sessile or short-petiolate, apex rounded or bluntly pointed; stem leaves alternate, 5.0–25.0 mm long, 4.0–27.0 mm wide, broadly ovate to kidney-shaped, sessile, base cordate, clasping stem, apex rounded to bluntly pointed, margins finely to sharply toothed, upper surface glabrous, lower surface scabrous or short soft trichomes. INFLORESCENCES axillary, sessile, dioecious, calyx tube somewhat inflated; lower flowers 1–3 per node, remain closed; upper flowers 1 per node, open, 8.0–13.0 mm long; corolla 5-lobed, showy, purplish blue to purple, rarely white. FRUITS capsule, 3.5–10.0 mm long, ovoid, 2–3 locules, glabrous or trichomes on veins, opening by pores near or below middle, numerous seeds. SEEDS elliptic to broadly oblong-elliptic, 0.4–0.7 mm long, slightly flattened or biconvex, minutely wrinkled to smooth, brown, dull or shiny.

Special Identifying Features

Erect to ascending pentangular stems; stem leaves sessile, bases cordate, clasping stem, broadly ovate, margins toothed; male and female flowers on different plants.

Toxic Properties

None reported.

Inflorescence

Leaf

Marijuana

Cannabis sativa L. · Cannabaceae · Hemp Family

Synonyms

Hemp

Habit, Habitat, and Origin

Erect annual herb; to 3.0 m tall; exposed sites, floodplains, pastures, roadsides, and waste areas; possibly native of Asia.

Seedling Characteristics

Cotyledons oval, glabrous; first leaves simple, opposite, margins serrate, stem pubescent.

Mature Plant Characteristics

ROOTS fibrous from highly branched taproot. STEMS erect, 0.5–3.0 m tall, much-branched, coarse, somewhat grooved, rough; pubescent, trichomes exuding sticky, odorous resin; inner bark fibrous. LEAVES opposite below, alternate above, palmately divided; leaflets 5–9, 4.0–15.0 cm long, 3.0–20.0 mm wide, middle leaflet longest and outer progressively shorter, apex acute, margins serrate, pubescent. INFLORESCENCES male

and female flowers on separate plants, staminate plants taller and less robust than female plants; male florets in axillary racemes, sepals 2.5–4.0 mm long, ovate to lanceolate, pedicels 0.5–3.0 mm long; female florets in spikelike clusters in leaf axils, without petals, enclosed by glandular bracteole, subtended by bract. FRUITS achene and persistent perianth, 4.0 mm long, ovoid to nearly round with obtuse edges, persistent perianth yellow to olive brown. SEEDS achene, 3.0 mm long, oval, somewhat compressed, mottled brown or purple.

Special Identifying Features

Erect annual herb; leaves palmately divided; male and female flowers on separate plants; characteristic odor.

Toxic Properties

Plant contains neurotoxins that impair reaction time, motor coordination, and visual perception.

TOP **Seeds**
MIDDLE **Seedling**
BOTTOM **Male flowers**

Flowers

TOP Seeds
MIDDLE Seedling
BOTTOM Inflorescence

African Spiderflower

Cleome gynandra L. · Capparaceae · Caper Family

Synonyms
Skunkweed

Habit, Habitat, and Origin
Erect, warm-season annual herb; 6.0 dm tall; open areas, woodland edges, and disturbed sites; native of tropical Africa.

Seedling Characteristics
Cotyledons oblong; first true leaves opposite, palmately compound.

Mature Plant Characteristics
ROOTS fibrous from weak taproot. STEMS erect, 4.0–6.0 dm tall, simple to branched, pubescent, trichomes simple or glandular, scattered. LEAVES alternate, attenuate, palmately compound; 3–5 leaflets, leaflet blades 2.0–8.0 cm long, obovate, oval-elliptic, or ovate, finely toothed, petiole 3.0–9.0 cm long. IN-FLORESCENCES arranged in open raceme; petals 4, pink to lavender or white, subtended by compound bracts, 3 obtuse divisions; stamens 6, blade 0.8–1.3 cm long, claw 3.0–5.0 mm long; filaments 1.5–2.5 cm long; pistil stipe 1.0–2.0 cm long. FRUITS linear capsule, 5.0–10.0 cm long, 3.0–5.0 mm wide, glabrous. SEEDS globose, 1.5–1.8 mm diameter, brown, reticulate, wrinkled.

Special Identifying Features
Erect warm-season annual; leaves compound, leaflets 3–5; leaves and stem with pungent odor; flowers pink, lavender, or white.

Toxic Properties
Plants reported to contain nitrate and to accumulate selenium.

Japanese Honeysuckle

Lonicera japonica Thunb. · Caprifoliaceae · Honeysuckle Family

Synonyms

None

Habit, Habitat, and Origin

Climbing, twining, or trailing woody perennial vine; stems to several meters

long; covering ground and bushes; native of eastern Asia.

Seedling Characteristics

Trailing or climbing; young stems densely pubescent, green to red, becoming gray and brittle with age, freely rooting at nodes and spreading by rhizomes.

Mature Plant Characteristics

ROOTS fibrous from taproot initially, then from extensive rhizomes. STEMS climbing, twining, or trailing vine; to several meters long, slender initially, pubescent, green or light reddish brown; developing scaly thin bark, hollow pith, older bark peels and shreds into long strips; tan or gray, rooting at nodes when in contact with soil. LEAVES opposite, 4.0–8.0 cm long, ovate to oblong, mostly evergreen, margins sometimes dentate or lobed, smooth or pubescent, base rounded, short-petioled. INFLORESCENCES in pairs, white fading to yellow, subtended by 2 leaflike sepals; corolla 2.0–10.0 mm long, 2-lipped, 3 short lobes on upper lip, lower lip narrow and unlobed, stamens long, curved; style exerted from corolla. FRUITS berry, 5.0–6.0 mm diameter, spherical or nearly so, glossy black. SEEDS oblong, 3.0–3.3 mm long, black.

Special Identifying Features

Climbing, twining, or trailing woody perennial vine; flowers in pairs, whitish, very fragrant.

Toxic Properties

Berries contain tetraterpenoids that cause nausea, vomiting, and diarrhea.

TOP **Flowers**
BOTTOM **Shoot**

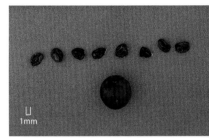

TOP **Seeds and fruit**
BOTTOM **Fruiting branch**

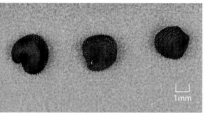

Corn Cockle

Agrostemma githago L. · Caryophyllaceae · Pink Family

Synonyms

Common corncockle, corn campion, corn rose, purple cockle

Habit, Habitat, and Origin

Erect, cool-season annual or biennial herb; to 1.0 m tall; grain fields, especially winter wheat, roadsides, fencerows, and waste sites; more common in the northern part of its range, occasionally cultivated in flower gardens; native of southern Europe.

Seedling Characteristics

Cotyledons broadly ovate to oblong; first true leaves initially compact, rosettelike before stem begins to elongate.

Mature Plant Characteristics

ROOTS fibrous from taproot. STEMS erect, 0.2–1.0 m tall, branching, often quadrangular, pubescent; trichomes soft, whitish. LEAVES opposite, 4.0–12.0 cm long, 5.0–10.0 mm wide, simple, sessile, entire, linear to lanceolate, acute tips, pubescent; trichomes soft, whitish. INFLORESCENCES solitary at end of branches; corolla 5-parted, regular, perfect; petals 2.4–3.6 cm long and pink, magenta, or purple; calyx swollen, 5.0–6.0 cm long, pubescent; pedicels 4.0–20.0 cm long. FRUITS capsule, 14.0–22.0 mm long, ascending, broadly ovate, 1-celled, dehiscent by 5 teeth. SEEDS tuberculate, 3.0–3.5 mm long, black, numerous per fruit.

Special Identifying Features

Erect annual or biennial; leaves opposite, linear, covered with long, silky, whitish pubescence; flowers showy, solitary at ends of branches, pink to purple, calyx swollen and pubescent, pedicels long.

Toxic Properties

Plants and seeds contain steroidal compounds that cause depression, anorexia, increased salivation, bloat, colic, diarrhea, and rarely tremors and paralysis.

Flower

Seedling

Mouseear Chickweed

Cerastium fontanum Baumg. ssp. *vulgare* (Hartman) Greuter & Burdet ·
Caryophyllaceae · Pink Family

Synonyms

Cerastium vulgatum L.

Habit, Habitat, and Origin

Erect to spreading, tufted, mat-forming
perennial herb; to 5.0 dm tall; lawns,
roadsides, pastures, meadows, prairies,
and open woodlands; native of Eurasia.

Seedling Characteristics

Hypocotyl light green, weak; adventitious roots arising from reclined seed
leaf node; cotyledons green, 2.0–7.0 mm
long, 0.5–2.0 mm wide, initially with
minute pubescence above.

Mature Plant Characteristics

ROOTS fibrous, shallow, frail, nearly
transparent, from slender taproot.
STEMS erect to spreading, 0.5–5.0 dm
tall, slender, weak, branches 1.5–5.0 cm
long, prostrate basally, often rooting at
lower nodes, pubescent; trichomes long,
sticky. LEAVES opposite, 1.0–2.0 cm long,
3.0–12.0 mm wide, dull green, oval to
elliptic, weakly nerved, upper surface
and principal veins beneath pubescent,
trichomes 0.5–1.0 mm long, petiole lacking. INFLORESCENCES usually in clusters
of 3 at ends of stems, open; petals white,
notched at tip; sepals 4.0–6.0 mm long,
pubescent. FRUITS capsule, 7.0–11.0 mm
long, 2.0–3.0 mm wide, cylindrical to
slightly curved, membranous. SEEDS
triangular or angular-ovate in outline,
0.5–0.8 mm in diameter, flattened,
chestnut brown, short irregular-knobby
tubercles.

Special Identifying Features

Erect to spreading, tufted perennial; fruit
capsule curved; sepals and leaves pubescent.

Toxic Properties

None reported.

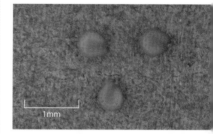

1mm

TOP Seeds
MIDDLE Fruit
BOTTOM Flower

Deptford Pink

Dianthus armeria L. · Caryophyllaceae · Pink Family

Synonyms

Grass pink

Habit, Habitat, and Origin

Erect annual or biennial herb; to 7.0 dm tall; ornamental, escaped to pastures, fields, roadsides, and disturbed areas; native of Europe.

Seedling Characteristics

Cotyledons club-shaped; first true leaves egg-shaped, petiole as long as leaf blades.

Mature Plant Characteristics

ROOTS fibrous from taproot. STEMS erect, 2.0–7.0 dm tall, branched; pubescent toward bottom, trichomes pointing in one direction. LEAVES opposite, 2.0–8.0 cm long, 1.0–5.0 cm wide, clasping at bases, finely pubescent, attenuate or blunt-tipped. INFLORESCENCES borne in terminal dense headlike cymes or sometimes solitary, sessile or with a very short pedicel; petals 5, long-clawed, dentate or crenate, pink or rose with whitish dots; calyx 10.0–15.0 mm long, 5-toothed, tubular; stamens 10; pistil with 2 styles; several bracts at base. FRUITS capsule 10.0–14.0 mm long, dehiscent by 4 teeth. SEEDS compressed and slightly concave, 1.0–1.5 mm long, laterally attached, numerous, dark red or brown.

Special Identifying Features

Erect annual or biennial; stems slender, upright; leaves cauline, sessile, linear; flowers pink or rose with white dots.

Toxic Properties

None reported.

Inflorescence

Flower

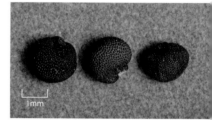

Bouncingbet

Saponaria officinalis L. · Caryophyllaceae · Pink Family

Synonyms

Soapwort, hedge-pink, wild sweet William

Habit, Habitat, and Origin

Erect perennial herb; to 9.0 dm tall; open woods, roadsides, disturbed areas, and old house sites; native of Europe.

Seedling Characteristics

Cotyledons elliptical with rounded tips, glabrous; first true leaves lanceolate to elliptic, glabrous, entire margins.

Mature Plant Characteristics

ROOTS fibrous from taproot, forming persistent short rhizomes. STEMS erect to reclining with ascending apices, 3.0–9.0 dm tall, stout, jointed, smooth, usually unbranched, glabrous. LEAVES opposite, 5.0–7.0 cm long, 2.0–3.0 cm wide, oval to lanceolate, simple, entire, glabrous, sessile. INFLORESCENCES conspicuous clusters of 5 pink or white blossoms at tops of stems, 2.0–3.0 cm diameter; petals 1.5–1.8 cm long, narrow triangular lobes with clawlike appendage at apex, base enclosed in tubular calyx; calyx 2.0–2.5 cm long, pedicellate. FRUITS capsule, 1.8–2.0 cm long, ellipsoid, many-seeded. SEEDS kidney-shaped, 1.5–1.8 mm long, flattened, dull black, with narrow rows of minute knobs covering surface.

Special Identifying Features

Erect perennial; stems typically unbranched; stems and leaves glabrous; leaves opposite; flowers showy, distinct.

Toxic Properties

Seeds rarely cause depression, anorexia, excessive salivation, bloat, colic, and diarrhea in livestock.

TOP Seeds
MIDDLE Cotyledons
BOTTOM Seedling

1mm

TOP Seeds
MIDDLE Seedling
BOTTOM Flowering plant

Flowers

White Campion

Silene latifolia Poir. · Caryophyllaceae · Pink Family

Synonyms

White cockle; *Silene alba* (P. Mill.) E. H. L. Krause

Habit, Habitat, and Origin

Erect to decumbent, short-lived perennial or biennial herb; to 8.0 dm tall; fields, roadsides, disturbed areas, and waste sites; native of Eurasia.

Seedling Characteristics

Cotyledons club-shaped, pubescent; first true leaves egg-shaped, pubescent.

Mature Plant Characteristics

ROOTS fibrous from stout, often laterally branched taproot. STEMS erect to decumbent, 6.0–8.0 dm tall, simple or branched from base, finely hirsute; trichomes glandular, multicellular, to 1.0 mm long. LEAVES opposite, 3.0–12.0 cm long, 6.0–30.0 mm wide, oblong-lanceolate to elliptic, apex acute, basal leaves sessile or short-petioled, pubescent on both surfaces. INFLORESCENCES terminal in axillary cymes, fragrant, dioecious; female flowers 20.0–30.0 mm long, 9.0–16.0 mm wide, pubescent; petals white, 7.0–10.0 mm long, 2-lobed; calyx 20-nerved; male flowers 10.0–24.0 mm long, 8.0–15.0 mm wide, subsessile to short-pedicellate. FRUITS capsule, unilocular, ovate, 10-toothed. SEEDS rounded, 0.8–1.5 mm diameter, grayish black, bluntly tuberculate.

Special Identifying Features

Erect to decumbent perennial or biennial; dioecious; flowers fragrant, calyx 20-nerved; fruit a capsule, 10-toothed.

Toxic Properties

None reported.

Flowers

Nightflowering Catchfly

Silene noctiflora L. · Caryophyllaceae · Pink Family

Synonyms

Night-flowering campion, sticky cockle

Habit, Habitat, and Origin

Erect annual herb; to 9.0 dm tall; fields, roadsides, disturbed areas, and waste sites; native of Europe.

Seedling Characteristics

Cotyledons club-shaped and fused together with trichomes on edges of base; young leaves egg-shaped.

Mature Plant Characteristics

ROOTS fibrous from unbranched taproot. STEMS erect, 3.0–9.0 dm tall, simple or few-branched from base, nodes swollen, pubescent, trichomes dense and sticky. LEAVES opposite, 3.0–12.0 cm long, 0.5–5.0 cm wide, simple, ovate to elliptic-oblanceolate or lanceolate, margins entire, sessile, pubescent; trichomes glandular, sticky, covering both surfaces. INFLORESCENCES terminal, loosely branched axillary cymes, most perfect, 1.9 cm diameter; petals 5, white shading into pink; calyx united, 1.5–2.3 cm long with prominent green ribs, glandular, freely cross-veined, with 5 awl-shaped teeth enclosing capsule. FRUITS capsule, trilocular, 6-toothed, sessile, ellipsoid. SEEDS rounded, 0.8–1.3 mm diameter, gray, uniformly wrinkled with minute projections.

Special Identifying Features

Erect annual; leaves opposite, covered with short, sticky trichomes; flowers open at night, close in morning; calyx united to form 10 prominent green ribs; seed much enlarged at maturity.

Toxic Properties

None reported.

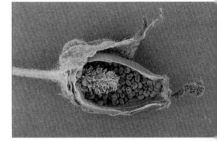

TOP Fruit and seeds
MIDDLE Two-leaf seedling
BOTTOM Four-leaf seedling

Common Chickweed

Stellaria media (L.) Vill. · Caryophyllaceae · Pink Family

Synonyms

Common starwort, starwort, winterweed

Habit, Habitat, and Origin

Erect, prostrate, or ascending, weakly tufted to matted annual or winter annual herb; to 8.0 dm tall; cultivated fields, gardens, lawns, roadsides, and waste areas; native of Eurasia.

Seedling Characteristics

Hypocotyl slender, often reddish, trichomes few, unequal, clear; cotyledons, 1.0–12.0 mm long, 0.25–2.0 mm wide, tender petiole, sparsely pubescent.

Mature Plant Characteristics

ROOTS shallow, fibrous and frail from weak taproot. **STEMS** erect, basally prostrate, or ascending; to 8.0 dm long, tender, freely branching, rooting at nodes, smooth or pubescent on older portions; trichomes soft, in vertical lines. **LEAVES** opposite, 1.0–3.0 cm long, 3.0–15.0 mm wide, oval or elliptic, tips acute or short-acuminate, smooth or pubescent toward base and on petioles; upper leaves sessile, lower leaves long-petioled. **INFLORESCENCES** solitary or in small clusters at ends of stems or leaf axils; petals 5, white, shorter than sepals; sepals 3.5–6.0 mm long, oblong-lanceolate; pedicels 3.0–30.0 mm long, slender, fragile. **FRUITS** capsule, 5.0–7.0 mm long, narrowly oval or elliptic, whitish, 5 segments, many-seeded. **SEEDS** subrotund, 1.0–1.3 mm diameter, margin notched, flattened, reddish brown, tubercles in curved rows.

Special Identifying Features

Erect, prostrate, or ascending, weakly tufted to matted annual or winter annual; stems pubescent in vertical lines.

Toxic Properties

None reported.

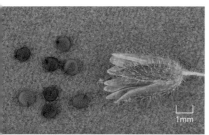

TOP Seeds
MIDDLE Seedling

Flower

Cowcockle

Vaccaria hispanica (P. Mill.) Rauschert · Caryophyllaceae · Pink Family

Synonyms

China cockle, cowbasil, cowherb, pint cockle, spring cockle, cow soapwort; *Vaccaria pyramidata* Medik.

Habit, Habitat, and Origin

Erect annual herb; to 9.0 dm tall; fields, roadsides, railroad beds, disturbed areas, and waste sites; native of Europe.

Seedling Characteristics

Cotyledons oblanceolate; cotyledons and young leaves glabrous, sessile.

Mature Plant Characteristics

ROOTS fibrous from taproot. STEMS erect, 3.0–9.0 dm tall, glabrous, whitish. LEAVES opposite, 5.0–10.0 cm long, 1.0–2.5 cm wide; lanceolate, ovate, to oblanceolate; 1-nerved, tip acute, margins entire, base cordate, sessile. INFLORESCENCES open cymes, numerous, showy, perfect; petals 1.8–2.2 cm long, pink or pale red; calyx tube ovoid, 5-ribbed, 8.0–14.0 mm long, 6.0–8.0 mm wide; lobes 1.5–3.0 mm long, margins scarious. FRUITS capsule, unilocular but rarely trilocular, dehiscent apically by 4 teeth or valves or a 1-seeded utricle. SEEDS globular, 2.0–2.7 mm diameter, black to bluish black, surface dull; covered with short, oval, pimplelike tubercles arranged in rows or a distinct pattern.

Special Identifying Features

Erect annual; flowers pink to pale red, calyx tube ovoid.

Toxic Properties

Grazing can cause depression, anorexia, excessive salivation, bloat, colic, and diarrhea in livestock.

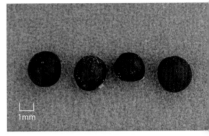

1mm

TOP Seeds
BOTTOM Seedling

TOP Flowering plant
BOTTOM Flower

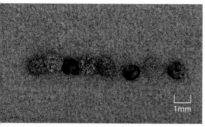

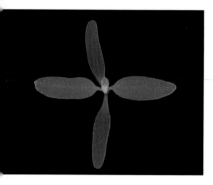

TOP Seeds
BOTTOM Two-leaf seedling

Common Lambsquarters

Chenopodium album L. · Chenopodiaceae · Goosefoot Family

Synonyms

Green pigweed, white goosefoot; *Dysphania ambrosioides* (L.) Mosyakin & Clemants

Habit, Habitat, and Origin

Erect annual herb; to 2.0 m tall; disturbed sites, old fields, cultivated areas, and gardens; native of Eurasia.

Seedling Characteristics

First leaves small, rounded, triangular, alternate, light green; petiole 2.0 cm long with a mealy gray case.

Mature Plant Characteristics

ROOTS fibrous from short-branched taproot, red. STEMS erect, 0.1–2.0 m tall, smooth, moderately branched, scruffy, grooved, light green with red coloration in various degrees. LEAVES alternate, 3.0–6.0 cm long, 2.0–4.0 cm wide, deltoid, simple, margins irregularly sinuate-dentate to entire, light green, with gray-mealy underside, smooth, petiole long. INFLORESCENCES clusters in dense paniculate spike at tips of branches and upper leaf axils, perfect; petals absent; stamens 6; pistil 1; styles 2 or 3; sepals 5, covered by mealy powder. FRUITS utricle, 1.1–1.5 mm diameter, star-shaped calyx of 5 sepals nearly covering seed. SEEDS lens-shaped with convex sides, 1.3 mm diameter, black, shiny, margin notched, fruit parts attached.

Special Identifying Features

Erect annual; leaves and flower calyx covered by mealy gray powder on lower surface.

Toxic Properties

In large quantities, leaves of *Chenopodium* species can cause intoxication leading to ataxia, depression, weakness, and tremors.

TOP Young plant
BOTTOM Flowering plant

Flowers

Mexicantea

Chenopodium ambrosioides L. · Chenopodiaceae · Goosefoot Family

Synonyms

Wormseed; *Dysphania ambrosioides* (L.) Mosyakin & Clemants

Habit, Habitat, and Origin

Erect or spreading annual, biennial, or perennial herb; to 1.0 m tall; disturbed areas, roadsides, fields, pastures, gardens, stockyards, and waste sites; native of tropical America.

Seedling Characteristics

Cotyledons oblong, glanded; first true leaves entire, oblong, with yellow glands.

Mature Plant Characteristics

ROOTS fibrous from taproot. STEMS erect or spreading, up to 0.2–1.0 m tall, numerous, ascending branches, glands yellow and resinous, pubescent, trichomes short. LEAVES alternate, 2.0–14.0 cm long, 1.0–6.0 cm wide, simple, lower leaves lanceolate, leaves reduced upward on branches, oblong to ovate, margins sinuate; glands minute, yellow, resinous. INFLORESCENCES glandular glomerules on bracted or bractless spikes; petals absent; sepals 5, glandular or glabrous; stamens 5; stigmas 2. FRUITS utricle, 0.7–1.0 mm diameter, pericarp readily separates from seed. SEEDS lenticular, 0.5–0.6 mm long, dark brown to black, shiny.

Special Identifying Features

Erect or spreading annual, biennial, or perennial; stems and leaves covered with yellow glands emitting an unpleasant chlorinelike odor when bruised or crushed.

Toxic Properties

See comments under *Chenopodium album*.

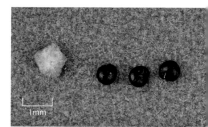

TOP Seeds
BOTTOM Seedling
BOTTOM Flowering branch

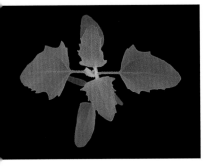

TOP **Fruit**
MIDDLE **Two-leaf seedling**
BOTTOM **Six-leaf seedling**

Flowering plant

Slimleaf Lambsquarters

Chenopodium leptophyllum (Moq.) Nutt. ex S. Wats. · Chenopodiaceae · Goosefoot Family

Synonyms

Narrowleaf goosefoot

Habit, Habitat, and Origin

Erect to semierect annual herb; to 9.0 dm tall; open areas, fields, and waste sites; native of North America.

Seedling Characteristics

First true leaves small, rounded, alternate, light green, with mealy gray cast.

Mature Plant Characteristics

ROOTS fibrous from short, red, much-branched taproot. STEMS erect or semierect, 4.0–9.0 dm tall, branched from base, branches erect, green to reddish. LEAVES alternate, 1.0–4.0 cm long, 2.0–3.0 mm wide, simple, linear-lanceolate to linear-oblong, 1-nerved, margins entire, fleshy, petiole short, mealy gray color. INFLORESCENCES open, widely spaced glomerules; petals absent; calyx lobes 1.0–2.0 mm wide; sepals 5, covered by mealy gray powder; stamens 5; stigmas 2. FRUITS utricle, 0.9–1.1 mm diameter, with star-shaped calyx of 5 sepals nearly covering seed. SEEDS roundish, 0.7–0.9 mm diameter, rugulate, black, pericarp adheres to seeds.

Special Identifying Features

Erect to semierect annual; leaves narrow, leaves and flowers mealy gray.

Toxic Properties

See comments under *Chenopodium album.*

Kochia

Kochia scoparia (L.) Schrad. · Chenopodiaceae · Goosefoot Family

Synonyms

Burning bush, Mexican fireweed, summer cypress

Habit, Habitat, and Origin

Erect, vigorous, early-germinating, drought-tolerant summer annual herb; to 2.0 m tall; fields, pastures, roadsides, and waste sites; native of Eurasia.

Seedling Characteristics

Rosette small, gray-green; first true leaves lanceolate, sessile, pubescent.

Mature Plant Characteristics

ROOTS fibrous from much-branched taproot; to 2.0 m deep and 2.0 m wide on either side of plant. STEMS erect, 0.5–2.0 m tall, bushy, spreading, with ascending branches from central stem forming a more or less round, bushy plant. LEAVES alternate, to 6.0 cm long, simple, lanceolate or narrowly linear, margins entire, sessile, yellowish green, may become reddish at maturity. INFLORESCENCES in axils of upper leaves and in terminal panicles, small, perfect, greenish, changing to reddish at maturity; stamens 5; styles 3 or rarely 2. FRUITS utricle enclosed in a 5–10-lobed cleft calyx, reddish. SEEDS ovate, 1.6–1.8 mm long, flattened, grooved on each side, end narrowed, surface dull, finely granular, brown with yellow markings, usually enclosed in a hull or membranous seed coat.

Special Identifying Features

Erect, vigorous annual; plants more or less round; mature plants break off at soil line and tumble in the wind.

Toxic Properties

Plants contain toxins that can cause weight loss, liver disease, and photosensitivity.

Flowers

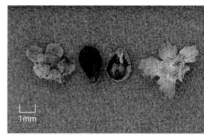

TOP **Seeds**
MIDDLE **Four-leaf seedling**
BOTTOM **Young plant**

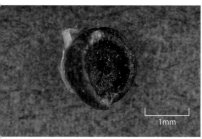

TOP **Seed**
MIDDLE **Seedling**
BOTTOM **Mature plant**

Russian-thistle

Salsola tragus L. · Chenopodiaceae · Goosefoot Family

Synonyms

Tumbleweed; *Salsola iberica* (Sennen & Pau) Botch. ex Czerepanoz, *S. kali* L.

Habit, Habitat, and Origin

Erect, much-branched, globular, early-germinating, drought-tolerant summer annual herb; to 1.5 m tall; fields, roadsides, railroad beds, disturbed and waste sites; native of Eurasia.

Seedling Characteristics

Cotyledons linear; first true leaves 2.0 cm long, cylindrical, succulent.

Mature Plant Characteristics

ROOTS fibrous from branched taproot; to 2.0 m deep and 2.0 m on either side of plant. STEMS erect, to 1.5 m tall, spiny, much-branched, forming a globular plant, at maturity turning reddish and breaking off from root to tumble in the wind. LEAVES alternate, simple, small; first leaves linear, succulent, later ones awl-shaped, ending in a spine. INFLO-RESCENCES inconspicuous in leaf and stem axils, mostly solitary, membranous saucers with pink or reddish centers; petals absent; calyx 2.5–3.5 mm long, wavy, veined, subtended by bracts; bracts 3, ovate, with acuminate tips; stamens 5, exserted beyond sepals; style 2-branched, exserted. FRUITS utricle, 4.0–10.0 mm diameter including sepal wings, becoming red when ripe, covered by calyx giving appearance of a small bladder forming a flat, circular wing at base. SEEDS coiled, round, 1.5–2.0 mm diameter, greenish brown to black, smooth, shiny.

Special Identifying Features

Erect, much-branched, globular summer annual; leaves resemble pine needles; plant breaks off from roots and tumbles freely.

Toxic Properties

Plants contain toxins that can cause diarrhea and abrupt onset of depression, weakness, labored respiration, prostration, seizures, and coma.

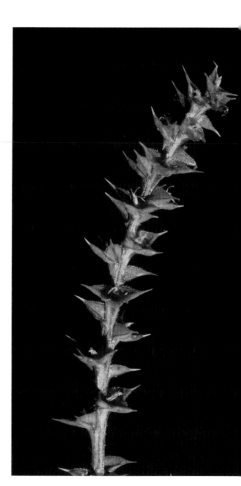

Flower

Flowers

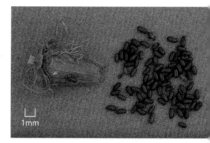

Common St. Johnswort

Hypericum perforatum L. · Clusiaceae · St. Johnswort Family

Synonyms

Goatweed, klamathweed

Habit, Habitat, and Origin

Erect, rhizomatous perennial; to 1.0 m tall; roadsides, pastures, prairies, open woods, and woodland edges; native of Europe.

Seedling Characteristics

Cotyledons glabrous, spatulate; first true leaves alternate.

Mature Plant Characteristics

ROOTS fibrous from branched taproot; branches extending to considerable depth; shallow rootstocks extending a few centimeters from crown. STEMS erect, 0.3–1.0 m tall, stout, often reddish, branched, erect from rootstock, woody, often winged, slender stolons with leafy basal offshoots. LEAVES opposite, 1.0–2.5 cm long, 2.0–5.0 mm wide, elliptic to oblong or linear, margins entire, sparsely black-dotted with translucent glands. INFLORESCENCES numerous, in leafy-bracted compound cymes, 0.5–3.0 dm diameter, individual florets 2.5 cm diameter, deep yellow; petals 8.0–12.0 mm long, narrow, black-dotted on margins; sepals 4.0–6.0 mm long, narrow, acute tip; stamens numerous. FRUITS capsule, 7.0–8.0 mm long, narrow-oblong. SEEDS 0.6–0.7 mm long, black, shiny, subcylindrical, reticulate.

Special Identifying Features

Erect rhizomatous perennial; stems numerous, small nonreproductive stems at base; leaves black-dotted with translucent glands.

Toxic Properties

Plants cause photosensitivity and some systemic effects in livestock.

TOP Seeds
MIDDLE Seedling
BOTTOM Flowering plant

Hedge Bindweed

Calystegia sepium (L.) R. Br. · Convolvulaceae · Morningglory Family

Synonyms

Convolvulus sepium L.

Habit, Habitat, and Origin

Climbing or trailing, perennial vining herb; to 3.0 m long; open areas, fencerows, thickets, and disturbed fields; native of Europe.

Seedling Characteristics

Cotyledons as long as broad, slightly indented at tip, glabrous.

Mature Plant Characteristics

ROOTS fibrous from taproot, long, branched; fleshy rootstock to 30.0 cm deep. STEMS climbing or trailing vines, to 3.0 m long, rounded or angular, glabrous, internodes to 7.0 cm long. LEAVES alternate, 2.0–15.0 cm long, 1.0–9.0 cm wide, simple, triangular or ovate to ovate-lanceolate, usually glabrous, margins entire or undulate, petiole 2.0–7.0 cm long. INFLORESCENCES solitary on pedicel, 3.0–14.0 cm long; corolla funnelform, 4.5–5.8 cm long, white or pink; stamens 2.3–2.9 cm long, subequal; style 2.0–2.3 mm long; bracts 1.4–2.6 cm long, 1.0–1.8 cm wide, surrounding and concealing sepals. FRUITS 1-celled capsule, 10.0–13.0 mm diameter, 2–4-seeded. SEEDS oblong, 5.0–6.0 mm long, 4.0 mm wide, dark reddish brown, dull, roughened by minute widely spaced tubercules.

Special Identifying Features

Climbing or trailing perennial vine; flowers white to pink, conspicuously funnelform; bracts larger than and concealing sepals.

Toxic Properties

Plants contain toxins causing diarrhea, vomiting, colic, depression, weakness, and drowsiness.

Flowers

Field Bindweed

Convolvulus arvensis L. · Convolvulaceae · Morningglory Family

Synonyms

Field morningglory

Habit, Habitat, and Origin

Climbing or trailing, perennial, vining herb; to 1.0 m long; cultivated areas, fields, roadsides, railroads, and waste sites; native of Europe.

Seedling Characteristics

Cotyledons as long as broad, barely indented at tip.

Mature Plant Characteristics

ROOTS fibrous from taproot; spirally twisting, branched rootstock sometimes exceeding 2.0 m in depth. STEMS climbing or trailing vine, up to 1.0 m long, forming dense mats, glabrous or pubescent. LEAVES alternate, 1.0–10.0 cm long, 0.3–6.0 cm wide, simple, variable in outline from triangular to oblong, glabrous or pubescent, base cordate to subtruncate, margins entire or somewhat undulate, petiole 0.3–4.0 mm long. INFLORESCENCES solitary or in cymes of 2–4, peduncles 1.0–9.0 cm long; corolla campanulate, 1.0–2.5 mm long, white to pink-tinged; sepals 3.0–4.5 mm long, 2.0–3.0 mm wide, obtuse to truncate; bracts 2.0–9.0 mm long, elliptic, ovate, or linear. FRUITS capsule, 5.0–7.0 mm diameter, globose-ovoid, bilocular, 1–4-seeded. SEEDS oblong, 4.0 mm long, 2.0–3.0 mm wide, dull, surface roughened by numerous blunt tubercules.

Special Identifying Features

Climbing or trailing perennial vine; flower white to pink-tinged; bracts stalked, small, distant from flower.

Toxic Properties

Plants contain toxins causing diarrhea, vomiting, colic, depression, weakness, and drowsiness.

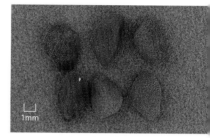

1mm

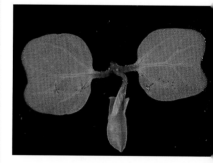

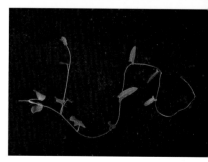

TOP Seeds
MIDDLE Cotyledons
BOTTOM Young plant

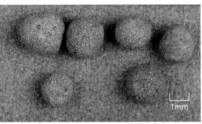

TOP Seeds
BOTTOM Flower

Carolina Dichondra

Dichondra carolinensis Michx. · Convolvulaceae · Morningglory Family

Synonyms

Kidney weed

Habit, Habitat, and Origin

Creeping, mat-forming perennial herb; to 7.0 cm tall; disturbed areas, fields, lawns, flowerbeds, roadsides, and waste sites; native of North America.

Seedling Characteristics

Cotyledons lanceolate, less than 1.0 cm long, dark green.

Mature Plant Characteristics

ROOTS fibrous from taproot. STEMS creeping, matting, and procumbent; 2.0–7.0 cm tall, round, green to whitish green, rooting at nodes. LEAVES alternate, 1.0–3.0 cm wide, simple, suborbicular to reniform, often retuse, margins entire, commonly pubescent on petioles 1.0–4.0 cm long. INFLORESCENCES inconspicuous in green-white clusters in leaf axils, solitary; corolla 1.0–1.5 cm wide, white, funnelform or funnelform-rotate, shallowly notched outwardly, shorter than calyx; sepals 2.0–3.0 mm long at anthesis; carpels nearly separate; stigmas capitate; styles separate, bractless pedicels from stem, 1.0–4.0 cm long. FRUITS 2-lobed capsule, each lobe 1-seeded, indehiscent or irregularly dehiscent. SEEDS pyriform, 1.8–2.5 mm long, brown, pilose.

Special Identifying Features

Creeping, mat-forming, procumbent perennial; stems round; leaves alternate, suborbicular to reniform, margins entire.

Toxic Properties

None reported.

TOP Seedling
BOTTOM Flowering plant

Flower

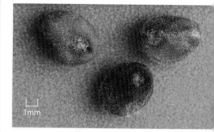

Swamp Morningglory

Ipomoea aquatica Forsk. · Convolvulaceae · Morningglory Family

Synonyms

Aquatic morningglory, water-spinach

Habit, Habitat, and Origin

Creeping or vining, aquatic perennial herb; to 3.0 m long; forming mats on surface of water or damp soil; native to Asia; listed as a Federal Noxious Weed.

Seedling Characteristics

Hypocotyl stout, green, glabrous; cotyledon lobes 2.0 cm long, 3.0–4.0 mm wide, pointed, glabrous.

Mature Plant Characteristics

ROOTS adventitious from nodes; with air cavities in moist environments, longer and brown in drier sites. STEMS creeping, to 3.0 m long, hollow or spongy, much shorter in dry sites, internodes 0.2–2.0 m long. LEAVES alternate, 3.0–15.0 cm long, 1.0–10.0 cm wide, simple, oblong-ovate or lanceolate; hastate, cordate, or truncate base; margins slightly sinuate; petiole 3.0–20.0 cm long, large and thin in water, small and thick when terrestrial. INFLORESCENCES axillary, 1–2 per axil, perfect; corolla 5-lobed, 3.0–5.0 cm long, funnelform; white, pink, pale lavender, or purple, with dark purple center; sepals 5, 8.0 mm long, partially fused, persistent, brownish pink; basal bracts small; pedicels 2.0–10.0 cm long. FRUITS capsule 6.0–10.0 mm long, spherical or ovoid, thin-walled, sepals clasped, to 5 fruits per node, to 4 seeds per fruit. SEEDS wedge-shaped, 4.0 mm long, 5.0–7.0 mm wide, grayish brown, glabrous or pubescent.

Special Identifying Features

Creeping or vining aquatic perennial; stems hollow, highly branched, forming mats; flowers showy, white, pink, pale lavender, or purple; new plants from cuttings or fragments.

Toxic Properties

None reported; edible to humans.

TOP Seeds
MIDDLE Four-leaf seedling
BOTTOM Three-leaf seedling

TOP Seeds
MIDDLE Fruit
BOTTOM Seedling

Red Morningglory

Ipomoea coccinea L. · Convolvulaceae · Morningglory Family

Synonyms

Scarlet morningglory; *Ipomoea coccinea* L. var. *hederifolia* (L.) House

Habit, Habitat, and Origin

Climbing or twining annual, vining herb; to 2.0 m long; cultivated areas, fields, pastures, fencerows, roadsides, disturbed and waste sites; native of tropical America.

Seedling Characteristics

Hypocotyl stout, green or maroon, glabrous; cotyledons often maroon-tinged, not deeply lobed, with rounded points.

Mature Plant Characteristics

ROOTS fibrous from vigorous taproot. **STEMS** climbing or twining prostrate vine, to 2.0 m long, many-branched, glabrous or rarely pubescent. **LEAVES** alternate, 2.0–14.0 cm long, 1.0–12.0 cm wide, simple, ovate, margins usually with 3–5 pointed projections or teeth in basal portion but may be entire or lobed, petiole 0.6–14.0 cm long. **INFLORESCENCES** in cymes, 2–6 or rarely solitary; corolla fused, funnelform, 2.2–3.0 cm long, orange-red; sepals subequal, 3.0–3.5 mm long, 2.0–3.0 mm wide, outer ones oblong to elliptic, tip mucronate; bracts 1.0–3.0 mm long, ovate to lanceolate; pedicels 5.0–15.0 mm long. **FRUITS** capsule, 6.0–7.0 mm diameter, rounded, glabrous, light brown, 4-seeded, pedicels reflexed. **SEEDS** wedge-shaped, 3.0–4.0 mm long, dark brown to black, finely tomentose.

Special Identifying Features

Climbing or twining annual vine; leaves with basal projections or teeth; flowers bright orange-red.

Toxic Properties

Seeds of *Ipomoea* species can contain hallucinatory alkaloids causing auditory and visual distortion, mood elevation, nausea, and sluggishness.

Flower

Tall Morningglory

Ipomoea purpurea (L.) Roth · Convolvulaceae · Morningglory Family

Synonyms

Common morningglory; *Pharbitis purpurea* (L.) Voigt

Habit, Habitat, and Origin

Climbing or trailing, vining annual herb; 2.0 m long or more; cultivated areas, fields, pastures, fencerows, roadsides, disturbed and waste sites; native of tropical America.

Seedling Characteristics

Hypocotyl stout, green; cotyledons moderately indented with rounded lobes.

Mature Plant Characteristics

ROOTS fibrous from taproot. STEMS climbing or trailing vine, to 2.0 m or more long, branched, loosely pubescent or tomentose; trichomes short, appressed. LEAVES alternate, 1.0–12.0 cm long, 1.0–12.0 cm wide, simple, heart-shaped, margins entire or rarely 5-lobed, loosely pubescent, trichomes flat on leaf surface, petiole 1.0–14.0 cm long. INFLORESCENCES axillary clusters of 3–5 in cymes, rarely solitary; corolla, 4.0–6.0 cm long, funnelform, purple or occasionally blue or white, with a white center; sepals subequal, 8.0–17.0 mm long, 1.5–4.5 mm wide, outer ovate-lanceolate or elliptical, tips blunt; style 14.0–24.0 mm long; stigmas 3. FRUITS capsule, 1.0 cm diameter, subglobose, with 3 locules, often 6-seeded. SEEDS wedge-shaped, 4.0–5.0 mm long, dark brown to black, scar smooth.

Special Identifying Features

Climbing or trailing annual vine; leaves heart-shaped, pubescent, trichomes flat on surface; flowers purple or occasionally white; sepals short, tips blunt, pubescent.

Toxic Properties

See comments under *Ipomoea coccinea*.

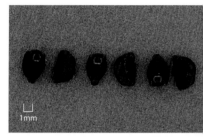
1mm

TOP Seeds
MIDDLE Fruit
BOTTOM Four-leaf seedling

Flower

Cypressvine Morningglory

Ipomoea quamoclit L. · Convolvulaceae · Morningglory Family

Synonyms

Quamoclit vulgaris Choisy

Habit, Habitat, and Origin

Climbing or twining, vining annual herb; to 3.0 m long; cultivated areas, fields, pastures, fencerows, roadsides, disturbed and waste sites; native of tropical America.

Seedling Characteristics

Hypocotyl stout; cotyledons long, angle between points much greater than 90 degrees.

Mature Plant Characteristics

ROOTS fibrous from taproot. STEMS climbing or twining vine, to 3.0 m long, glabrous, branched. LEAVES alternate, 1.0–9.0 cm long, 0.8–7.0 cm wide, simple, deeply pinnately divided with 9–19 pairs of opposite linear segments 0.2–1.5 mm wide, glabrous, petiole 0.2–4.5 cm long. INFLORESCENCES axillary, solitary or 2–5 in cymes; corolla 2.5–4.0 cm long, funnelform, deep red to rarely scarlet or white; sepals subequal, 4.0–6.0 mm long, 2.0–3.0 mm wide, elliptic to oblong; style exserted, 2.3–3.0 cm long; stigmas 2; bracts 0.6–1.0 mm long; pedicels 8.0–25.0 mm long. FRUITS capsule, 7.0–9.0 mm long, ovoid, glabrous, with 4 locules. SEEDS elongate, wedge-shaped, 4.5–5.5 mm long, reddish brown, pubescent; trichomes sparse, short, scaly.

Special Identifying Features

Climbing or twining annual vine; cotyledons long, wide-angled; leaves deeply divided; flowers funnelform, dark red.

Toxic Properties

See comments under *Ipomoea coccinea*.

TOP Seeds
MIDDLE Fruit
BOTTOM Seedling

Flower

Flower

Purple Morningglory

Ipomoea turbinata Lag. · Convolvulaceae · Morningglory Family

Synonyms

Purple moonflower; *Calonyction muri-catum* (L.) G. Don., *Ipomoea muricata* (L.) Jacq.

Habit, Habitat, and Origin

Climbing or twining, vining annual herb; to 2.0 m long or more; cultivated areas, fields, pastures, fencerows, road-sides, disturbed and waste sites; native of tropical America.

Seedling Characteristics

Hypocotyl stout, stem glabrous with fleshy prickles; cotyledons large, butterfly-shaped.

Mature Plant Characteristics

ROOTS fibrous from taproot. STEMS climbing or twining vine, to more than 2.0 m long, green, glabrous, with oc-casional fleshy prickles. LEAVES alternate, 7.0–18.0 cm long, 6.5–15.0 cm wide, simple, heart-shaped, glabrous, peti-ole 4.0–12.0 cm long. INFLORESCENCES axillary, solitary or 2–5 in cymes; corolla 6.0–7.5 cm wide, 7.0–9.0 cm long, fun-nelform, light purple to lavender with dark center; sepals 3, 7.0–8.0 mm long; stigma 2-lobed; anthers large, base cordate; bracts 8.0 mm long; pedicels 1.0–2.0 cm long. FRUITS capsule, 1.8–2.0 cm long, subglobose, glabrous, 3–4-seeded. SEEDS wedge-shaped, 1.0–1.2 cm long, dark brown, glabrous.

Special Identifying Features

Climbing or twining vine; cotyledons large, butterfly-shaped; stems glabrous, with occasional fleshy prickles; flowers light purple or lavender, opening at night.

Toxic Properties

See comments under *Ipomoea coccinea*.

TOP **Seeds**
MIDDLE **Seedling**
BOTTOM **Prickles**

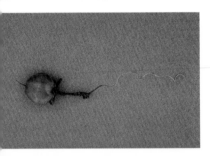

TOP Seeds
BOTTOM Fruit
BOTTOM Seedling

Flower

Palmleaf Morningglory

Ipomoea wrightii Gray · Convolvulaceae · Morningglory Family

Synonyms

Willowleaf morningglory

Habit, Habitat, and Origin

Climbing or trailing, vining annual herb; to 2.0 m long; alluvial or clay soils, cultivated areas, ditch banks, fields, and roadsides; probably native of India.

Seedling Characteristics

Hypocotyl stout; cotyledons deeply indented with pointed ends, glabrous.

Mature Plant Characteristics

ROOTS fibrous from taproot. STEMS low-climbing or trailing vine, to 2.0 m long, branched, slender, glabrous. LEAVES alternate, 2.0–6.0 cm long, 3.0–6.0 cm wide, palmately compound, 3–7-lobed; lobes subequal to conspicuously unequal, lanceolate to linear-lanceolate, 1.5–6.0 cm long, margins entire, petiole 1.5–5.0 cm long. INFLORESCENCES axillary, soli-tary; corolla 1.5–2.3 cm long, funnelform, lavender to lavender pink or rose, with dark center; sepals subequal, 4.0–6.0 mm long, ovate or oblong-ovate, tips rounded or blunt; stamens and style not exerted; bracts minute; pedicel 1.0–5.0 cm long, spirally twisted. FRUITS capsule, 6.0–7.0 mm diameter, subglobose, brown, glabrous; long, spirally coiled stalk. SEEDS wedge-shaped, 3.0–5.0 mm long, with a few long trichomes.

Special Identifying Features

Climbing or trailing annual vine; leaves palmately compound; flowers lavender, pedicel spiral-coiled; seed with long pubescence.

Toxic Properties

See comments under *Ipomoea coccinea*.

Smallflower Morningglory

Jacquemontia tamnifolia (L.) Griseb. · Convolvulaceae · Morningglory Family

Synonyms

Hairy clustervine

Habit, Habitat, and Origin

Erect, becoming climbing or twining, vining annual herb; to 2.0 m tall; cultivated areas, fields, pastures, fencerows, gardens, roadsides, disturbed and waste sites; probably native of the American tropics.

Seedling Characteristics

Hypocotyl stout, green, glabrous; cotyledons slightly indented with rounded points.

Mature Plant Characteristics

ROOTS fibrous from vigorous taproot. **STEMS** erect initially, becoming climbing or twining, to 2.0 m long, many branches, sparsely to densely pubescent, trichomes spreading. **LEAVES** alternate, 3.0–12.0 cm long, 2.0–9.0 cm wide, simple, margins entire and pubescent, sparsely pubescent on both surfaces, petiole 3.0–12.0 cm long. **INFLORESCENCES** axillary, cymose clusters of 20 or more; corolla 1.0–2.0 cm wide, funnelform, blue or rarely white; sepals 5, 8.0–15.0 mm long, lanceolate; stamens 5, unequal, not exserted; stigmas 2; bracts and calyx with long, conspicuous pubescence; pedicels long-stalked. **FRUITS** capsule, 4.0–6.0 mm long, subglobose, shorter than sepals, 4-seeded, in clusters. **SEEDS** wedge-shaped, 3.0 mm long, 2.0 mm wide, light brown, surface dull, roughened by blisterlike protuberances.

Special Identifying Features

Erect, becoming climbing or twining, annual vine; cotyledons rounded; unlike *Ipomoea* species, flowers are in clusters, characteristically with fuzzy pubescence.

Toxic Properties

None reported.

Flowers

1mm

TOP Seeds
BOTTOM Seedling

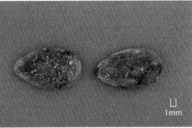

1mm

Flowers and fruit

Citronmelon

Citrullus lanatus (Thunb.) Matsumura & Nakai var. *citroides* (Bailey) Mansf. · Cucurbitaceae · Cucumber Family

Synonyms

Watermelon

Habit, Habitat, and Origin

Procumbent, trailing annual vine; to 3.0 m long; fields, cultivated areas, pastures, disturbed sites; native of Africa, introduced and escaped from cultivation.

Seedling Characteristics

Cotyledons oval to oblong, petioled; first true leaves pinnately lobed, pubescent.

Mature Plant Characteristics

ROOTS fibrous from shallow taproot. **STEMS** procumbent trailing vine, to 3.0 m long, pubescent, leafy, branched. **LEAVES** alternate, 3.0–8.0 cm long, deeply bipinnatifid, lobe tips rounded. **INFLO-**RESCENCES solitary in leaf axils, separate male and female flowers, light yellow, 2.0–3.0 cm diameter; corolla in 5 parts nearly to base. **FRUITS** hard, spherical melon with green skin and white spots or stripes; hard outward, white to pinkish flesh inward, containing many seeds; wild types inedible. **SEEDS** ovoid, 5.0–15.0 mm long, flattened, white; turning dark reddish brown, tan, or blackish brown.

Special Identifying Features

Procumbent, trailing, annual vine; fruit with hard rind, inedible white flesh; seeds white and turning dark reddish brown, tan, or blackish brown.

Toxic Properties

None reported.

Fruit

Burgherkin

Cucumis anguria L. · Cucurbitaceae · Cucumber Family

Synonyms

Wild Indian gherkin, wild spiny cucumber

Habit, Habitat, and Origin

Slender, trailing, vining annual; to 3.0 m; disturbed areas, fields, pastures, and waste sites; native of South America.

Seedling Characteristics

Cotyledons oval to oblong, entire, about half as wide as long, petiole as long as leaf blade.

Mature Plant Characteristics

ROOTS fibrous from taproot. STEMS trailing and branching vine, to 3.0 m, rough, pubescent, somewhat angled, tendrils small. LEAVES alternate, 9.0 cm long, divided into 3 main obtuse lobes with rounded sinuses, outer lobes sometimes divided. INFLORESCENCES solitary, yellow, separate male and female flowers, 9.0–13.0 mm wide; male flowers larger than female flowers, peduncles slender. FRUITS oval or oblong on crooked peduncle, 5.0 cm long, furrowed and prickly. SEEDS ovoid, 4.0–5.0 mm long, flattened, white, numerous.

Special Identifying Features

Slender, trailing, annual vine; stems rough, pubescent; leaves lobed; fruit with scattered short prickles, peduncle elongated and crooked.

Toxic Properties

Foliage and fruits may cause digestive irritation, diarrhea, and mild liver and kidney necrosis.

TOP Seeds
MIDDLE Seedling
BOTTOM Flower

Smellmelon

Cucumis melo L. · Cucurbitaceae · Cucumber Family

Synonyms

Dudaim melon, wild muskmelon
Cucumis melo L. var. *dudaim* (L.) Naud.

Habit, Habitat, and Origin

Trailing or climbing, annual, herbaceous
vine; to 3.0 m long; disturbed areas,
fields, pastures, fencerows, waste sites,
picnic areas, and trash heaps; native of
Asia.

Seedling Characteristics

Cotyledons broadly ovate to oblong-
ovate with slight notch at tip, glabrous,
stem and petioles softly pubescent;
first true leaves scabrous, orbicular, and
angled but not lobed.

Mature Plant Characteristics

ROOTS fibrous from shallow taproot.
STEMS trailing or climbing, to 3.0 m
long, branched, angled, striate, softly
pubescent. LEAVES alternate, to 13.0
cm wide, simple, angled but not lobed,
margins sinuate-dentate, pubescent,
scabrous. INFLORESCENCES separate
male and female flowers on same plant,
yellow, 25.0 mm across, 1 per leaf axil;
corolla tube of male flower 0.8–2.0
mm long, corolla tube of female flower
0.8–2.8 mm long. FRUITS broadly ellipsoid, 6.0–8.0 cm long, 4.0–6.0 cm wide,
golden yellow, brown-marbled, pubescent, extremely aromatic; flesh yellowish,
orange, or green. SEEDS ovoid, 10.0–12.0
mm long, flattened, white, slender,
numerous per fruit.

Special Identifying Features

Trailing or climbing, annual herbaceous
vine; flowers yellow; fruit extremely aromatic, otherwise resembling muskmelon.

Toxic Properties

None reported.

Flower and fruit

Buffalo Gourd

Cucurbita foetidissima Kunth · Cucurbitaceae · Cucumber Family

Synonyms

Missouri gourd

Habit, Habitat, and Origin

Trailing, coarse, perennial herb; numerous stems to 6.0 m or more long; disturbed areas, fields, pastures, fencerows, and waste sites; native of North America.

Seedling Characteristics

Cotyledons oblong, squashlike; first true leaves pubescent, coarse, triangular-ovate.

Mature Plant Characteristics

ROOTS fibrous from taproot; becoming a huge, fusiform overwintering structure producing numerous stems. **STEMS** trailing, numerous, each to 6.0 m or more long, coarse, widely running, pubescent. **LEAVES** alternate, to at least 3.0 dm long, coarse, thick, scabrous, pubescent, gray-green, triangular-ovate, unpleasant odor when bruised. **INFLORESCENCES** monoecious, solitary in axis; male flowers on long pedicels; corolla to 1.0 dm long, flaring above middle, yellow, 5-lobed; anthers united; stigmas 3–5, bilobed.

FRUITS subglobose, 5.0–10.0 cm diameter, smooth, green with lighter stripes, lemon yellow to greenish orange when ripe. **SEEDS** ovate, 6.0–12.0 mm long, flattened, smooth, light straw-colored.

Special Identifying Features

Trailing, coarse perennial herb; rootstock fusiform; leaves covered with scabrous pubescence, gray-green, unpleasant odor when bruised; fruit roundish, green-striped.

Toxic Properties

Fruits may cause digestive irritation including diarrhea, vomiting, and cramps.

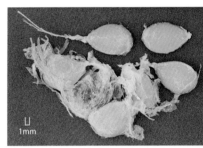

TOP **Seeds**
MIDDLE **Flowers**
BOTTOM **Mature plant**

Fruit

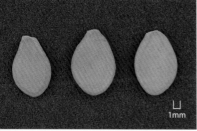

TOP Seeds
MIDDLE Seedling
BOTTOM Fruit

Flower

Texas Gourd

Cucurbita pepo L. var. *texana* (Scheele) D. Decker ·
Cucurbitaceae · Cucumber Family

Synonyms

Egg gourd; *Cucurbita texana*
(Scheele) Gray

Habit, Habitat, and Origin

Trailing or climbing, herbaceous annual
vine; stems to 10.0 m; along streams,
floodplains, fields, pastures, fencerows,
disturbed and waste sites; native of
North America.

Seedling Characteristics

Cotyledons oblong, thick, squashlike;
first true leaf roundish; older leaves
triangular to heart-shaped.

Mature Plant Characteristics

ROOTS fibrous from taproot. STEMS
trailing or climbing vine, to 10.0 m
long, slender, much-branched, abundant
tendrils, pubescent, scabrous. LEAVES
alternate, 10.0–16.0 cm long, 9.0–15.0
cm wide, simple, broadly ovate, margins
serrate and sometimes lobed, pubescent,
scabrous, cordate at base, stout-petioled,
green to grayish. INFLORESCENCES male
and female flowers separate, in leaf axils;
petals, 7.0 cm wide, orange-yellow or
greenish yellow. FRUITS pepo, 4.0–9.0
cm diameter, globular or sometimes
oblong-ovoid, light green or light green
with dark green stripes, nearly white
when ripe, rind hard and smooth. SEEDS
flattened, 8.0–10.0 mm long, light
straw-colored, numerous per fruit.

Special Identifying Features

Trailing or climbing, herbaceous annual
vine; leaves large and triangular, wedge-
or heart-shaped; flowers orange-yellow
or greenish yellow; fruit green with
lighter green stripes to nearly white
when mature.

Toxic Properties

None reported.

Burcucumber

Sicyos angulatus L. · Cucurbitaceae · Cucumber Family

Synonyms

None

Habit, Habitat, and Origin

Climbing, summer annual vine; stems to several meters long; moist sites, cultivated fields, lowlands, and fencerows, often climbing high into trees; native of North America.

Flowers and fruit

Seedling Characteristics

Cotyledons oblong, thick, green, slightly roughened with spreading trichomes on upper and lower surfaces.

Mature Plant Characteristics

ROOTS fibrous from taproot. STEMS climbing vine, to several meters long, branched, tendrils 3-forked from sides of leaves, pubescent at leaf nodes, longitudinally ridged. LEAVES alternate, 6.0–20.0 cm wide, palmate, broadly heart-shaped with 5 pointed lobes, pubescent; petiole 2.0–5.0 cm long, pubescent. INFLORESCENCES unisexual, male and female on separate stalks from same leaf axil; petals 5, uniting at base; female flowers in dense clusters of 10–20, whitish to green, sessile at end of stalk; male flowers in short clusters, stalks as long as or longer than leaves. FRUITS firm, not inflated, 1.0–2.0 cm long, indehiscent, clusters of 3–5, covered with barbed bristles. SEEDS broadly oval, 1.0 cm long, flattened, light brown to gray, smooth, dull surface, 1 per fruit.

Special Identifying Features

Climbing annual vine; stems to several meters long; leaves 5-lobed; flowers whitish to green; fruit spiny.

Toxic Properties

None reported.

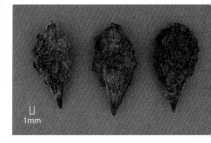

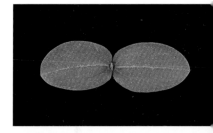

TOP Seeds
MIDDLE Cotyledons
BOTTOM Seedling

TOP **Seeds**
MIDDLE **Seedling**
BOTTOM **Rosette**

Common Teasel

Dipsacus fullonum L. · Dipsacaceae · Teasel Family

Synonyms

Wild teasel; *Dipsacus sylvestris* Huds.

Habit, Habitat, and Origin

Erect, robust, biennial herb; to 3.0 m tall; pastures, fields, wet ditches, cemeteries, roadsides, and waste areas; native of Europe.

Seedling Characteristics

Cotyledons oval to rounded, smooth, short-stalked; first true leaves paired, oval to ovate, margins dented.

Mature Plant Characteristics

ROOTS fibrous from shallow taproot. STEMS erect, 0.5–3.0 m tall, stout, striate, angled, covered with numerous prickles. LEAVES robust rosette, then opposite apically, 20.0–60.0 cm long, 4.0–10.0 cm wide; rosette leaves oblanceolate, auriculate, margins scalloped; stem leaves sessile, connate-perfoliate, lanceolate, margins entire; midribs and often margins with short prickles. INFLORESCENCES heads at first ovoid, becoming cylindrical, 3.0–10.0 cm tall, 3.0–5.0 cm wide; corolla 8.0–14.0 mm long, white to lilac; bracts leaflike, linear, curving upward; scales of elongated receptacle with straight, flexible awns. FRUITS achene surrounded by persistent calyx lobes. SEEDS achene, 2.0–3.0 mm long, quadrangular, rigid, grayish brown.

Special Identifying Features

Erect, robust biennial; rosette robust; stems, branches, peduncles, and leaf midribs armed with numerous short prickles.

Toxic Properties

None reported; may cause mechanical irritation.

Inflorescence

Flowers and fruit

Hophornbeam Copperleaf

Acalypha ostryifolia Riddell · Euphorbiaceae · Spurge Family

Synonyms

Virginia copperleaf

Habit, Habitat, and Origin

Erect annual herb; to 8.0 dm tall; fields, cultivated areas, roadsides, and waste sites; native of North America.

Seedling Characteristics

Cotyledons pubescent; stems with short pubescence; first leaves toothed, sparsely pubescent.

Mature Plant Characteristics

ROOTS fibrous from taproot. STEMS erect, 3.0–8.0 dm tall, simple to usually branched, pubescent; trichomes sparse to dense, glandular or short and recurved. LEAVES alternate, 3.0–10.0 cm long, 1.5–5.0 cm wide, simple, ovate to rhombic-ovate, margins serrate, sparsely pubescent, petiole 2.0–7.0 cm long, stipules to 1.5 mm long. INFLORESCENCES terminal and axillary on separate spikes, monoecious; female flowers terminal, 3.0–11.0 cm long, subtending bracts deeply cleft; male florets on axillary spikes, 1.0–3.0 cm long, crowded, subtending bracts 2.0–7.0 mm long, 5.0–12.0 mm wide. FRUITS capsule, 2.0–3.5 mm long, with tuberculate projections, 3-seeded. SEEDS ovoid, 1.4–2.3 mm long, silvery gray, tuberculate.

Special Identifying Features

Erect annual; leaf blades somewhat heart-shaped at base; flowers with female spike terminal; fruit a pubescent capsule.

Toxic Properties

Plants may cause digestive irritation, colic, and diarrhea, but to a lesser extent than *A. virginica*.

TOP Seeds
MIDDLE Four-leaf seedling
BOTTOM Young plant

TOP Seeds
BOTTOM Flowers

Seedling

Virginia Copperleaf

Acalypha virginica L. · Euphorbiaceae · Spurge Family

Synonyms

Three-seeded Mercury

Habit, Habitat, and Origin

Erect, rarely branched annual herb; to 7.0 dm tall; open woodlands, fields, roadsides, cultivated areas, and waste sites; native of North America.

Seedling Characteristics

Cotyledons smooth, rounded; first true leaves pubescent with crenate margins.

Mature Plant Characteristics

ROOTS fibrous from taproot. **STEMS** erect, 2.0–7.0 dm tall, simple or rarely branched, pubescent; trichomes long and spreading, sparse to dense. **LEAVES** alternate, 2.0–12.0 cm long, 4.0–36.0 mm wide, simple, ovate, narrowly rhombic to broadly lanceolate, margins shallowly crenate, petiole 2.0–60.0 mm long, exceeding floral bract in axil, stipules to 1.0 mm long. **INFLORESCENCES** axillary spikelets, monoecious; female flowers 2–8, subtended by bracts, bracts 1–3 at base of spike, 4.6–7.6 mm long, 5.0–20.0 mm wide; male flowers few to many on upper portion of spike, to 2.0 cm long, bracts with 9–15 deeply cleft lobes. **FRUITS** capsule, 2.0–2.4 mm long, pubescent, 3-lobed, 3-seeded. **SEEDS** ovoid, 1.2–1.8 mm long, reddish to gray or brown, pitted in shallow rows, rarely mottled.

Special Identifying Features

Erect, rarely branched annual; leaf blade longer than petiole; flowering bracts with 9–15 deeply cleft lobes.

Toxic Properties

Plants cause digestive irritation, colic, and diarrhea.

Flowers and fruit

Texasweed

Caperonia palustris (L.) St. Hil. · Euphorbiaceae · Spurge Family

Synonyms

Sacatropo

Habit, Habitat, and Origin

Erect annual herb; to 10.0 dm tall; open wet soils, roadsides, rice fields, cultivated soils, open woodlands, and waste sites; native of warmer parts of America; introduced into Texas circa 1920.

Seedling Characteristics

Cotyledons smooth; stems and petioles coarsely pubescent.

Mature Plant Characteristics

ROOTS fibrous from taproot. STEMS erect, 3.0–10.0 dm tall, stout, simple or with few ascending branches above, coarsely pubescent, trichomes simple and spreading or glandular-tipped, lower portions becoming glabrous as trichomes slough off. LEAVES alternate, 3.0–15.0 cm long, simple; broadly lanceolate, lanceolate, or lance-oblong; bases rounded, apices acute, margins serrate; petiole 3.0–25.0 mm long, pubescent. INFLORESCENCES axillary, monoecious; female flowers 1–6 per spike, petals shorter than calyx; male flowers 6–20 per spike, petals longer than calyx, stamen short. FRUITS capsule, about 3.0 mm long, 6.0 mm wide basally, subtended by persistent calyx, pubescent; trichomes broad-based, flat, glandular-tipped. SEEDS globose, 2.5 mm diameter, dark brown, surface minutely pitted; trichomes few, flat, plain.

Special Identifying Features

Erect annual; stems, leaves, petioles, and capsules coarsely pubescent; unique fruiting structure; plants in wet habitats.

Toxic Properties

None reported.

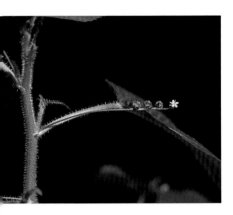

TOP Flower
BOTTOM Fruit

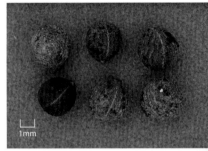

1mm

TOP Seeds
BOTTOM Seedling

Prostrate Spurge

Chamaesyce humistrata (Engelm. ex Gray) Small · Euphorbiaceae · Spurge Family

Synonyms

Euphorbia humistrata Engelm.

Habit, Habitat, and Origin

Prostrate, highly branched annual herb; stems to 5.0 dm long; stream and pond edges, low fields, lawns, cultivated areas, and waste sites; native of North America.

Seedling Characteristics

Cotyledon and hypocotyl glabrous, red.

Mature Plant Characteristics

ROOTS fibrous from taproot. **STEMS** prostrate, 1.0–5.0 dm long, highly branched, pubescent, rooting at nodes, milky sap. **LEAVES** opposite, 5.0–15.0 mm long, 2.0–8.0 mm wide, simple, oblong to oblong-ovate, variable in size, margins entire or irregularly and minutely toothed, milky sap, petiole short. **INFLORESCENCES** cyathium, 1 or 2 per leaf axil; male and female flowers separate; male flowers 2–5 per cyathium; female flowers with 3 styles, 0.5–0.7 mm long, bifid their length. **FRUITS** capsule, 1.4 mm long, ovate in outline, pubescent, trichomes appressed. **SEEDS** irregularly quadrangular, 1.0 mm long, pale brown to reddish silvery, faces papillose, wrinkled, surface not ridged.

Special Identifying Features

Prostrate, highly branched annual; stem and leaves with milky sap; stems and capsule pubescent; stems rooting at nodes; fruiting styles 2, as long as capsule; seed wrinkled, angled, not ridged.

Toxic Properties

Plants produce an irritant sap when wounded.

TOP Seeds
MIDDLE Seedling
BOTTOM Fruit and flower

Flowering branch

Flowers

Hyssop Spurge

Chamaesyce hyssopifolia (L.) Small · Euphorbiaceae · Spurge Family

Synonyms

Euphorbia hyssopifolia L.

Habit, Habitat, and Origin

Erect, branched annual herb; to 8.0 dm tall; cultivated areas, fields, pastures, roadsides; native of subtropical America.

Seedling Characteristics

Cotyledons and stem glabrous, reddish; first true leaves glabrous, opposite, irregularly toothed.

Mature Plant Characteristics

ROOTS fibrous from taproot. **STEMS** erect, 1.0–8.0 dm tall, branches arching outward then ascending, green and red, glabrous, milky sap. **LEAVES** opposite, 5.0–40.0 mm long, 3.0–15.0 mm wide, simple, oblong or oblong-elliptic, variable in size, margins irregularly toothed, occasionally long trichomes near base, otherwise glabrous, petiole short. **INFLORESCENCES** cyathium in stalked, compound cymes, appearing clustered at node, male and female flowers separate; male flowers 4–5 per cyathium; female flowers with 3 styles, 0.5–0.9 mm long, bifid more than half their length. **FRUITS** capsule, 1.5–2.1 mm long, 1.4–2.0 mm wide, glabrous. **SEEDS** oblong, 1.3–1.6 mm long, 1.0–1.4 mm wide, pale brown, angled, 2 facets with 2–4 transverse low rounded ridges.

Special Identifying Features

Erect, branched annual; stems and leaves with milky sap; stems erect, branches arching outward then ascending; fruiting capsule glabrous; seed with low transverse ridges.

Toxic Properties

See comment under *Chamaesyce humistrata*.

1mm

TOP Seeds
MIDDLE Seedling
BOTTOM Stem

TOP Seeds
MIDDLE Seedling
BOTTOM Fruit

Flowers

Spotted Spurge

Chamaesyce maculata (L.) Small · Euphorbiaceae · Spurge Family

Synonyms

Euphorbia maculata L., *E. supina* Raf. ex Boiss.

Habit, Habitat, and Origin

Prostrate or decumbent, highly branched, annual herb; stems to 5.0 dm long; cultivated areas, fields, roadsides, lawns, turf, and pastures; native of North America.

Seedling Characteristics

Cotyledons glabrous; stem red, pubescent; first true leaves opposite, irregularly toothed.

Mature Plant Characteristics

ROOTS fibrous from taproot; prostrate stems seldom rooting at nodes. **STEMS** prostrate or decumbent, 0.5–5.0 dm long, highly branched, radiating from a central point, red, with milky sap, pubescent. **LEAVES** opposite, 4.0–17.0 mm long, 1.0–10.0 mm wide, simple; oblong-ovate, ovate-elliptic, to linear-oblong; variable in size, margins minutely serrate, milky sap, few to numerous long trichomes or glabrous, petiole 1.0–1.5 mm long. **INFLO-RESCENCES** cyathium, solitary at a node but appearing clustered, involucre 1.0 mm long, glands 4, unequal, male and female flowers separate; male flowers 2–5 per cyathium; female flowers with 3 styles, 0.3–0.4 mm long, cleft into lobes one-fourth to one-third their length. **FRUITS** capsule, 1.4 mm long, ovoid-triangular, pubescent. **SEEDS** oblong-quadrangular, 1.0 mm long, pale brown, with irregularly low transverse ridges.

Special Identifying Features

Prostrate or decumbent, highly branched annual; not rooting at nodes; stems and leaves with milky sap; stems and fruiting capsules pubescent; seeds with low rounded ridges.

Toxic Properties

See comment under *Chamaesyce humistrata*.

Nodding Spurge

Chamaesyce nutans (Lag.) Small · Euphorbiaceae · Spurge Family

Synonyms

Spotted spurge; *Euphorbia nutans* Lag.

Habit, Habitat, and Origin

Erect summer annual herb; to 8.0 dm tall; cultivated areas, fields, pastures, roadsides; native of North America.

TOP **Flowers and fruit**
BOTTOM **Stem**

Seedling Characteristics

Cotyledons glabrous; stem red, glabrous; first true leaves opposite, margins irregularly toothed.

Mature Plant Characteristics

ROOTS fibrous from taproot. STEMS erect, 3.0–8.0 dm tall, dichotomously branched, green and red, milky sap, glabrous except for stem tips, upper nodes with white pubescence, trichomes crispy. LEAVES opposite, 0.5–4.0 cm long, 3.0–15.0 mm wide, simple, oblong or oblong-elliptic, bases inequilateral, milky sap, margins minutely serrate, glabrous or a few long trichomes at base, petiole 1.0–2.0 mm long or essentially absent. INFLORESCENCES cyathium, solitary or in forks or in cymose clusters, glands 4–5, 0.3–0.5 mm diameter, petaloid appendages smaller than or equal to gland length, male and female flowers separate; male flowers 5–29 per cyathium, single stamen attached to involucre side; female flowers center of involucre, exserted. FRUITS capsule, 2.0–2.5 mm long, 1.8–2.2 mm wide, glabrous, 3-lobed. SEEDS ovoid, 1.2 mm long, dark gray, irregularly wrinkled.

Special Identifying Features

Erect summer annual; stems and leaves with milky sap; leaves opposite; stem tips pubescent; fruiting capsule glabrous; seed wrinkled.

Toxic Properties

See comment under *Chamaesyce humistrata.*

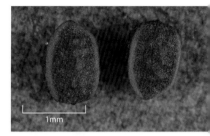

TOP **Seeds**
BOTTOM **Seedling**

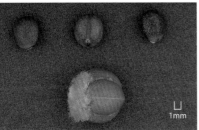

Flowering plant

TOP Seeds and fruit
MIDDLE One-leaf seedling
BOTTOM Four-leaf seedling

Woolly Croton

Croton capitatus Michx. · Euphorbiaceae · Spurge Family

Synonyms

Goatweed, hogwort

Habit, Habitat, and Origin

Erect annual herb; to 1.2 m tall; fields, pastures, roadsides, woodland edges, and waste sites; native of North America.

Seedling Characteristics

Cotyledons and stem pubescent, trichomes star-shaped, stem orange-brown.

Mature Plant Characteristics

ROOTS fibrous from taproot. STEMS erect, 0.2–1.3 m tall, dichotomous or umbel-branched above, orange or orange-brown, pubescent, trichomes stellate. LEAVES alternate, 4.0–15.0 cm long, 1.5–6.0 cm wide, simple, lanceolate, pubescent, appearing white or tawny, margins entire, petiole 2.0–9.0 cm long, stipules subulate, deciduous. INFLORESCENCES monoecious, terminal; female petals absent, styles 3, deeply dichotomous, ovary subglobose, calyx 6–9-lobed, 2.0–3.0 mm long; male flowers with 5 petals, 1.0 mm long, stamens 7–12, sepals 5, 1.0 mm long, subulate to deltoid. FRUITS capsule, 6.0–9.0 mm long, nearly globose, 3-seeded, pubescent, trichomes stellate. SEEDS ovoid, 3.5–5.2 mm long, glossy, brown.

Special Identifying Features

Erect annual; entire plant pubescent with star-shaped trichomes; stems orange to orange-brown.

Toxic Properties

Croton species contain irritants and if ingested can cause diarrhea and colic, and can adversely affect nursing.

Tropic Croton

Croton glandulosus L. var. *septentrionalis* Muell.-Arg. · Euphorbiaceae · Spurge Family

Synonyms

Vente conmigo

Habit, Habitat, and Origin

Erect, summer annual herb; to 5.0 dm tall; prairies, pastures, fields, road-sides, and waste sites; native of North America.

Seedling Characteristics

Cotyledons heart-shaped, 5.0–7.0 mm long, 7.0–10.0 mm wide; first true leaves rounded to oval, distinctly toothed, generally longer than wide.

Mature Plant Characteristics

ROOTS fibrous from shallow taproot. **STEMS** erect, 1.0–5.0 dm tall, rough, green-ish, somewhat umbel-branched, pubes-cent but not woolly. **LEAVES** alternate, 1.0–6.5 cm long, 0.5–3.0 mm wide, simple, oblong to egg-shaped, margins sharply serrate; trichomes glandular, whitish cartilaginous, saucer-shaped, on each side of petiolar attachment; petioles short. **INFLORESCENCES** monoecious, terminal racemes 1.0 cm long; female flowers essentially without petals, styles 3.0–4.0 mm long, grayish, sepals 5, 1.5 mm long, whitish; male flower petals 4–5, stamens 6–13, sepals 5. **FRUITS** capsule, 4.0–6.0 mm long, solitary or in small clusters at base of upper leaves. **SEEDS** ovate or nearly elliptic, about 3.0 mm long.

Special Identifying Features

Erect summer annual; leaves serrate, pubescent; trichomes glandular, white, saucer-shaped.

Toxic Properties

See comments under *Croton capitatus.*

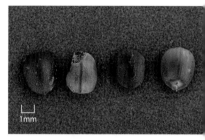

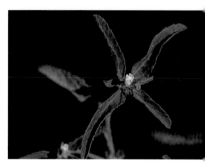

TOP **Seeds**
BOTTOM **Flower**

Flowering plant

Seedling

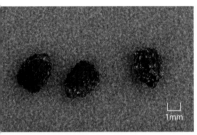

TOP Seeds
MIDDLE Seedling
BOTTOM Fruit

Painted Poinsettia

Euphorbia cyathophora Murr. · Euphorbiaceae · Spurge Family

Synonyms

Wild poinsettia, painted leaf; *Poinsettia cyathophora* (Murray) Klotzsch & Garcke

Habit, Habitat, and Origin

Erect annual herb; to 7.0 dm tall; moist soils, open woodlands, roadsides, fields, and waste sites; native of North America.

Seedling Characteristics

Hypocotyl smooth, glabrous; cotyledons oblong, shorter than first leaf; seedling leaves elliptic.

Mature Plant Characteristics

ROOTS fibrous from slender taproot. STEMS erect, 4.0–7.0 dm tall, alternately branched, light green, smooth, glabrous or pubescent, trichomes few and short, milky sap. LEAVES alternate, 4.0–9.5 cm long, 2.0–4.2 mm wide, simple; oblanceolate, lanceolate, or linear; margins entire, slightly toothed, or with several large lobes; uppermost bracteal leaves with distinct red patch; milky sap. INFLORESCENCES cyathium lacking petalous lobes, involucre elongate to elongate-turbinate, 3.0–4.5 mm long; female flowers with ovary glabrous, style and stigma 1.0–1.5 mm long; male flowers 35–60, anthers yellow; bracteal leaves usually red or rarely yellow at base. FRUITS capsule, 4.1–5.2 mm long, 3.5–4.0 mm wide, smooth, distinctly 3-chambered.

SEEDS ovate-elliptical, 2.5–3.0 mm long, 2.0–2.5 mm wide, brown or black, rough, tuberculate.

Special Identifying Features

Erect annual; stem alternately branched; leaves alternate; upper bracteal leaves with showy red patch on upper surface, stems and leaves with milky sap.

Toxic Properties

Plants produce an irritant sap when wounded.

Flowering plant

Flowering plant

Toothed Spurge

Euphorbia dentata Michx. · Euphorbiaceae · Spurge Family

Synonyms

Poinsettia dentata (Michx.) Klotzsch & Garcke

Habit, Habitat, and Origin

Erect, spreading, and branched annual herb; to 8.0 dm tall; sandy levees, prairies, fields, roadsides, fencerows, and waste sites; native of North America.

Seedling Characteristics

Cotyledons oval, petioled, leaves opposite, serrate margins, pubescent.

Mature Plant Characteristics

ROOTS fibrous from taproot. **STEMS** erect, spreading, or curved upward; 2.0–8.0 dm tall, often branched, pubescent. **LEAVES** mostly opposite, 4.0–9.0 cm long, 1.0–3.0 cm wide, simple, variously shaped from lanceolate to ovate, margins coarsely toothed, slender petioles, pubescent on both sides. **INFLORESCENCES** cyathium; male flowers small, 25–40 per cyathium; female flowers with 3 styles, cleft into lobes about one-half their length; involucres in terminal clusters, nearly sessile, subtended by opposite leaves pale at base, 5 oblong lobes, 1 or a few short-stalked glands, subtended by opposite leaves. **FRUITS** capsule, 5.0 mm diameter, smooth, trilocular, each locule with 1 seed. **SEEDS** oblong, 2.5–3.0 mm long; dark brown, gray, or black; sharply angled, roughly tuberculate.

Special Identifying Features

Erect or spreading annual; plants with milky sap; leaves opposite, margins coarsely toothed, pubescent; fruiting capsule smooth; seeds ovoid, roughly tuberculate.

Toxic Properties

See comments under *Euphorbia cyathophora*.

TOP **Fruit and seeds**
MIDDLE **Seedling**
BOTTOM **Flowers**

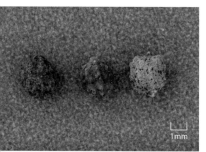

Inflorescence

TOP Seeds
MIDDLE Seedling
BOTTOM Flowering plant

Wild Poinsettia

Euphorbia heterophylla L. · Euphorbiaceae · Spurge Family

Synonyms

Various-leaved spurge; *Poinsettia hetero-phylla* (L.) Klotzsch & Garcke

Habit, Habitat, and Origin

Erect, ascending, or spreading branched annual herb; to 9.0 dm tall; cultivated areas, fields, roadsides, shallow ditches, pastures, and waste sites; native of North America.

Seedling Characteristics

Hypocotyl smooth, glabrous; cotyledons linear, at least as long as first true leaf.

Mature Plant Characteristics

ROOTS fibrous from slender taproot. **STEMS** erect, ascending, or spreading; 6.0–9.0 dm tall, pale green, smooth, occasional pubescence at nodes, branches opposite. **LEAVES** opposite, 2.0–20.0 cm long, simple, 0.5–6.0 cm wide; linear, lanceolate, or elliptic; margins 2-lobed, gray-green with small red blotches, petioles pubescent. **INFLORESCENCES** cyathium, without petal-like lobes, glandular, 1.0 mm wide, single, depression disklike. **FRUITS** capsule, smooth, distinctly trilocular, each locule with 1 seed. **SEEDS** oblong-ovoid, 2.1–2.5 mm long, brownish black, wrinkled, and rough.

Special Identifying Features

Erect, ascending, or spreading plants; cotyledons linear; leaves variable; red blotches present but not as extensive as red coloration of cultivated poinsettia.

Toxic Properties

See comments under *Euphorbia cyathophora.*

Snow-on-the-Mountain

Euphorbia marginata Pursh · Euphorbiaceae · Spurge Family

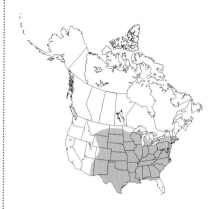

Synonyms

Snowy spurge

Habit, Habitat, and Origin

Erect annual herb; to 7.0 dm tall; roadsides, pastures, disturbed fields, and waste sites; native of North America.

Seedling Characteristics

Leaves 10–20; stem broad, rounded 15.0 cm prior to elongation.

Mature Plant Characteristics

ROOTS fibrous from simple taproot. **STEMS** erect, 0.5–7.0 dm tall, lower part unbranched, upper stem branched first into 3 branches and then whorled, divergent, densely villous. **LEAVES** lower leaves alternate, 3.0–8.0 cm long, simple, sessile, somewhat clasping, oblong to ovate, mostly green, margins glabrous to ciliate; upper leaves whorled at base of first branches and opposite at forks, narrower than lower leaves, with wide white margins. **INFLORESCENCES** solitary in forks of branches; peduncles slender, erect, 7.0–15.0 mm long, densely villous; glands 5, oblong, 1.0 mm long, cupped, yellow with red tinge; appendages white, 2.0–3.0 mm, obtuse, margins irregular; male flowers about 35; female flowers exserted, reflexed. **FRUITS** capsule, 5.0 mm wide, flat-globose, with 3 rounded lobes, usually pubescent, columella well-developed. **SEEDS** ovoid, 4.0 mm long, dark gray, tubercles in reticulate pattern.

Special Identifying Features

Erect herb; flowering late spring to fall; top, flowering portion branching like a compound cyme having 3 or more branches at each division; leaves with white margins above branching part of stem.

Toxic Properties

See comments under *Euphorbia cyathophora*.

TOP Seeds and fruit
MIDDLE Flowering plant
BOTTOM Mature plant

Flowers

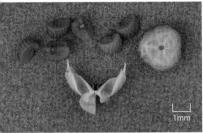

TOP Seeds and fruit
MIDDLE Seedlings
BOTTOM Mature plant

Chamber Bitter

Phyllanthus urinaria L. · Euphorbiaceae · Spurge Family

Synonyms

Gale of wind, leafflower, niruri

Habit, Habitat, and Origin

Erect, spreading, or procumbent annual herb; to 5.0 dm tall; shaded disturbed areas including lawns, flowerbeds, and container pots; native of tropical East Africa.

Seedling Characteristics

Cotyledons reddish, glabrous, spatulate; first 2 leaves apparently opposite.

Mature Plant Characteristics

ROOTS fibrous from long taproot. STEMS erect, spreading, or procumbent; 0.5–5.0 dm tall, main stem narrow, single or sparsely branching, smooth, green, primary axis angled, with spirally arranged scale leaves. LEAVES alternate, 6.0–25.0 mm long, 2.0–9.0 mm wide, simple, oblong or lanceolate, on side branches only, appearing pinnately compound. IN-FLORESCENCES monoecious; male flowers in groups on distal nodes; female flowers solitary on proximal nodes, small. FRUITS capsule, 2.0–2.2 mm wide, tuberculate to nearly smooth. SEEDS oval, 1.1–1.2 mm long, light brown, with sharp transverse ridges and 1–3 lateral pits.

Special Identifying Features

Erect, spreading, or decumbent annual; flower-bearing branches and leaves appear to be a compound leaf, but leaves are simple and flowers are borne on lower surface of branches; pinnately compound leaves.

Toxic Properties

Plants produce an irritant sap when wounded.

Inflorescence

Flower

Indian Jointvetch

Aeschynomene indica L. · Fabaceae · Bean Family

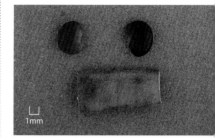

Synonyms

Sensitive vetch

Habit, Habitat, and Origin

Erect, often bushy, branched annual herb; to 2.5 m tall; moist soil, shallow water, and cultivated areas, especially rice fields; native of North America.

Seedling Characteristics

Cotyledons ovate in outline, green, and slightly furrowed at midrib, margins entire; first true leaves compound.

Mature Plant Characteristics

ROOTS fibrous from taproot. STEMS erect, 0.5–2.5 m tall, often bushy-branched, fine longitudinal lines or ridges furrowed, glabrous or glabrate below, pubescent above, trichomes glandular. LEAVES alternate, 5.0–12.0 cm long, evenly pinnately compound; leaflets 50–70, 5.0–8.0 mm long, 0.8–2.5 mm wide, glabrous; stipules lance-shaped, smooth, or minutely toothed; leaflets sensitive, folding when touched. INFLORESCENCES borne on axillary racemes, 2.0–10.0 cm long; pedicel 2.0–5.0 mm long, subtended by toothed bract 3.0–5.0 mm long; petals yellowish to reddish purple, standard 0.8–1.0 cm long. FRUITS legume, 2.0–5.0 cm long, oblong-linear, compressed, with disarticulating segments. SEEDS kidney-shaped, 4.0–5.0 mm long, light brown to black.

Special Identifying Features

Erect, bushy annual; leaves evenly pinnately compound; leaflets sensitive, folding when touched.

Toxic Properties

None reported.

TOP Seeds and fruit segment
MIDDLE One-leaf seedling
BOTTOM Four-leaf seedling

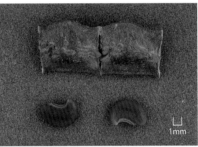

TOP Seeds
MIDDLE Seedling
BOTTOM Fruit

Flower

Rough Jointvetch

Aeschynomene rudis Benth. · Fabaceae · Bean Family

Synonyms

Zigzag jointvetch

Habit, Habitat, and Origin

Erect, robust annual herb; to 2.0 m tall; moist soil or shallow water in cultivated areas, especially rice fields; native of tropical Americas.

Seedling Characteristics

Cotyledons ovate in outline, green, furrow near midrib not conspicuous, margins entire; first true leaves compound.

Mature Plant Characteristics

ROOTS fibrous from taproot. STEMS erect, 1.0–2.0 m tall, with fine longitudinal furrows or ridges, glabrous or pubescent, trichomes stiff and pustular. LEAVES alternate, 4.0–15.0 cm long, evenly pinnate; leaflets 20 to numerous, 5.0–20.0 mm long, oblong, margins entire, glabrous, 1-nerved, stipules retrorsely lobed. INFLORESCENCES slightly foliate short raceme, 1–3 flowers, pedicels generally glandular-pustulate and 2.0–5.0 mm long, bracts finely toothed; corolla 9.0–15.0 mm long, yellow to copper red; calyx 5.0–9.0 mm, bilabiate, margin toothed. FRUITS straight legume, 3.0–6.0 cm long, 4.0–6.0 mm wide, upper margin entire to slightly undulate, lower margin undulate, 7–12 disarticulating loments 4.0–6.0 mm long and wide, broad isthmi between loments, surface covered with stiff, pustulate pubescence, medially muricate. SEEDS kidney-shaped, 4.0–5.0 mm long, light brown to black.

Special Identifying Features

Erect annual; stems often with stiff, pustular trichomes; fruit stalk 4.0–7.0 mm long, 7–12-lomentaceous, rough, and generally covered by stiff, pustulate trichomes.

Toxic Properties

None reported.

Showy Crotalaria

Crotalaria spectabilis Roth · Fabaceae · Bean Family

Synonyms

Rattlebox

Habit, Habitat, and Origin

Erect, summer annual herb; to 2.0 m tall; cultivated areas, fields, pastures, roadsides, and waste sites; native of India.

Seedling Characteristics

Cotyledons bean-shaped, thick, green above, light green below; first true leaves smooth above, densely covered with appressed trichomes below.

Mature Plant Characteristics

ROOTS fibrous from taproot. STEMS erect, 0.5–2.0 m tall, stout, green or purplish, usually ribbed. LEAVES alternate, 5.0–15.0 cm long, simple, broadest at apex, wedge-shaped at base, smooth above, densely pubescent below; stipules 5.0–8.0 mm long, 4.0–6.0 mm wide, ovate to lance-shaped, persistent.

INFLORESCENCES borne on few to several racemes clustered near top of plant; large, showy, spirally arranged, petals yellow, stalk subtended by ovate to ovate-lanceolate persistent bract 7.0–12.0 mm long, 5.0–9.0 mm wide. FRUITS legume, 3.0–5.0 cm long, cylindrical pod with inflated appearance, turning black at maturity, splitting lengthwise; seed dissemination projectile, up to several meters. SEEDS kidney-shaped, brown to black at maturity, distinctly notched at base.

Special Identifying Features

Erect annual; flowers showy, bright yellow; pod cylindrical with inflated appearance; projectile dissemination of seeds.

Toxic Properties

Plants contain alkaloids that damage the liver and lungs.

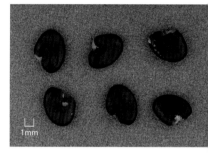

TOP Seeds
BOTTOM Seedling

Flowers

Fruit

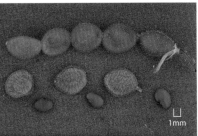

TOP **Seeds and fruit**
BOTTOM **Seedling**

Florida Beggarweed

Desmodium tortuosum (Sw.) DC. · Fabaceae · Bean Family

Synonyms

Beggar lice, beggar tick, stick-tight

Habit, Habitat, and Origin

Erect, much-branched summer annual; to 3.0 m tall; cultivated areas, fields, pastures, roadsides, open woodlands, and waste sites; native of North America, West Indies, and South America.

Seedling Characteristics

Cotyledons round to oval in outline, margins smooth, blue-green; first true leaf usually unifoliate.

Mature Plant Characteristics

ROOTS fibrous from taproot. STEMS erect, 0.5–1.5 (rarely 3.0) m tall, much-branched, rounded, with fine longitudinal lines; reddish purple; usually covered with dense, short, stiff pubescence. LEAVES alternate; lower leaves may be unifoliate, others pinnately trifoliate; leaflets elliptic to oblong-elliptic, 2.0–8.0 cm long, 1.0–3.0 cm wide; stipules persistent, ovate to lance-shaped, 0.5–1.0 cm long. INFLO-RESCENCES usually borne on terminal or axillary simple racemes or branched panicles near top of plant; petals 5.0–7.0 mm long, purple to bluish purple; calyx short-pilose to pilose. FRUITS loment, 1.0–4.0 cm long, flattened, constricted, 2–7 oval to suborbicular segments between seeds, densely covered with short, stiff trichomes. SEEDS kidney-shaped, 2.0–4.0 mm long, light brown to brown.

Special Identifying Features

Erect, many-branched annual; fruit with distinctive shape, segmented, with rough texture that adheres to clothing, hair, fur, and other objects.

Toxic Properties

None reported.

Flowers

Leaf axil

Bagpod

Glottidium vesicarium (Jacq.) Harper · Fabaceae · Bean Family

Synonyms

Bagpod sesbania; *Sesbania vesicaria* (Jacq.) Elliott

Habit, Habitat, and Origin

Erect annual herb; to 3.0 (rarely 5.0) m tall; open, usually wet, disturbed pastures, roadsides, swamps, or agricultural soils; native of North America.

Seedling Characteristics

Cotyledon oblong; first true leaf elliptic to ovate.

Mature Plant Characteristics

ROOTS fibrous from much-branching taproot. STEMS erect, 1.0–3.0 (rarely 5.0) m tall, branching, green toward top, brownish toward base, smooth. LEAVES alternate, 0.8–2.0 dm long, evenly compound along a central axis; 16–52 leaflets, short-petioled, oblong to elliptic, 1.5–4.0 cm long, 2.5–6.0 mm wide, margins entire, rounded at base, rounded or pointed at tip, silky-pubescent becoming smooth at maturity; stipules 7.0–10.0 mm long, linear, pointed at tip, early deciduous. INFLORESCENCES borne on 3–12 loosely disposed axillary racemes; petals 8.0–9.0 mm long, orange to yellowish orange, often maroon-tipped or red- or maroon-tinged; calyx minimally lobed to nonlobed. FRUITS legume, 2.0–8.0 cm long, 1.5–2.0 cm wide, oblong to elliptical, flattened, strongly beaked, 1 or 2 (usually 2) seeds per seedpod, persistent in winter. SEEDS oblong, 1.0 cm long, 0.5 cm wide, plump, reddish brown.

Special Identifying Features

Erect, branching annual 3.0–5.0 m tall; flower calyx minimally lobed to nonlobed; pods 1–2-seeded, flattened, butterbean-like.

Toxic Properties

Plants contain toxins causing depression, colic, diarrhea, and weakness.

TOP Mature plant
BOTTOM Flowers

1mm

TOP Seeds
BOTTOM Seedling

Everlasting Peavine

Lathyrus latifolius L. · Fabaceae · Bean Family

Synonyms

Perennial peavine, perennial sweetpea, sweetpea

Habit, Habitat, and Origin

Ascending, climbing, or twining perennial herb; stems to 2.0 m long; disturbed fields, cultivated areas, pastures, roadsides, and waste sites; native of Europe.

Seedling Characteristics

Cotyledons glabrous, linear; first true leaves with 2 leaflets.

Mature Plant Characteristics

ROOTS fibrous from weak taproot. STEMS ascending, climbing, or twining, and somewhat vinelike; 1.0–2.0 m long, slightly flattened, glabrous, slightly glaucous, broad-winged; wing segments thin, green, somewhat membranous. LEAVES compound and bearing branched tendrils distally at base with 2 leaflets in a single pair per leaf; leaflets 5.0–10.0 cm long, oblong-lanceolate to elliptic, tips acute, strongly nerved; stipules 3.0–5.0 cm long, usually single. INFLORESCENCES borne in loose raceme atop long, curved axillary peduncle, 5–15 per inflorescence, 2.5 cm long; petals usually purple, less commonly red, pink, or white. FRUITS legume, 6.0–10.0 cm long, 7.0–10.0 mm diameter, linear, cylindrical, glabrous, dehiscent, splitting and coiling lengthwise when dry, 10–25 seeds per pod. SEEDS subspherical to cuboidal, black to brown, surface texture slightly rough.

Special Identifying Features

Ascending, climbing, or twining perennial; stems dramatically winged; leaves compound, with basally protruding tendrils; leaflets 2 per leaf; flowers purple, pink, red, or white.

Toxic Properties

None reported.

Mature plant

Sericea Lespedeza

Lespedeza cuneata (Dumont) G. Don · Fabaceae · Bean Family

Synonyms

Chinese bush clover, Chinese lespedeza, pubescent lespedeza, Himalayan bush clover, silky bush clover; *Lespedeza juncea* var. *sericea* Forbes and Hemsl., *L. sericea* Miq.

Habit, Habitat, and Origin

Erect, branching, semiwoody perennial herb; to 1.8 m tall; fields, pastures, roadsides, and waste sites; native of eastern Asia, introduced as a forage crop and for erosion control and soil improvement.

Seedling Characteristics

Cotyledons broadly ovate; first 2 true leaves simple, opposite, broadly ovate to suborbicular with notched tip.

Mature Plant Characteristics

ROOTS fibrous from taproot; lateral branches to more than 1.0 m long. **STEMS** erect, 0.9–1.8 m tall, branching, arising from root crown buds, pubescent; trichomes sharp, stiff, flattened. **LEAVES** alternate, compound; 3 leaflets, each 5.0–25 mm long, 1.5–6.0 mm wide; leaf base wedge-shaped, apex rounded with short sharp-pointed tip, pubescent, short trichomes giving gray-green to silvery appearance; stipules narrow, linear, 3.0–11.0 mm long. **INFLORESCENCES** single or in groups of 2–4 in axils of median to upper leaves; petaliferous flowers, 3.0–4.0 mm long, creamy white with pink to purple throat, especially in veins; apetalus flowers 1.5–2.0 mm long, without showy petals. **FRUITS** legume, 3.0–5.0 mm long, oval, flattened, with appressed pubescence. **SEEDS** shiny, flattened and oval with a shallow notch near one end, tan to olive.

Special Identifying Features

Erect, branching, semiwoody perennial; leaves wedge-shaped at base and rounded with sharp point at tip, pubescent, gray-green to silvery.

Toxic Properties

None reported.

TOP Seeds
MIDDLE Seedling
BOTTOM Flowers

TOP Seeds
MIDDLE Seedling
BOTTOM Flowering plants

Perennial Lupine

Lupinus perennis L. · Fabaceae · Bean Family

Synonyms

Blue pea, Indian bean, old maid's bonnets, Quaker's bonnets, wild lupine

Habit, Habitat, and Origin

Erect or ascending perennial herb; to 6.0 dm tall; sandy soils in clearings, roadsides, fields, and pastures; native of North America.

Seedling Characteristics

Cotyledons ovate; first true leaves palmately foliate, pubescent.

Mature Plant Characteristics

ROOTS fibrous from taproot; producing perennial spreading rootstocks. STEMS erect, 2.0–6.0 dm tall, stout, branched, succulent, glabrous or pubescent, trichomes long and shaggy. LEAVES alternate, palmately foliate, 7–11 oblanceolate leaflets; leaflets 1.5–5.0 cm long, 3.0–12.0 mm wide, apex mucronate, pubescent; petiole 2.0–15.0 cm long, slender; stipules filiform, 8.0–12.0 mm long. INFLORESCENCES terminal raceme, 1.0–3.0 dm long, on a short peduncle, densely few-flowered; petals showy, 1.0–1.8 cm long, blue, purplish blue, or rarely pinkish, 2-lipped; calyx tube 2.0 mm long. FRUITS legume, 3.0–5.0 cm long, 8.0–10.0 mm wide, oblong, 4–7-seeded, densely short-pubescent to villous. SEEDS oblong, yellow to white with brown or black markings.

Special Identifying Features

Erect or ascending perennial from spreading rootstocks; flowers showy, blue to purplish blue; leaflets 7–11, usually 8, pubescent, petiole slender.

Toxic Properties

None reported, but many *Lupinus* species are cyanogenic.

Flowers

Flowering
plant

Yellow Sweetclover

Melilotus officinalis (L.) Pall. · Fabaceae · Bean Family

Synonyms

Hubam, meli, white sweetclover; *Melilotus alba* L.

Habit, Habitat, and Origin

Erect, annual, occasionally biennial herb; to 2.0 m tall; disturbed habitats; native of Europe, imported and planted to increase soil nitrogen.

Seedling Characteristics

Cotyledons oblong; first true leaf simple, remaining leaves compound.

Mature Plant Characteristics

ROOTS fibrous from taproot. STEMS erect, 0.2–2.0 m tall, branching, green, glabrous to slightly pubescent apically. LEAVES alternate, compound, pinnately trifoliate; leaflets 1.0–2.5 cm long, 5.0–20.0 mm wide, oblanceolate to obovate, margins minutely serrate, stipules fused toward base. INFLORESCENCES in slender, elongate raceme at least four to eight times as long as wide, 35–75 per raceme; petals 3.0–5.0 mm long, white or yellow; stamens 10, united; sepals papilionaceous, green. FRUITS legume, 1.8–3.4 mm long and wide, globose, nearly indehiscent, very short-stalked, usually dark brown to blackish at maturity, 1-seeded, glabrous, reticulate-veined, calyx not inflated below pod. SEEDS yellowish or greenish yellow, not mottled, one side flat, the other smoothly rounded.

Special Identifying Features

Erect annual or occasionally biennial; flowers white or yellow; pod covered with network of coarse nerves; seed not mottled.

Toxic Properties

Improperly cured yellow sweetclover hay can interfere with postsurgical blood coagulation in livestock.

TOP Seeds
MIDDLE Three-leaf seedling
BOTTOM Cotyledons

TOP **Seeds**
BOTTOM **Seedling**

Kudzu

Pueraria montana (Lour.) Merr. var. *lobata* (Willd.) Maesen & S. M. Almeida ·
Fabaceae · Bean Family

Synonyms

Pueraria lobata (Willd.) Ohwi

Habit, Habitat, and Origin

Trailing or climbing perennial vine commonly known as "the weed that took over the South"; up to 30.0 m long; roadsides, forests, fields, hillsides, gullies, and waste areas, climbing on almost anything; native of Asia.

Seedling Characteristics

Vinelike young stems covered with long, fine trichomes.

Mature Plant Characteristics

ROOTS fibrous from taproot; then enlarged, long, deep, mealy, tuberous rhizomes. STEMS high-climbing and twining, up to 30.0 m long, herbaceous but may become woody; young stems pubescent, up to 2.5 cm thick. LEAVES alternate, pinnately trifoliate, leaflets broadly ovate in outline, 5.0–12.0 cm long, margins entire or 2–3-lobed, pubescent beneath, petiole as long as leaf, stipules ovate to lance-shaped, 8.0–12.0 mm long. INFLORESCENCES showy, borne in axillary racemes up to 20.0 cm long, largest petal violet purple to reddish purple with yellow spot at base, odor of grape drink. FRUITS legume, 4.0–5.0 cm long, linear-oblong, somewhat flattened, reddish brown, pubescent, trichomes long and fine. SEEDS round to oval in outline, somewhat compressed.

Special Identifying Features

Perennial with trailing or climbing vines up to 30.0 m long; flowers have odor of grape drink; most infestations are on highly erodible soils on roadsides or in forests.

Toxic Properties

None reported.

Fruit

Flowers

Flower

Hemp Sesbania

Sesbania herbacea (P. Mill.) McVaugh · Fabaceae · Bean Family

Synonyms

Coffeeweed; *Sesbania exaltata* (Raf.) Rydb. ex A. W. Hill, *S. macrocarpa* Muhl.

Habit, Habitat, and Origin

Erect annual herb; to 4.0 m tall; cultivated areas, pastures, fields, ditches, roadsides, and waste sites; native of North America.

Seedling Characteristics

Cotyledons lance- to spoon-shaped, two times longer than wide; first true leaf simple.

Mature Plant Characteristics

ROOTS fibrous from taproot. STEMS erect, 1.0–4.0 m tall, branching, green, glabrous, may become woody with age, round to slightly angled. LEAVES alternate, up to 30.0 cm long, pinnately compound; leaflets 20–70, opposite, 1.0–3.0 cm long, 2.0–6 mm wide, glabrous above and somewhat pubescent below; stipules linear-lanceolate, soon deciduous, 1.0 cm long. INFLORESCENCES usually only 2–6 on racemes borne in leaf axils; largest petal 1.2–1.5 cm long, yellow, may be streaked or spotted with purple. FRUITS legume, 10.0–20.0 cm long, linear,

somewhat compressed, curved, beaked, containing 30–40 seeds. SEEDS brown, mottled with black, two times longer than wide.

Special Identifying Features

Erect, branching herb; pod distinctive, curved, often tipped with beak, 0.5–1.0 cm long; largest flower petal showy, yellow.

Toxic Properties

Many *Sesbania* species contain toxins causing depression, colic, diarrhea, and weakness.

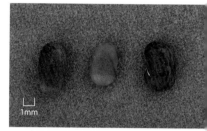

1mm

TOP Seeds
BOTTOM Seedling

Fruit

Crimson Clover

Trifolium incarnatum L. · Fabaceae · Bean Family

Synonyms

Italian clover

Habit, Habitat, and Origin

Ascending or erect annual or winter annual herb; to 20.0 cm tall; fields, roadsides, and pastures; native of southern Europe.

Seedling Characteristics

Cotyledons oblong; first true leaf simple, remainder trifoliate.

Mature Plant Characteristics

ROOTS fibrous from taproot. STEMS ascending or erect, 5.0–20.0 cm long, elongate, pubescent. LEAVES alternate, palmately trifoliate; leaflets fewer than 3, usually one to two times as long as broad, broadly obovate, somewhat cuneate to nearly orbicular, toothed near apex, sessile or nearly so; stipules mostly 1.0–2.0 cm long, fused to petiole base. INFLORESCENCES nearly sessile and very numerous on head, spikelike, at least twice as long as wide, 2.0–7.0 cm long, peduncle 4.0–12.0 cm, terminal; corolla 8.0–12.0 mm long; sepals united; calyx tube cylindrical to campanulate, 3.5–5.0 mm long, 10-nerved, lobes subulate; upper petal much longer than wing or keel, scarlet or red, rarely white. FRUITS legume, sessile, ovoid, shorter than calyx, 1 seed per pod. SEEDS oval, 2.5–3.0 mm long, rotund, yellow, smooth, glossy; hilum on side, scarcely indented, thick rim around depressed hilar area.

Special Identifying Features

Ascending or erect annual or winter annual; leaves palmately trifoliate, rarely more than twice as long as wide; flower heads twice as long as wide, on peduncles, corolla red.

Toxic Properties

None reported.

Four-leaf seedling

Inflorescence

Flowers

Red Clover

Trifolium pratense L. · Fabaceae · Bean Family

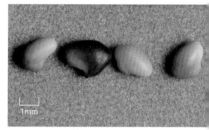

Synonyms

None

Habit, Habitat, and Origin

Erect or ascending, cool-season, short-lived perennial or biennial herb; up to 6.0 dm tall; fields, roadsides, and pastures; native of Europe.

Seedling Characteristics

Cotyledons spatula-shaped, 6.0–7.0 mm long, glabrous; first true leaf unifoliate, oval, truncate; subsequent leaves trifoliate.

Mature Plant Characteristics

ROOTS fibrous from taproot and stolons; nitrogen-fixing nodules present. **STEMS** prostrate or occasionally erect, up to 6.0 dm tall, glabrous or usually densely pubescent, rooting at nodes. **LEAVES** trifoliate; leaflets 2.0–3.0 cm long, 1.0–1.5 cm wide, elliptic to ovate, margins entire, each with a prominent white

V; leaves usually pubescent on both sides but sometimes only on lower surface; stipules 1.0–3.0 cm long, ovate-lanceolate or ovate. **INFLORESCENCES** nearly globose cluster, 1.0–3.0 cm long, sessile or peduncle less than 5.0 mm long; corolla 1.2–1.8 cm long, pink to purplish; calyx tube 10-nerved, pubescent. **FRUITS** legume, 4.0–5.0 mm long, ovoid or oblong, smooth above, wrinkled below, 1 seed per pod. **SEEDS** mitten-shaped, 2.0–3.0 mm long, yellow to brown to purple.

Special Identifying Features

Erect or ascending, cool-season perennial or biennial; leaves trifoliate, each leaflet with a prominent white V; flowers pink to purple; plants larger than white clover.

Toxic Properties

None reported.

TOP Seeds
MIDDLE Seedling
BOTTOM Flowering plant

White Clover

Trifolium repens L. · Fabaceae · Bean Family

Synonyms

Dutchclover, honeysuckle clover, ladino clover, purple-grass, purplewort, white trefoil

Habit, Habitat, and Origin

Erect or ascending, mat-forming perennial herb; to 4.0 dm tall; fields, lawns, roadsides, and pastures; native of northern Eurasia.

Seedling Characteristics

Cotyledons spatulate; first true leaves simple, broadly oblong to ovate.

Mature Plant Characteristics

ROOTS fibrous from taproot and stolons. STEMS erect or ascending, 0.3–4.0 dm tall, glabrous; stolons prostrate, creeping, often forming mats, rooting at most nodes. LEAVES palmately trifoliate or rarely quadrifoliate (four-leaf clover), from stolons, 0.5–4.0 cm long; leaflets obovate to obcordate with small-toothed margins, white-membranous, lanceolate; stipules tubular; petiole 5.0–20.0 cm long, glabrous. INFLORESCENCES nearly globose, 40–85-flowered, 1.0–3.0 cm diameter; corolla white turning pink and then brown, 7.0–12.0 mm long; calyx tube cylindrical, 2.0–3.0 mm long, lobes lanceolate, 5–10 nerves to each sinus between lobes, ending in a purple spot or dark purplish area; peduncle 0.3–3.0 dm long, erect. FRUITS legume, 3.0–5.0 mm long, oblong, sessile. SEEDS globose to reniform, 1.5 mm long, yellowish, 3–4 seeds per pod.

Special Identifying Features

Erect or ascending stoloniferous perennial; leaves and flowering stems arising from stolons; flower head full, white, globose, with up to 85 flowers; sinuses between each calyx lobe purplish or purple-spotted.

Toxic Properties

Plants can be infected with a fungus that causes profuse salivation, lacrimation, diarrhea, bloat, stiffness, and anorexia in livestock.

TOP Seeds
BOTTOM Four-leaf seedling

Flowering plant

Flowers

Narrowleaf Vetch

Vicia sativa L. ssp. *nigra* (L.) Ehrh. · Fabaceae · Bean Family

Synonyms

Blackpod vetch, common vetch; *Vicia angustifolia* L.

Habit, Habitat, and Origin

Decumbent to ascending, more or less climbing, annual herb; to 8.0 dm long; fields, cultivated areas, roadsides, and pastures; native of Europe and the Middle East.

Seedling Characteristics

Glabrous to short-pubescent, stout with small cotyledons; first true leaflets paired.

Mature Plant Characteristics

ROOTS fibrous from shallow taproot. **STEMS** decumbent to ascending, tending to climb, 1.0–8.0 dm long, glabrous or pubescent, trichomes short. **LEAVES** alternate, compound, leaflets 6–12, narrow, 1.5–2.0 cm long, all but lowermost with branched tendrils; stipules 2.0–7.0 mm long, toothed, with nectaries. **INFLORESCENCES** pealike, 2 in subsessile cluster in upper axils; petals rose-purple, 1.0–1.8 cm long. **FRUITS** legume, 3.4–5.0 cm long, 4.0–6.0 mm wide, almost round, very dark brown to black at maturity, initially with short pubescence,

8–10-seeded. **SEEDS** rounded, about 2.0 mm diameter, dark brown mottled with black spots.

Special Identifying Features

Decumbent to ascending, more or less climbing annual; flowers paired, pealike, in leaf axils; seedpods black, nearly round.

Toxic Properties

Ingested seeds of *Vicia* species can cause neuropathy, dermatopathy, and favism.

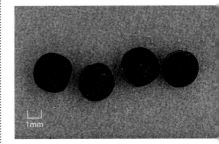

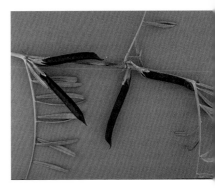

TOP Seeds
BOTTOM Fruit

Seedling

Hairy Vetch

Vicia villosa Roth · Fabaceae · Bean Family

Synonyms

Smooth vetch, winter vetch; *Vicia dasy-carpa* Ten.

Habit, Habitat, and Origin

Erect or ascending herbaceous perennial, often acting as a winter annual herb; to 1.0 m tall; fields, roadsides, pastures, and waste sites; native of Europe.

Seedling Characteristics

Villous to almost smooth; cotyledons linear, small.

Mature Plant Characteristics

ROOTS fibrous from shallow taproot. STEMS erect or ascending, 0.5–1.0 m long, tending to climb, villous to nearly smooth, slender. LEAVES alternate, compound, with 8–20 pairs of narrowly oblong leaflets; leaflets 1.0–2.5 cm long, pubescent; with terminal, branched tendril, stipules persistent, pubescent. INFLORESCENCES long-peduncled raceme with 10–40 flowers; calyx irregular, villous; corolla slender, 12.0–20.0 mm long, purple or rarely white. FRUITS legume, 20.0–40.0 mm long, dark to light straw color, glabrous, oblong, flattened, 4–5-seeded. SEEDS spherical to nearly sublenticular, 3.4–5.0 mm diameter, smooth, greenish to reddish brown, may be mottled.

Special Identifying Features

Erect or ascending herbaceous perennial, often acting as a winter annual; stems smooth or pubescent; leaves compound,

8–20 pairs of leaflets, terminal branched tendril on upper leaves; flowers in raceme.

Toxic Properties

See comment under *Vicia sativa* ssp. *nigra*.

TOP **Flowers**
BOTTOM **Fruit**

TOP **Seeds**
BOTTOM **Seedling**

Fruit

Cowpea

Vigna unguiculata (L.) Walpers · Fabaceae · Bean Family

Synonyms

Black-eyed pea, China-bean, southern pea, field pea

Habit, Habitat, and Origin

Trailing, climbing, or vining annual herb; to 3.0 m long; fields, roadsides, pastures, and waste sites; native of Asia.

Seedling Characteristics

First pair of true leaves unifoliate, ovate, opposite; subsequent leaves trifoliate.

Mature Plant Characteristics

ROOTS fibrous from taproot. **STEMS** trailing, climbing, or vining; 0.5–3.0 m long, smooth or slightly pubescent, green. **LEAVES** alternate, compound, trifoliate; leaflets 4.0–15.0 cm long, ovate to ovate-hastate, glabrous or nearly so on both surfaces; stipules lanceolate, to 1.5

cm long. **INFLORESCENCES** papilionaceous raceme; most petals purple but may be white or violet, upper petal 1.5–2.5 cm long, peduncle 1.5–2.5 dm long. **FRUITS** legume, linear, 1.0–3.0 dm long, 1.0 cm wide, usually glabrous. **SEEDS** kidney-shaped, 8.0 mm long, 5.0 mm wide, flattened, light brown or tan, black, or red-maroon depending on cultivar.

Special Identifying Features

Trailing, climbing, or vining annual; leaves trifoliate, glabrous; flowers purple; seedpod linear.

Toxic Properties

None reported.

1mm

TOP Seeds
MIDDLE Seedling
BOTTOM Flower

1mm

Flowers

Redstem Filaree

Erodium cicutarium (L.) L'Hér. ex Ait. · Geraniaceae · Geranium Family

Synonyms

Alfileria, common storksbill, filaria, heron's-bill, pin-clover, stork's bill

Habit, Habitat, and Origin

Prostrate or ascending winter annual herb; to 5.0 dm tall; fields, lawns, roadsides, and waste sites; native of Europe.

Seedling Characteristics

Stem and petiole coarsely pubescent; first leaves deeply lobed.

Mature Plant Characteristics

ROOTS fibrous from shallow taproot. STEMS prostrate or ascending, 4.0–5.0 dm tall, short until flowering, diffusely branched, pubescent. LEAVES basal rosette, alternate or opposite above, 1.0–2.5 cm long, elongate-oblanceolate in general outline, pinnately compound with several sessile, ovate or oblong, deeply pinnate, cleft lobes. INFLORESCENCES long-peduncled cyme, 2–8 flowers per cyme, each 1.0 cm wide; corolla pink to purple; anther bearing filaments without teeth; pedicels 1.0–2.0 cm long. FRUITS carpel, 2.0–4.0 cm long, stork's-bill-like beak, becoming spirally twisted when dry. SEEDS ellipsoid, 2.0–3.0 mm long, brown.

Special Identifying Features

Prostrate or ascending winter annual; leaves pinnately compound with deeply cleft lobes; fruit with stork's-bill beak.

Toxic Properties

Fruit causes contact irritation to livestock, and some *Erodium* species cause photosensitization in calves and are cyanogenic after light frost.

Carolina Geranium

Geranium carolinianum L. · Geraniaceae · Geranium Family

Synonyms

Carolina cranesbill, Carolina wild geranium, cranesbill, wild geranium

Habit, Habitat, and Origin

Erect, widely branching, winter annual or biennial herb; to 7.0 dm tall; fields, lawns, roadsides, and waste sites; native of North America.

Seedling Characteristics

Cotyledons apically broadly truncate or slightly indented, pubescent, 6.0 mm wide and broad; first leaves deeply lobed, forming a basal rosette, pubescent.

Mature Plant Characteristics

ROOTS fibrous from shallow taproot. STEMS erect, 3.0–7.0 dm tall, freely branching near base, densely pubescent. LEAVES alternate near base, opposite above, 2.0–6.0 cm wide, deeply cut into 5–9 fingerlike lobed or toothed segments, pubescent on both surfaces. INFLORESCENCES at tips of stems and branches, 2 to several in compact clusters; petals whitish pink to pale purple, 4.0–6.0 mm long. FRUITS carpel, body less than 3.0 mm long, with elongated beak 8.0–12.0 mm long, splitting at maturity into 5 curled sections, covered with short pubescence, trichomes nonglandular or few longest ones glandular. SEEDS oval to slightly oblong, 1.5–2.0 mm long, light to dark brown, surface prominently reticulate, 1 per fruit segment.

Special Identifying Features

Erect, widely branching winter annual or biennial; leaves divided; flowers pink to pale purple; fruit with distinctive elongated beak.

Toxic Properties

None reported.

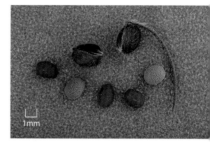

TOP Seeds
MIDDLE Seedling
BOTTOM Fruit

Flower

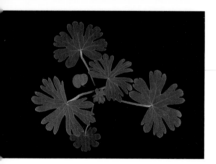

TOP **Seeds**
BOTTOM **Seedling**

Flower and fruit

Cutleaf Geranium

Geranium dissectum L. · Geraniaceae · Geranium Family

Synonyms

Cutleaf cranesbill, cut-leaved geranium

Habit, Habitat, and Origin

Erect or spreading, widely branching, winter annual or biennial herb; to 7.0 dm tall; fields, lawns, roadsides, and waste sites; native of Eurasia.

Seedling Characteristics

Cotyledons broadly truncate or slightly indented apically, 6.0 mm long and wide; first leaves deeply lobed, forming a basal rosette.

Mature Plant Characteristics

ROOTS fibrous from shallow taproot. **STEMS** erect, 2.0–7.0 dm tall, freely branching near base, densely pubescent. **LEAVES** basal rosette, then alternate near base, opposite upward, 2.0–6.0 cm wide, deeply palmate, divided into 5–7 sec-

tions, sections again divided into linear segments; pubescent on both surfaces. **INFLORESCENCES** paired, 5 petals, purple to deep purplish pink, 4.0–6.0 mm long. **FRUITS** carpel, body less than 3.0 mm long excluding elongate beak, densely pubescent, trichomes glandular. **SEEDS** oval to slightly oblong, 1.5–2.0 mm long, light to dark brown, surface prominently reticulate, 1 per fruit segment.

Special Identifying Features

Erect or spreading, widely branching, winter annual or biennial; leaves divided into 5–7 sections, each section again divided into linear segments; flowers paired, purple to deep purplish pink; fruit with distinctive elongated beak.

Toxic Properties

None reported.

Ground Ivy

Glechoma hederacea L. · Lamiaceae · Mint Family

Synonyms

Creeping Charlie, gill-over-the-ground, run-away-robin

Habit, Habitat, and Origin

Creeping perennial herb; stems to 1.0 m long; in lawns, yards, flowerbeds, and other disturbed areas; native of Eurasia.

Seedling Characteristics

Cotyledons kidney-shaped, small notch at tip, glabrous; first leaves opposite, kidney- to heart-shaped, pubescent, margins crenate.

Mature Plant Characteristics

ROOTS fibrous from taproot, stems, stolons, and rhizomes. STEMS creeping, to 1.0 m long, slender, square, with sparse pubescence; flowering stems 0.5–2.0 dm tall, erect, rooting at nodes. LEAVES opposite, 1.5–5.0 cm wide, round to cordate or reniform, pubescent, margins crenate, ciliate, green or purplish green on one or both sides; petioles on flowering stem 0.5–4.0 cm long, and on prostrate stem up to 10.0 cm long. INFLORESCENCES terminal or axillary cyme, clusters of 2–7 flowers; corolla blue-violet, speckled reddish purple, bilaterally symmetrical, 1.0–2.0 cm long, or in pistillate plants only 8.0–15.0 mm long, upper lip smaller, lower lip larger, 3-lobed, bearded in throat with club-shaped trichomes; calyx slightly bilaterally symmetrical, 4.0–8.0 mm long, 15-nerved, lobes sharp-pointed, triangular. FRUITS dry indehiscent fruit or mericarp, 1.3–2.0 mm long, 1.0 mm wide, reddish brown, ovate, smooth except for slight bumps at apex. SEEDS nutlet, 1.0–2.0 mm long, 1.0 mm wide, flat on two sides and round on the third, dark brown.

Special Identifying Features

Creeping perennial herb; leaves opposite; stems square; flowers reddish purple, speckled, terminal or in leaf axils.

Toxic Properties

Plants are reported to cause colic, excessive salivation, and sweating in livestock.

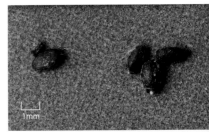

TOP **Seeds**
BOTTOM **Seedling**

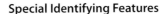

TOP **Flowers**
BOTTOM **Flowering plant**

TOP Seeds
MIDDLE Two-leaf seedling
BOTTOM Four-leaf seedling

Henbit

Lamium amplexicaule L. · Lamiaceae · Mint Family

Synonyms

Dead-nettle, henbit dead-nettle, henbit-nettle

Habit, Habitat, and Origin

Erect or decumbent winter annual or biennial herb; to 4.0 dm tall; cultivated areas, fields, pastures, lawns, and waste sites; introduced from Eurasia.

Seedling Characteristics

Cotyledons oval, 3.0–12.0 mm long, 1.0–4.0 mm wide, smooth, midvein ending in gland at leaf tip; petiole with spreading trichomes; hypocotyl green becoming dull purple, smooth.

Mature Plant Characteristics

ROOTS fibrous from taproot. STEMS green- or purple-tinged, decumbent with numerous ascending branches, 1.0–4.0 dm tall, nearly square in cross section, rooting at lower nodes, pubescent, trichomes retrorsely spreading. LEAVES opposite, 0.6–3.5 cm long, triangular to circular, margins with rounded teeth, venation palmate, dark green above, light green below; trichomes above soft, minute, and on prominent veins below; middle and upper leaves sessile, lower with petioles. INFLORESCENCES whorls in axis of upper leaves; corolla fused, 1.5–1.8 cm long, 2-lipped, reddish purple. FRUITS mericarp, 1.5–2.4 mm long. SEEDS obovate, blunt apically, triangular in cross section, mottled light and dark brown or occasionally with white granules, 4 per mericarp, retained in mericarp.

Special Identifying Features

Stems square; leaves reflexed; flowers reddish purple, without pedicels.

Toxic Properties

Plants reported to contain neurotoxins causing stilted gait, tremors, and collapse in sheep.

Flowers

Flowering plant

TOP Seeds
MIDDLE Two-leaf seedling
BOTTOM Four-leaf seedling

Lanceleaf Sage

Salvia reflexa Hornem. · Lamiaceae · Mint Family

Synonyms

Blue sage, lambsleaf sage, Rocky Mountain sage, sage mint

Habit, Habitat, and Origin

Erect annual herb; to 8.0 dm tall; disturbed areas, pastures, fields, and roadsides; native of North America.

Seedling Characteristics

Glabrous or pubescent, cotyledon spade-shaped; first leaves opposite and oblong.

Mature Plant Characteristics

ROOTS fibrous from taproot. STEMS erect, 1.0–8.0 dm tall, branched above, square, pubescent or glabrous. LEAVES opposite, 3.0–6.5 cm long, 4.0–15.0 mm wide, simple, lanceolate to narrowly oblong or elliptic, glabrous above, pubescent or glabrous beneath, tip rounded or obtuse, fragrant, petiole 3.0–20.0 mm long. IN-FLORESCENCES whorls of flowers around stem in interrupted spike, to 10.0 cm long; corolla blue to white, with distinct upper and lower lips; bracts lanceolate to ovate-lanceolate, 2.0–6.0 mm long, bell-shaped. FRUITS mericarp, aggregate of 4 nutlets. SEEDS nutlet, 2.0–2.5 mm long, triangular, outer face elliptic, convex, smooth, light tan or buff and mottled with dark brown.

Special Identifying Features

Erect, bushy-branched annual; stems square; leaves opposite, lanceolate, fragrant; flowering summer and fall, flowers blue to white.

Toxic Properties

Plants can produce acute nitrate intoxication; contaminated hay can poison cattle and sheep.

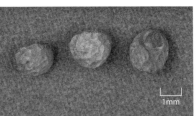

TOP Seeds
MIDDLE Seedling
BOTTOM Flowering plant

American Germander

Teucrium canadense L. · Lamiaceae · Mint Family

Synonyms

Germander, wood sage

Habit, Habitat, and Origin

Erect, simple or sparsely branched, perennial herb; to 1.5 m tall; moist soils in open or shaded areas; native of North America.

Seedling Characteristics

Cotyledons linear-lanceolate, smooth, yellow to yellowish green; hypocotyl brownish; true leaves linear to linear-lanceolate, smooth.

Mature Plant Characteristics

ROOTS fibrous from taproot and rhizomes, sometimes producing whitish tubers. **STEMS** erect, 0.3–1.5 m tall, simple or sparsely branched, square, retrorsely pubescent; trichomes spreading, soft or sometimes villous. **LEAVES** opposite, 3.0–16.0 cm long, 1.0–6.0 cm wide; blades ovate, elliptic, ovate-lanceolate, or lanceolate; margins serrate or crenate; upper surface lightly pilose to glabrate, lower surface sparsely to densely canescent, base cuneate, apex acute or short-acuminate, petiole 0.3–2.0 cm long. **INFLORESCENCES** compact cylindrical or ovate panicle with axillary cymes (thryse), rarely leafy; zygomorphic; corolla pink, light rose, lavender, or purple, rarely white, 1.0–1.5 cm long; calyx 5-lobed, lower lip 3-lobed with middle lobe longest, upper lobes shorter than lower, pubescent. **FRUITS** mericarp of 4 nutlets. **SEEDS** nutlet, 1.5–2.4 mm long; ovoid, obovoid, or ellipsoid; reddish or yellowish brown, glabrous, reticulate, wrinkled toward tip.

Special Identifying Features

Erect perennial forming colonies from rhizomes; stems square; leaves opposite; flowers zygomorphic, pink to purple; pubescent throughout.

Toxic Properties

None reported.

Flowers

Flowers

Redstem

Ammannia coccinea Rottb. · Lythraceae · Loosestrife Family

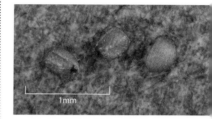

1mm

Synonyms

Purple ammannia, toothcup; *Ammannia teres* Raf.

Habit, Habitat, and Origin

Erect annual herb; to 10.0 dm tall; in or along edges of shallow water in rice fields, wet ditches, ponds, and lakes; native of North America.

Seedling Characteristics

Cotyledons reddish, glabrous, linear to linear-lanceolate; first leaves reddish, glabrous, linear to linear-lanceolate, sessile.

Mature Plant Characteristics

ROOTS fibrous from taproot. STEMS erect, 2.0–10.0 dm, usually simple to freely branching, reddish, square, angles slightly winged. LEAVES opposite, 2.0–10.0 cm long, 2.0–16.0 mm wide, linear to linear-lanceolate or linear-oblong, sessile, auriculate to clasping. INFLORESCENCES in tight axillary cymes, 3–5 per cyme, sessile, including style 1.5–3.0 mm long; corollas 4–5, 1.5–2.0 mm long, 1.5–2.0 mm wide, purple to rose-pink; stamens 4–7, exserted; anthers dark yellow. FRUITS capsule, 3.5–5.0 mm diameter, reddish, style persistent. SEEDS more or less triangular in outline with one face rounded and granular, the others flat and reticulate, olive brown, minute, numerous per capsule.

Special Identifying Features

Erect annual; stems square, slightly winged, reddish; leaves auriculate to clasping; flower purple, not stalked, style 1.5–3.0 mm long.

Toxic Properties

None reported.

TOP **Seeds**
MIDDLE **Seedling**
BOTTOM **Flowering plant**

TOP Seeds
MIDDLE Seedling
BOTTOM Flowers

Purple Loosestrife

Lythrum salicaria L. · Lythraceae · Loosestrife Family

Synonyms

Spiked loosestrife

Habit, Habitat, and Origin

Erect, aquatic perennial herb; to 2.3 m tall; reproducing from seeds or root crowns and invading vast areas of wetlands; native of Eurasia.

Seedling Characteristics

Cotyledons up to 5.0 mm long, orbicular becoming oblong to broadly oblong, almost triangular at base; petioles flattened; first true leaves up to 2.0 cm long, opposite, oblong, sessile, prominent venation, glabrous or nearly so.

Mature Plant Characteristics

ROOTS fibrous from single or branched taproot. STEMS erect, 0.6–2.3 m tall, glabrous to finely pubescent, round to angular, up to 50 arising from large crown. LEAVES opposite or whorled, alternate and smaller near inflorescence, 2.0–15.0 cm long, 5.0–15.0 mm wide, simple, lanceolate to narrowly oblong or nearly linear, margins entire, sessile, larger leaves cordate at base, glabrous to finely pubescent. INFLORESCENCES terminal spike, 10.0–40.0 cm long, whorl-like cymes numerous, small, pubescent to subtomentose; petals 5–7, free, red-purple, 7.0–10.0 mm long; styles 3; stamens twice as many as petals; sepals united into column with 8–12 prominent green veins ending in several long, thin, pointed lobes. FRUITS capsule, oblong-ovoid, enclosed by persistent hypanthium. SEEDS broadly ovoid, less than 1.0 mm, red-brown, numerous per capsule.

Special Identifying Features

Escaped perennial ornamental with regrowth from crown; flowers numerous, very showy, red-purple.

Toxic Properties

None reported.

Flowering plants

Flowering plant

Toothcup

Rotala ramosior (L.) Koehne · Lythraceae · Loosestrife Family

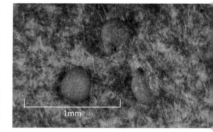

Synonyms

Rowland rotala

Habit, Habitat, and Origin

Erect, glabrous annual; to 4.0 dm tall; marshes, ditches, and wet depressions; native of North America.

Seedling Characteristics

Cotyledons fleshy; first true leaves fleshy and shallowly toothed.

Mature Plant Characteristics

ROOTS fibrous from shallow taproot. **STEMS** erect, 1.0–4.0 dm tall but usually less than 2.0 dm, usually branched, glabrous. **LEAVES** opposite, 1.0–5.0 cm long, 1.0–10.0 mm wide, linear to lanceolate or oblanceolate, bases tapering, sessile or short-petiolate. **INFLORESCENCES** axillary, usually single, sessile, quadrimerous; floral tubes 2.0–5.0 mm long; petals 4–6, white, pink, or purple; stamens 4–6; calyx tube 3.0–4.5 mm long, lobed. **FRUITS** capsule, 2.0–5.0 mm diameter, pyramidal or subglobose, yellow, 2–4-locular. **SEEDS** ovoid, 1.0 mm long, yellow or red, numerous per capsule.

Special Identifying Features

Erect glabrous annual; leaves opposite; flowers single, quadrimerous, axillary; distinguished from *Ammannia coccinea* by usually having a single flower per leaf axil.

Toxic Properties

None reported.

TOP **Seeds**
MIDDLE **Seedling**
BOTTOM **Flower**

TOP **Seeds**
BOTTOM **Seedling**

Fruit

Flower

Okra

Abelmoschus esculentus (L.) Moench · Malvaceae · Mallow Family

Synonyms

Gumbo, ladyfingers

Habit, Habitat, and Origin

Erect, stout, branched annual herb; to 2.0 m tall; fields, disturbed areas, and waste sites; native of India or Africa.

Seedling Characteristics

Cotyledons round, finely pubescent; first leaves not lobed, with serrate margins; stem pubescent.

Mature Plant Characteristics

ROOTS fibrous from thick taproot. **STEMS** erect, 0.5–2.0 m tall, stout, branched, pubescent. **LEAVES** alternate, 10.0–40.0 cm long, 5.0–40.0 cm wide, variable in shape but usually deeply lobed, pubescent, trichomes stinging. **INFLORESCENCES** solitary, axillary or terminal; corolla with 5 cream to light yellow petals, about 5.0 cm tall and wide, each petal with spot in center. **FRUITS** columnar capsule, 8.0–30.0 cm long, ribbed, beaked, containing mucilage, sparsely pubescent, trichomes stinging. **SEEDS** round, 1.5–4.0 mm diameter, grayish to brown, 200 or more per capsule.

Special Identifying Features

Erect, stout, branched annual; flowering summer and fall, flowers showy, cream to light yellow; fruit and seedpods distinctive, containing mucilage; weedy, edible variant escaped from cultivation.

Toxic Properties

None reported.

Velvetleaf

Abutilon theophrasti Medik. · Malvaceae · Mallow Family

Synonyms

Butterprint, buttonweed, Indian mallow, piemarker, wild cotton

Habit, Habitat, and Origin

Erect, sparingly branched, annual herb; to 2.1 m tall; cultivated areas, fields, pastures, roadsides, and waste sites; native of Asia.

Seedling Characteristics

Cotyledons: one rounded and one heart-shaped with rounded apices, margins entire, covered on both surfaces and along margins with short, nonglandular trichomes; true leaves alternate, ovate to heart-shaped with round to acute apices, serrate margins, velvety pubescent surfaces; stem densely pubescent.

Mature Plant Characteristics

ROOTS fibrous from taproot. **STEMS** erect, 0.5–2.1 m tall, sparingly branched on upper portion, smooth, pubescent, trichomes short and velvety. **LEAVES** alternate, 5.0–15.0 cm long and wide, simple, heart-shaped, with long, tapered tips, fine-toothed margins, pubescent, trichomes velvety. **INFLORESCENCES** solitary, borne on short peduncles in leaf axils of upper portion of stem; petals 5, orange-yellow, to 2.5 cm diameter, peduncle jointed near middle. **FRUITS** capsule, 2.5 cm diameter, 10–15 beaked carpels with ring of prickles around upper edge, pubescent. **SEEDS** ovate, 3.0 mm long, strongly notched, flattened, dull grayish brown.

Special Identifying Features

Erect, sparingly branched annual; cotyledons: one rounded, one heart-shaped; leaves densely pubescent, trichomes velvety; strong, unpleasant odor when crushed.

Toxic Properties

Plants contain toxins that cause visible organ damage, depression, fluid accumulation, and mild liver necrosis.

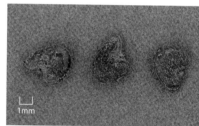

TOP Seeds
MIDDLE Two-leaf seedling
BOTTOM Four-leaf seedling

Flower and fruit

Spurred Anoda

Anoda cristata (L.) Schlecht. · Malvaceae · Mallow Family

Synonyms

Crested anoda

Habit, Habitat, and Origin

Erect, freely branched annual herb; to 1.0 m tall; cultivated areas, fields, pastures, roadsides, and waste sites; native of southwestern North America and South America.

Seedling Characteristics

Cotyledons ovate to heart-shaped with rounded apices, generally one rounded and one heart-shaped; margins entire with short, gland-tipped trichomes; first leaves alternate, pubescent on both surfaces and along margins, irregularly and slightly toothed, stems densely pubescent.

Mature Plant Characteristics

ROOTS fibrous from taproot. STEMS erect, 0.4–1.0 m tall, stout, freely branching especially from base, vertically ridged apically, densely pubescent. LEAVES alternate, 5.0–10.0 cm long, 3-lobed, triangular-ovate or hastate, margins coarsely toothed, often with purplish veins, pubescent on both surfaces. INFLO-RESCENCES solitary in leaf axils, 7.0–12.0 mm wide; petals pale blue or lavender; calyx 2.0 cm wide, much exceeding the 10–20 carpels. FRUITS capsule consisting of 10–20 beaked, 1-seeded carpels. SEEDS kidney-shaped, 2.8–3.2 mm long, dark brown to black, surface with short trichomes or fine bumps or both.

Special Identifying Features

Erect, freely branched annual; cotyledons: one rounded and one heart-shaped, veins purplish, trichomes along margins short and gland-tipped; calyx star-shaped and much exceeding carpels.

Toxic Properties

None reported.

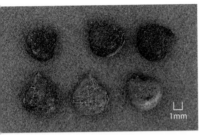

TOP Seeds
MIDDLE Seedling
BOTTOM Fruit

Flower

Flower

Venice Mallow

Hibiscus trionum L. · Malvaceae · Mallow Family

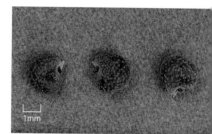

Synonyms

Bladder-ketmia, flower-of-an-hour, shoo-fly

Habit, Habitat, and Origin

Erect or spreading, branched annual herb; to 5.0 dm tall; cultivated areas, fields, pastures, roadsides, railroad beds, and waste sites; native of southern Europe.

Seedling Characteristics

Cotyledons round or heart-shaped with pubescent petiole; leaves alternate, dull green, irregularly shaped, pubescent; first leaf with powdery bloom, second leaf toothed, subsequent leaves deeply lobed.

Mature Plant Characteristics

ROOTS fibrous from shallow taproot. STEMS erect, 3.0–5.0 dm tall, spreading or branching from base, pubescent.

LEAVES alternate, 2.0–6.0 cm long, palmately lobed, divided into 3–7 lobes, margins coarsely toothed, long petiole. INFLORESCENCES solitary in leaf axils; corolla 2.5–5.0 cm diameter; petals 5, yellow or whitish, purple petal spot, often with purplish margins, remaining open 1–2 hours. FRUITS capsule 1.0–1.5 cm long, round, pubescent, enclosed by membranous sepals. SEEDS kidney-shaped, 2.0–2.5 mm long, dark brown to black, rough.

Special Identifying Features

Erect or spreading, branched annual; leaves divided into 3–7 lobes, margins coarsely toothed; flowers yellow or whitish with purple base, petals often with purple margins.

Toxic Properties

None reported.

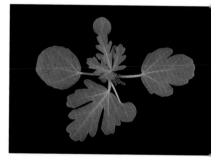

TOP Seeds
MIDDLE Seedling
BOTTOM Fruit

Common Mallow

Malva neglecta Wallr. · Malvaceae · Mallow Family

Synonyms

Button weed, cheeses, dwarf mallow, malice, running mallow, round dock

Habit, Habitat, and Origin

Spreading or nearly erect winter annual or biennial; to 3.0 dm tall; cultivated areas and waste sites; native of northern Africa and Europe.

Seedling Characteristics

Cotyledons heart-shaped, smooth, 5.0–7.0 mm long, 3.0–4.0 mm wide; first leaves simple, circular with toothed to shallowly lobed margins, pubescent.

Mature Plant Characteristics

ROOTS short, straight, fibrous from taproot. STEMS nearly erect or spreading, 1.0–3.0 dm long, freely branched at base, tips generally turning upward. LEAVES alternate, 2.0–6.0 cm long and wide, simple, round to heart-shaped, shallowly 5–9-lobed or nearly unlobed, margins toothed, short pubescence on both sur-faces and margins; petiole 5.0–20.0 cm long, with short pubescence. INFLORES-CENCES borne singly or in clusters of 2–4 in leaf axils; pedicels 1.0–4.0 cm long; petals 5, 0.6–1.3 cm long, white or tinged pink or purple. FRUITS capsule, 5.0–8.0 mm diameter, flattened, disklike, 10–20 (usually 12–15) small 1-seeded segments, pubescent. SEEDS kidney-shaped, 1.3–1.8 mm long, flattened, circular with deep marginal notch, reddish brown to black, surface finely rough.

Special Identifying Features

Spreading or nearly erect winter annual or biennial; leaves rounded, margins toothed and obscurely 5–9-lobed, long petiole; alternate leaves distinguish mallow from ground ivy; flowers borne singly or in clusters in leaf axils, pedicels long; fruit flattened, disklike, with 10–20 sections.

Toxic Properties

None reported.

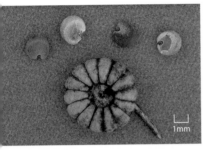

TOP Seeds and fruit
MIDDLE Two-leaf seedling
BOTTOM Four-leaf seedling

Flower

Flower

1mm

Arrowleaf Sida

Sida rhombifolia L. · Malvaceae · Mallow Family

Synonyms

Ironweed

Habit, Habitat, and Origin

Erect, much-branched annual herb; to 1.0 m tall; cultivated areas, fields, pastures, roadsides, and waste sites; native of North America.

Seedling Characteristics

Cotyledons heart-shaped with small notch at tip; first leaves alternate, rhombic, widest at or slightly above middle, serrate on distal half of leaf margins.

Mature Plant Characteristics

ROOTS fibrous from slender, branching, long taproot. STEMS erect, 0.2–1.0 m tall, much branched, densely pubescent, commonly with a short spinelike projection at base of each leaf. LEAVES alternate, 1.5–8.0 cm long, 1.0–3.0 cm wide, simple, rhombic, widest at or slightly above middle, margins serrate only on upper one-half to three-fourths of blade, petiole less than one-third as long as blade. INFLORESCENCES solitary in leaf axils; petals 5, 4.0–8.0 mm long, pale yellow; peduncles 1.0–5.0 cm long,

articulate above middle, longer than subtending leaf petiole. FRUITS capsule, 1.0–1.5 cm long; 8–12 segments, each 1-seeded and separating at maturity; 2 sharp, spreading spines at top. SEEDS triangular, 2.0–3.0 mm long, brown, 8–12 per capsule.

Special Identifying Features

Erect, much-branched annual; leaf margins serrate only on upper one-half to three-fourths of blade; peduncles long; petioles short.

Toxic Properties

Plants suspected to produce substances toxic to poultry.

Fruit

TOP Seeds and fruit
BOTTOM Seedling

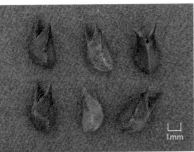

Prickly Sida

Sida spinosa L. · Malvaceae · Mallow Family

Synonyms

False-mallow, Indian mallow, spiny sida, teaweed

Habit, Habitat, and Origin

Erect, much-branched annual herb; to 1.0 m tall; cultivated areas, fields, pastures, roadsides, and waste sites; native of North America.

Seedling Characteristics

Cotyledons smooth to rough, heart-shaped with small notch at tip; true leaves alternate, ovate to triangular, margins serrate throughout.

Mature Plant Characteristics

ROOTS fibrous from slender, branching, long taproot. STEMS erect, 0.2–1.0 m tall, much-branched, softly pubescent, commonly with short spinelike projection at base of each blade. LEAVES alternate, 2.0–5.0 cm long, 1.0–2.0 cm wide, simple, lanceolate to oval, margins serrate, petiole 1.0–3.0 cm long, one-third or more times as long as leaf blade. INFLORESCENCES solitary or clustered in leaf axils, 1-flowered; petals 5, 4.0–6.0 mm long, pale yellow, peduncles shorter than subtending leaf petiole. FRUITS capsule with 5 carpels, splitting into 5 one-seeded segments at maturity, with 2 sharp, spreading spines at top. SEEDS triangular, 1.0–3.0 mm long, somewhat egg-shaped, brown or reddish brown, 5 per capsule.

Special Identifying Features

Erect, much-branched annual; cotyledons heart-shaped with small notch at tip; short, spinelike projection at base of each leaf; leaf margins serrate throughout; peduncles short; petioles long.

Toxic Properties

None reported.

Flower

Fruit

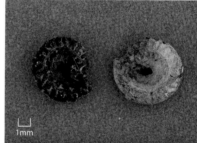

1mm

Redberry Moonseed

Cocculus carolinus (L.) DC. · Menispermaceae · Moonseed Family

Synonyms

Cocculus, coralbeads, snailseed

Habit, Habitat, and Origin

Dioecious, prostrate or scandent, twining, perennial vine; to 3.0 m long; cultivated areas, fields, thickets, and open woodlands; native of North America.

Seedling Characteristics

Seldom seen; emerging sprout with alternate, hastately lobed leaves, usually in clumps.

Mature Plant Characteristics

ROOTS fibrous from perennial woody rootstock. **STEMS** prostrate or scandent, twining vine to 3.0 m long, green, pubescent. **LEAVES** alternate, 2.5–12.0 cm long and wide, ovate to hastately lobed, margins entire, glabrous above, pubescent beneath, petioles to 10.0 cm long. **INFLORESCENCES** dioecious, numerous, borne on axillary and terminal panicles; small, green, and short-pedicellate; staminate panicle to 1.5 dm long, pistillate panicle much shorter; sepals and petals 1.0–2.0 mm long; stamens 6; anthers quadrilocular; carpels usually 3. **FRUITS** drupe, 5.0–8.0 mm diameter, bright red, round, in clusters. **SEEDS** ridged, depressed in center.

Special Identifying Features

Dioecious, prostrate or scandent, twining vine; leaves glabrous above, pubescent below; fruit bright red at maturity in fall; seed distinctively ridged with depressed center.

Toxic Properties

Plants contain biologically active alkaloids, but few incidences of poisoning have been reported.

Flowers

TOP Seeds
BOTTOM Shoot

Flowers

TOP Seeds
MIDDLE Seedling
BOTTOM Fruit

Illinois Bundleflower

Desmanthus illinoensis (Michx.) MacMill. ex B. L. Robins & Fern. · Mimosaceae · Mimosa Family

Synonyms

None

Habit, Habitat, and Origin

Erect or spreading, deciduous, perennial herb; to 1.0 m tall; fields, pastures, roadsides, ditches, and waste sites; native of North America.

Seedling Characteristics

Typical legume; cotyledons large; first leaf once-compound.

Mature Plant Characteristics

ROOTS fibrous taproot initially, forming a thickened overwintering rootstock. STEMS erect or spreading, 0.4–1.0 m long, herbaceous, strongly angled, glabrous to hirsute, unarmed. LEAVES alternate, 2.0–10.0 cm long, evenly bipinnate, 6–16 pairs of pinnae; 20–30 pairs of linear leaflets, 2.0–3.5 mm long, depressed; saucer-shaped gland between lowermost pinnae. INFLORESCENCES dense head on tips of upper branches, greenish white and globular, perfect or the lowest staminate, 5-parted, sessile; petals 5, distinct. FRUITS legume, 1.5–2.5 cm long, 4.0–6.0 mm wide, oblong, moderately to strongly curved, glabrous, numerous in dense heads. SEEDS oblong or slightly triangular, 3.0–5.0 mm long, brown, flattened.

Special Identifying Features

Erect or spreading, deciduous perennial; leaves bipinnately compound, not extremely sensitive; saucer-shaped gland between lowermost pinnae.

Toxic Properties

None reported.

Flowering plant

1mm

TOP Seeds
BOTTOM Seedling

Carpetweed

Mollugo verticillata L. · Molluginaceae · Carpetweed Family

Synonyms

Common carpetweed, devil's-grip, Indian chickweed

Habit, Habitat, and Origin

Prostrate summer annual herb; to 0.6 m diameter; cultivated areas, fields, pastures, roadsides, gardens, lawns, and waste sites; native of tropical America.

Seedling Characteristics

Cotyledons oblong, thickened, smooth, persistent, 1.5–3.0 mm long; hypocotyl short, green early, brown later, not evident after second leaf.

Mature Plant Characteristics

ROOTS fibrous, little-branched, from taproot. STEMS prostrate, 5.0–30.0 cm long, green, smooth, glabrous, branched at base and branching in all directions, creating circular mat on soil surface. LEAVES whorled, 1.0–3.0 cm long, 0.8–1.0 cm wide, 5–6 or occasionally 3–8 at each node, spatulate to linear-oblanceolate, apex obtuse to acute, petiole short. INFLORESCENCES in clusters of 2–5 on slender stalks 3.0–14.0 mm long from leaf axils; sepals 5, small, pale green to white; stamens 3. FRUITS capsule, 1.5–2.5 mm long, thin-walled, 3-parted, partitions breaking to form many-seeded axis; seeds visible through wall. SEEDS kidney-shaped, 0.5 mm long, orange-red, somewhat flattened, surface glossy, marked by parallel curved lines.

Special Identifying Features

Prostrate, late-germinating summer annual; circular matlike growth quickly covering soil; stems and leaves light green; leaves in whorled arrangement.

Toxic Properties

None reported.

Flowers

TOP Seeds
MIDDLE Two-leaf seedling
BOTTOM Four-leaf seedling

Mulberryweed

Fatoua villosa (Thunb.) Nakai · Moraceae · Mulberry Family

Synonyms

Fatoua japonica (Thunb.) Blume, *F. pilosa* Gaud.

Habit, Habitat, and Origin

Erect annual herb; to 1.0 m tall; shaded areas, vegetable gardens, lawns, flowerbeds, container plants, and greenhouses; native of eastern Asia, introduced into the New Orleans area in the 1950s.

Seedling Characteristics

Cotyledons ovate, dropping early; leaves ovate with serrate margins, short appressed pubescence.

Mature Plant Characteristics

ROOTS fibrous from taproot, light brown. STEMS erect, 0.3–1.0 m tall, branched, thin, weak, tending to fall over or lean on surrounding vegetation, pubescence irregular length, trichomes hooked. LEAVES first pair opposite, then alternate, 5.0–15.0 cm long, 3.0–8.0 cm wide, broadly ovate, tip acute, margins obtusely toothed; coarsely pubescent, trichomes appressed, hooked on both surfaces; petiole slender, about as long as blade. INFLORESCENCES male and female flowers mixed together in dense axillary cymes, small, light green, numerous; male flower a 4-lobed perianth, 4 stamens opposite perianth lobes; female flower a 6-lobed boat-shaped perianth, ovary depressed-globose, style oblique and filiform with one arm elongate and the other reduced, purple, stigmas purple. FRUITS axillary pendulous glomerule of achenes. SEEDS achene, 1.1 mm long, 1.0 mm wide, triangular, granular, face sharply angled, light to dark brown, papillose, enclosed in a persistent perianth.

Special Identifying Features

Erect herbaceous annual; leaves alternate; leaves and stems pubescent; male and female flowers together in dense, light green axillary cymes.

Toxic Properties

None reported.

Flowering plant

Flowers

Seedling

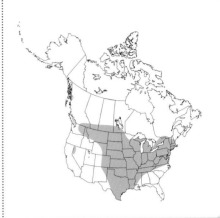

TOP Seeds
BOTTOM Mature plant

Wild Four-o'clock

Mirabilis nyctaginea (Michx.) MacMill. · Nyctaginaceae · Four o'clock Family

Synonyms

Heartleaf four o'clock

Habit, Habitat, and Origin

Erect, freely branching perennial herb; to 1.5 m tall; prairies, fields, pastures, roadsides, and waste sites; native of North America.

Seedling Characteristics

Cotyledons broadly ovate to orbicular, glabrous; first true leaves lanceolate, margins entire.

Mature Plant Characteristics

ROOTS large, thick, tough, fleshy taproot, extending to 1.0 m. **STEMS** erect, 0.3–1.5 m tall, freely branching, repeatedly forked, nodes thickened, square, glabrous or sparsely pubescent. **LEAVES** opposite, 3.0–15.0 cm long, 1.0–9.0 cm wide, broadly ovate to ovate-lanceolate, base sometimes heart-shaped, tip often pointed, margins entire, upper and lower surfaces glabrous, lower and mid-stem petioles 1.0–7.0 cm long, petioles absent apically. **INFLORESCENCES** slightly pubescent terminal umbels; corolla absent; calyx 5-lobed, 2.0 mm long, pink to purple bell-shaped tube, subtended by slender pedicels 1.0 cm long. **FRUITS** cylindrical to narrowly elliptic, 4.0–7.0 mm long, 5-ribbed, gray to brown, hard, warty or wrinkled. **SEEDS** cylindrical to club-shaped, 3.0–4.0 mm long, brown to yellow.

Special Identifying Features

Erect, freely branching perennial; flowers open in late afternoon and close next morning; stems square, smooth, fleshy; nodes swollen, resembling ball-and-socket joints.

Toxic Properties

Plants irritate skin and digestive tract.

Winged Waterprimrose

Ludwigia decurrens Walter · Onagraceae · Evening Primrose Family

Synonyms

Primrose-willow, water primrose, wing-stem waterprimrose

Habit, Habitat, and Origin

Erect, freely branching annual; to 2.0 m tall; swamps and shallow water; native of North America.

Seedling Characteristics

Cotyledons linear, smooth; first leaves lanceolate, glabrous, margins serrate, quadrangular.

Mature Plant Characteristics

ROOTS fibrous from taproot. STEMS erect, 0.2–2.0 m tall, quadrangular, 4-winged, smooth, glabrous. LEAVES alternate, 5.0–18.0 cm long, 1.0–4.0 cm wide, lanceolate to linear, sessile or subsessile, lanceolate bases cuneate to rounded, decurrent, apices acuminate, surfaces glabrous or sparsely short-pubescent, margins finely scabrous. INFLORESCENCES solitary in axils of reduced leaves or bracteal leaves, sessile or on stalks 1.0–5.0 mm long; 2 minute, scalelike bractlets at base of floral tube or on flower stalk; calyx segments 4, mostly triangular-subulate, 7.0–10.0 mm long; petals 4, yellow, 8.0–12.0 mm long, suborbicular to obovate, broadly rounded apically, 8.0–12.0 mm long; stamens 8; capsule narrowly obconic, angled or winged, glabrous, 1.0–2.0 cm long; pedicels with irregular locules. FRUITS capsule, 1.0–2.0 cm long,

Flowering plant

slender, pyramidal, quadrangular or narrowly 4-winged. SEEDS ellipsoid, 0.3–0.4 mm long, several series per locule.

Special Identifying Features

Erect, freely branching annual; stems quadrangular, 4-winged; leaf bases decurrent.

Toxic Properties

None reported.

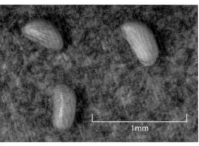

TOP **Seeds**
MIDDLE **Seedling**
BOTTOM **Flower**

Creeping Waterprimrose

Ludwigia peploides (Kunth) Raven · Onagraceae · Evening Primrose Family

Synonyms

Floating evening primrose, floating primrose-willow

Habit, Habitat, and Origin

Creeping or floating aquatic perennial herb; stems to 0.6 m long; damp soils, wetlands, streams, ponds, and ditches; native of North America.

Seedling Characteristics

Cotyledons ovate-lanceolate, with short pubescence; first leaves ovate-lanceolate, petioled, with short pubescence.

Mature Plant Characteristics

ROOTS fibrous, much-branching from taproot. STEMS creeping and rooting at nodes over wet substrate or in shallow water, 0.2–0.6 m long, sometimes forming floating mats. LEAVES alternate, 1.0–8.0 cm long, 0.4–3.5 cm wide, elliptic to obovate on young shoots, oblanceolate to spatulate or sometimes suborbicular on lower branches, lanceolate to narrowly elliptic on distal portions and usually larger, glabrous, margins glabrous or sparsely short-pubescent, petioles to 2.5 cm long. INFLORESCENCES solitary in axils of upper leaves, stalks 1.0–6.0 cm long, branches usually ascending, glabrous or occasionally sparsely pubescent; petals 5, 10.0–15.0 mm long, yellow, obovate; calyx 5-segmented, 8.0–12.0 mm long, usually glabrous or with a few long trichomes, subulate, lanceolate, or ovate. FRUITS capsule, 2.0–4.0 cm long, 3.0–4.0 mm wide, hard, cylindrical, glabrous, tardily dehiscent. SEEDS ellipsoid, 0.3–0.5 mm long, embedded in woody endocarp, single vertical series in capsule.

Special Identifying Features

Creeping or floating aquatic perennial; rooting at nodes; forming mats in shallow water; flowers yellow.

Toxic Properties

None reported.

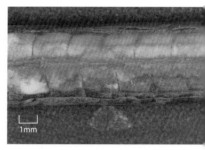

1mm

TOP **Fruit**
MIDDLE **Seedling**
BOTTOM **Flowering plant**

Flower

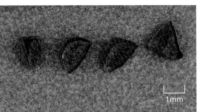

TOP Seeds
MIDDLE Seedling
BOTTOM Flowering plant

Flower

Common Evening-primrose

Oenothera biennis L. · Onagraceae · Evening Primrose Family

Synonyms

Candlestick, evening primrose, Victoria's evening primrose, yellow evening-primrose

Habit, Habitat, and Origin

Erect biennial herb; to 2.0 m tall; fields, roadsides, and ditch banks; native of North America.

Seedling Characteristics

Cotyledons pubescent, lanceolate; first leaves alternate, petioled, forming a rosette.

Mature Plant Characteristics

ROOTS fibrous from taproot. STEMS erect, 0.5–2.0 m tall, branched, short appressed trichomes. LEAVES basal rosette, alternate upward, 0.5–3.0 dm long, 1.0–7.0 cm wide, lanceolate, margins entire to lobed, long petioles, cauline leaves sessile or with short petioles, sparsely pubescent. INFLORESCENCES terminal spike with bracts; mature buds erect; floral tube 2.0–5.0 cm long, slender, greenish yellow, pubescent, trichomes glandular or plain; petals yellow, fading to pale reddish, 1.0–2.5 cm long, opening near sunset; stamen equal to or longer than stigma and style. FRUITS capsule, 1.4–3.5 cm long, 3.5–6.0 mm wide, cylindrical near base, strigose to subglabrous. SEEDS sharply angled, 1.3–1.6 mm long, reddish brown, in 2 rows in each locule.

Special Identifying Features

Erect biennial; plants usually 0.6 m or more tall; flowers in terminal spike, yellow.

Toxic Properties

None reported.

Cutleaf Evening-primrose

Oenothera laciniata Hill · Onagraceae · Evening Primrose Family

Synonyms

None

Habit, Habitat, and Origin

Erect, prostrate, or weakly ascending winter annual or biennial herb; stems to 8.0 dm tall; cultivated areas, fields, pastures, roadsides, gardens, and waste sites; native of eastern North America.

Seedling Characteristics

Cotyledons kidney-shaped; petioles flat on upper surface; hypocotyl short, becoming magenta, smooth, not evident above soil after second leaf develops.

Mature Plant Characteristics

ROOTS fibrous from taproot. STEMS erect, prostrate, or weakly ascending; 1.0–8.0 dm long, simple or many-branched from base, pubescent. LEAVES alternate, 3.0–8.0 cm long, 5.0–20.0 mm wide, oblong to lanceolate, margins coarsely toothed to irregularly lobed or deeply incised, dull green, glabrous or pubescent; trichomes short, appressed, or ascending. INFLORESCENCES single, sessile in leaf axis; corolla yellow to reddish, 8.0–25.0 mm long, basally fused into a long, narrow tube; anthers 3.0–6.0 mm long. FRUITS capsule, 2.0–4.0 cm long, 3.0–4.0 mm wide, cylindrical, straight or curved, 4-lobed, pubescent becoming smooth with age. SEEDS thick-ellipsoid, 1.2–1.4 mm long, 0.8 mm wide, variably shaped but most are sharply angular, pale brown, strongly pitted.

Special Identifying Features

Erect, prostrate, or weakly ascending winter annual or biennial; flowers and fruits in axils of upper pinnatifid leaves or bracts; flowers yellow to reddish; seeds deeply pitted.

Toxic Properties

None reported.

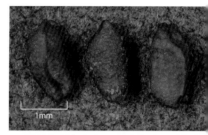

TOP Seeds
MIDDLE Fruit
BOTTOM Seedling

Flower

TOP **Seeds**
MIDDLE **Seedling**
BOTTOM **Flowers**

Showy Evening-primrose

Oenothera speciosa Nutt. · Onagraceae · Evening Primrose Family

Synonyms

White evening primrose

Habit, Habitat, and Origin

Ascending to erect, branched perennial; to 7.0 dm tall; fields, prairies, open waste sites, and roadsides; native of North America.

Seedling Characteristics

Forms rosettes when young.

Mature Plant Characteristics

ROOTS fibrous, adventitious from lateral roots and taproot. **STEMS** ascending to erect, 1.0–7.0 dm tall, simple or branched, base semiwoody, sparsely to densely pubescent. **LEAVES** alternate, 0.6–3.0 dm long, 1.0–7.0 cm wide, elliptic to oblanceolate; margins nearly entire to irregularly lobed or toothed, especially on basal half; petioles absent or up to 3.0 cm long. **INFLORESCENCES** solitary in upper axils, buds nodding, white or pink, self-incompatible, sessile; floral tube slender, 1.0–2.0 cm long; petals inversely heart-shaped, 2.5–4.0 cm long; stamens two-thirds length of petals; anthers 10.0–12.0 mm long; style as long as petals; stigmas 3.0–6.0 mm long. **FRUITS** capsule, 6.0–18.0 mm long, obovoid to club-shaped, angled, ribbed, sessile or on pedicels 2.0–8.0 mm long. **SEEDS** spindle-shaped, 0.9–1.2 mm long, brown, in rows.

Special Identifying Features

Ascending to erect, branched perennial; flowers showy; white flowers of diploid plants open in evening and are more common northward; pink flowers of tetraploid plants open in morning and are more common southward.

Toxic Properties

None reported.

Flowering plant

Flower

Yellow Woodsorrel

Oxalis stricta L. · Oxalidaceae · Wood Sorrel Family

Synonyms

European woodsorrel, lady's sorrel, sheep's clover, sheep sorrel, sheep sour, upright yellow oxalis, yellow sorrel, yellow wood-sorrel;

Oxalis corniculata L., *O. dillenii* Jacq., *O. europaea* Jord., *O. florida* Salisb.

Habit, Habitat, and Origin

Prostrate to erect annual or perennial; to 0.5 m tall; lawns, gardens, flowerbeds, cultivated areas, open woodlands, and waste sites; native of North America.

Seedling Characteristics

Cotyledons oblong, green, pubescent.

Mature Plant Characteristics

ROOTS fibrous from taproot or rhizomes. STEMS prostrate to erect, to 0.5 m tall, branched at base, weak, glabrous to densely pubescent, trichomes septate. LEAVES alternate, compound, leaflets 3, heart-shaped, green or purplish green, petiole long, sometimes pubescent. INFLORESCENCES clusters, 2.0 cm wide, unequally branched umbel, regular, perfect; petals 5, yellow; sepals 5, green; stamens 10, two lengths. FRUITS capsule, 5.0–15.0 mm long, slender, 5-ridged, pointed, glabrous or pubescent, trichomes septate. SEEDS flat, small, brown with white markings on transverse ridges, dehiscent when capsules burst.

Special Identifying Features

Prostrate to erect annual or perennial, spring or summer; leaflets 3, heart-shaped; flowers in unequally branched umbel; septate pubescence; sour taste.

Toxic Properties

Plants contain toxins that cause renal disease, metabolic tetany, depression, weakness, labored respiration, weight loss, and collapse.

TOP Seeds
MIDDLE Seedling
BOTTOM Fruit

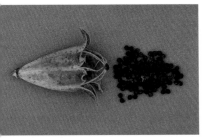

TOP Seeds and fruit
MIDDLE Three-leaf seedling
BOTTOM Four-leaf seedling

Flower and fruit

Mexican Pricklepoppy

Argemone mexicana L. · Papaveraceae · Poppy Family

Synonyms

Devil's fig, Mexican thistle, yellow thistle, yellow prickly poppy

Habit, Habitat, and Origin

Erect, branched annual or biennial herb; to 8.0 dm tall; prairies, fields, pastures, roadsides, and waste sites; native of the Americas.

Seedling Characteristics

Cotyledons sessile, linear, 15.0–20.0 mm long; first leaves sessile, coarsely dentate, spiny.

Mature Plant Characteristics

ROOTS fibrous from taproot. **STEMS** erect, 2.0–8.0 dm tall, mostly solitary, spiny, often branching from near base, bright yellow latex sap. **LEAVES** alternate, sessile, clasping-auriculate, pinnatifid and irregularly dentate, spiny, glaucous with grayish white veins, 2.5–20.0 cm long, 3.0–9.0 cm wide. **INFLORESCENCES** conspicuous, solitary; petals bright yellow to orange, 2.0–4.0 cm long, free, entire, obovate; sepals 5.0–10.0 mm long, free, spiny. **FRUITS** oblong capsule, 25.0–50.0 mm long, spiny, longest spines 4.0–7.0 mm long, many-seeded. **SEEDS** 1.5–2.5 mm diameter, gray to brown-black.

Special Identifying Features

Erect, branched annual or biennial; stems spiny with yellow latex sap; flowers bright yellow; capsules spiny.

Toxic Properties

Argemone species contain toxic alkaloids, especially in the seeds, and intoxication impairs growth and causes sluggishness, passiveness, sedation, and either constipation or diarrhea.

Flower

Maypop Passionflower

Passiflora incarnata L. · Passifloraceae · Passionflower Family

Synonyms

Maypop

Habit, Habitat, and Origin

Deciduous, herbaceous perennial vine; to 2.0 m long; cultivated areas, fields, pastures, roadsides, and waste sites; native of North America.

Seedling Characteristics

Emerging sprout with palmately lobed leaves and axillary tendrils.

Mature Plant Characteristics

ROOTS initially fibrous from vigorous taproot, developing large rootstock. STEMS erect, repent, or climbing; to 2.0 m long, glabrous or minutely pubescent. LEAVES alternate, 6.0–15.0 cm long, 6.0–15.0 cm wide, palmately 3-lobed or rarely 5-lobed, margins serrate, finely or often remotely pubescent, paired nectar glands at junction of petiole and leaf blade, long petiole. INFLORESCENCES axillary, solitary, peduncle 5.0–10.0 cm long, stout; petals bluish to white, 3.0–4.0 cm long with corona of segments white or lavender banded with purple, to 3.0 cm long; sepals green on black, white or whitish above, 1.5–3.5 cm long. FRUITS berry, 4.0–7.0 cm long, green or yellowish, ellipsoid. SEEDS flattened, 4.0–6.0 mm long, dark brown, prominently reticulate.

Special Identifying Features

Deciduous, herbaceous perennial vine with tendrils; flowers colorful, bluish purple and white; leaves palmately 3-lobed.

Toxic Properties

None reported; fruit edible by humans and wildlife.

TOP Seeds
MIDDLE Seedling
BOTTOM Fruit

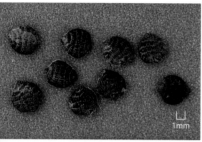

TOP Seeds
BOTTOM Shoot

Yellow Passionflower

Passiflora lutea L. · Passifloraceae · Passionflower Family

Synonyms

Passionflower

Habit, Habitat, and Origin

Climbing or trailing perennial vine; to 5.0 m long; gardens, flowerbeds, open woodlands, pastures, roadsides, and waste sites; native of North America.

Seedling Characteristics

First true leaves palmately 3-lobed, whitish along veins.

Mature Plant Characteristics

ROOTS initially fibrous from taproot, developing enlarged rootstock. STEMS perennial climbing or trailing vine, to 5.0 m long, smooth to sparingly pilose. LEAVES alternate, 3.0–7.0 cm long, 4.0–10.0 cm wide, palmately 3-lobed, lobes blunt or rounded, bases truncate or subcordate, margin entire, glandless, deciduous, petiole to 5.0 cm long. INFLO-RESCENCES solitary or in pairs, greenish yellow, 0.8–1.2 cm long, 1.5–2.5 cm wide, complex; yellow exterior corona without bracts, peduncles very slender, calyx tube cup-shaped, sepals linear-oblong, petals linear and whitish; corona filaments in 2 series, outer ones about 30, narrowly linear or filiform, inner ones narrowly liguliform, slightly thickened toward apex, white above, pink-tinged toward base. FRUITS berry, 0.8–1.5 cm diameter, nearly round or slightly apressed, green turning black or purple. SEEDS ovoid, 3.0–4.5 mm long, dark brown, prominently cross-ribbed, ribs broken into small segments.

Special Identifying Features

Perennial climbing or trailing vine; leaves blunt or rounded, margins entire; flowers greenish yellow.

Toxic Properties

None reported.

Flower and tendrils

Flowers

1mm

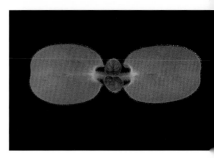

Devil's-claw

Proboscidea louisianica (P. Mill.) Thellung · Pedaliaceae · Unicorn-plant Family

Synonyms

Unicorn-plant

Habit, Habitat, and Origin

Erect or decumbent, branching annual herb; 0.5 m tall; cultivated areas, fields, pastures, and waste sites; native of North America.

Seedling Characteristics

Cotyledons oblong, entire, short-petioled, 2.0 cm long; true leaves short-petioled, light green, cordate, entire, pubescent.

Mature Plant Characteristics

ROOTS deep fibrous from taproot. **STEMS** erect or decumbent, thick, fleshy, low-branching with clammy pubescence, up to 0.5 m tall, at maturity stems break at axils and plants lie flat on soil surface. **LEAVES** alternate, round to heart-shaped, with undulating edges, petioled, pubescent, clammy, heavy odor. **INFLORESCENCES** terminal raceme, pinkish white, showy, spotted with yellow and purple, 3.0 cm across. **FRUITS** capsule, 8.0–15.0 cm long, fleshy, clammy; 2-valved with long, arched beak that sheds flesh and exposes 2 woody horned, curved, beaked pods at maturity; up to 50 seeds per pod in 2 compartments. **SEEDS** ovoid to somewhat obpyramidal, 8.0–10.0 mm long, blunt at each end, flattened; seed coat thick, rough, black.

Special Identifying Features

Erect or decumbent, fleshy, branching annual; leaves and stems filled with slimy, sticky, smelly mucus; flowers attractive; fruit long-beaked.

Toxic Properties

None reported.

Fruit

TOP Seeds
MIDDLE Seedling
BOTTOM Fruit

Common Pokeweed

Phytolacca americana L. · Phytolaccaceae · Pokeweed Family

Synonyms

American pokeweed, garget, inkberry, pigeonberry, poke, pokeberry, pokeweed, poke salad, scoke

Habit, Habitat, and Origin

Erect, freely branched perennial herb; to 3.0 m tall; cultivated areas, fields, pastures, roadsides, open woodlands, gardens, and waste sites; native of North America, introduced into Europe.

Seedling Characteristics

Cotyledons unequal, 7.0–33.0 mm long, 6.0–11.0 mm wide, pale beneath, becoming leaflike; hypocotyl smooth, succulent, magenta-tinged.

Mature Plant Characteristics

ROOTS well-developed, large, white, fleshy, poisonous, taproot system up to 15.0 cm diameter. STEMS erect, 0.3–3.0 m tall, coarse, stout, freely branching from main stem and root crown, smooth, succulent, bright magenta-tinged. LEAVES alternate, 9.0–30.0 cm long, 3.0–11.0 cm wide, broadly lanceolate or oval, petiole 1.0–5.0 cm long, smooth. INFLORESCENCES stalked raceme, 5.0–30.0 cm long, nodding or erect; petals 5, 2.0–3.0 mm long and wide, whitish to pink-tinged, rounded, persistent; stamens 10; pistil 1; ovary superior. FRUITS berry, 7.0–12.0 mm, dark purple to black at maturity. SEEDS lens-shaped, 7.0–10.0 mm diameter, black, glossy, 10 per fruit.

Inflorescence

Special Identifying Features

Erect, freely branched perennial; leaf margins magenta, continuous down stem; mature berries dark purple to black.

Toxic Properties

Plants contain several toxins causing digestive irritation, diarrhea, sedation, seizures, and decreased weight gain; young plants can be consumed if cooking water is changed several times; older plants are very poisonous.

Flowers

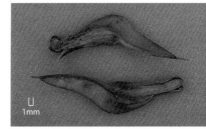

Redvine

Brunnichia ovata (Walt.) Shinners · Polygonaceae · Buckwheat Family

Synonyms

Eardrops, ladies' eardrops; *Brunnichia cirrhosa* Gaertn.

Habit, Habitat, and Origin

Decumbent, climbing, or somewhat shrubby perennial vine; to several meters long; cultivated areas, fields, pastures, roadsides, open woodlands, fencerows, ditches, and waste sites; native of North America.

Seedling Characteristics

Usually not seen, but emerging from perennial rootstock; leaves alternate, entire, glabrous; reddish tint to stems and tips of sprout.

Mature Plant Characteristics

ROOTS overwintering, large, extensive rootstock, deeply rooted. **STEMS** partly woody vine climbing by tendrils, to several meters long, terminating in short lateral branches, perennial, deciduous. **LEAVES** alternate, 5.0–15.0 cm long, simple, ovate, base truncate to slightly cordate. **INFLORESCENCES** paniculate raceme, perfect, whitish or greenish; perianth lobes 5; stamens 8; pistil 3-parted; calyx enlarging to 2.0–3.0 cm long, pink and showy, on a long, flattened, winglike base. **FRUITS** nutlet, 7.0 mm long, obscurely triangular, enclosed in well-developed hypanthium. **SEEDS** oblong, 1.0–2.0 mm long, dark brown, tightly enclosed in fruit.

Special Identifying Features

Decumbent, climbing, or somewhat shrubby perennial vine; tendrils on terminal branches; leaves alternate, entire; live plant and dead parts have reddish tint.

Toxic Properties

None reported.

TOP Seeds
MIDDLE Seedling
BOTTOM Shoots

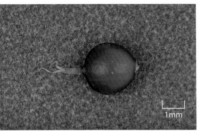

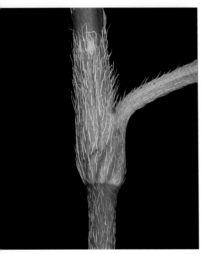

TOP Seed
MIDDLE Sprout
BOTTOM Ocrea

Flowers

Swamp Smartweed

Polygonum amphibium L. var. *emersum* Michx. · Polygonaceae ·
Buckwheat Family

Synonyms

Devil's shoestring, shoestring smartweed, water smartweed; *Polygonum
coccineum* Muhl. ex Willd., *Persicaria
amphibia* (L.) Gray

Habit, Habitat, and Origin

Erect perennial from long, creeping,
woody rootstocks; to 1.0 m tall; shallow
water, shorelines, marshes, and roadside
ditches; native of North America.

Seedling Characteristics

Usually not seen; cotyledons oblong
or lance-shaped; shoots erect, enlarged
at nodes with sheath covering base of
leaves.

Mature Plant Characteristics

ROOTS stout, woody, creeping rootstocks.
STEMS erect, 0.3–1.0 m tall, enlarged at
nodes, green, glabrous or pubescent in
drier environments, rooting at lower
nodes. **LEAVES** alternate, 4.0–25.0 cm
long, 1.0–6.0 cm wide, lanceolate, base
round, tip pointed, veins prominent,
glabrous or pubescent in drier environments, ocrea membranous and surrounding stem at nodes. **INFLORESCENCES**
compact spike at tip of stem, erect,
2.0–15.0 cm long, often absent in dry
or northern areas; perianth 4.0–5.0 mm
long, rose to pink; sepals 5-parted. **FRUITS**
includes adhering calyx containing
achene. **SEEDS** achene, 2.5–3.0 mm long,
oval, flattened on one side, black, shiny,
slightly rough.

Special Identifying Features

Perennial from woody, creeping rootstock; nodes swollen; ocrea membranous, surrounding stem at nodes.

Toxic Properties

Polygonum species may cause acute
photosensitization and death.

Prostrate Knotweed

Polygonum aviculare L. · Polygonaceae · Buckwheat Family

Synonyms

Wireweed, knotweed

Habit, Habitat, and Origin

Prostrate to loosely ascending annual herb; stems to 1.0 m long; cultivated areas, fields, moist soils, waste areas; native of North America.

Seedling Characteristics

Cotyledons small, linear; stem reddish; leaves blue-green, glabrous.

Mature Plant Characteristics

ROOTS slender from taproot, commonly not rooting at nodes. STEMS prostrate, to 1.0 m long, highly branched, conspicuous nodes, glabrous. LEAVES alternate, 5.0–30.0 mm long, 1.0–8.0 mm wide, simple, linear, margins entire, blue-green, petiole short, ocrea fringeless and veinless. INFLORESCENCES axillary clusters, small, perfect, included in or barely exserted from leaf axil; sepals green with white or pinkish margins, 1.0–1.6 mm long. FRUITS includes achene and adhering calyx, 2.0–3.0 mm long, 5–6 lobes, green with pink or white margins. SEEDS achene, 2.0–25.0 mm long, ovoid, reddish brown to dark brown, lustrous.

Special Identifying Features

Prostrate to loosely ascending annual; flowers in leaf axil, inconspicuous; ocrea not fringed, veinless; ocrea distinguishes *Polygonum* species from bindweeds and morningglories.

Toxic Properties

See comments under *Polygonum amphibium* var. *emersum*.

1mm

TOP Seeds
BOTTOM Seedling

Flower

Flowers

Wild Buckwheat

Polygonum convolvulus L. · Polygonaceae · Buckwheat Family

Synonyms

Black bindweed, climbing buckwheat; *Fallopia convolvulus* (L.) A. Löve

Habit, Habitat, and Origin

Twining or trailing annual vine; to 1.0 m tall; moist soils in fields, fencerows, and ditches; native of Europe.

Seedling Characteristics

Hypocotyl glabrous, reddish; cotyledons linear, glabrous; first leaves round, glabrous, petioled.

Mature Plant Characteristics

ROOTS fibrous from taproot. STEMS initially erect, become twining or creeping and branched at base, to 1.0 m long. LEAVES alternate, 3.0–6.0 cm long, 2.0–5.0 cm wide, simple, sagittate or cordate, heart-shaped at base, pointed at tip, petiole to 6.0 cm long, ocrea at leaf base. INFLORESCENCES axillary fascicle or raceme; perianth greenish, often purple-spotted, styles 3, united; calyx, 5.0 mm long, strongly triangular, green to white or pinkish, farinose or slightly rough. FRUITS includes achene and adhering calyx, rarely keeled. SEEDS achene, 4.0–5.0 mm long, dull black, ovate, trigonous, faces convex.

Special Identifying Features

Twining or trailing annual vine; leaves sagittate or cordate, alternate arrangement distinguishing from honeyvine swallow-wort; ocrea at leaf base distinguishes from bindweeds and morningglories; flowers greenish to white or pinkish.

Toxic Properties

See comments under *Polygonum amphibium* var. *emersum*.

Pale Smartweed

Polygonum lapathifolium L. · Polygonaceae · Buckwheat Family

Synonyms

Nodding smartweed, willowleaf, willow weed, smartweed; *Persicaria lapathifolia* (L.) Gray

Habit, Habitat, and Origin

Erect, branching annual herb; to 2.5 m tall; damp soils, ditches, fields, cultivated areas, wet meadows, and shorelines; native of North America.

Seedling Characteristics

Hypocotyl erect or prostrate; cotyledons lanceolate.

Mature Plant Characteristics

ROOTS fibrous from taproot. **STEMS** erect, 0.5–2.5 m tall, glabrous, reddish, nodes swollen. **LEAVES** alternate, 4.0–20.0 cm long, 3.5–5.0 cm wide, lanceolate, pointed at tip, pubescent on lower surface, trichomes stalkless and glanded, ocrea surrounding stem at nodes, smooth, striate-nerved, membrane shattering distally to give a bristly appearance. **INFLORESCENCES** numerous racemes in slender arching or nodding spike, 2.0–8.0 cm long; perianth white, green, or light pink, 2.0–4.0 mm long; sepals 3-nerved; styles 2; bracts entire, glabrous. **FRUITS** includes adhering calyx containing achene. **SEEDS** achene, 1.8–2.2 mm long, 1.6–2.0 mm wide, oval or round, flattened, pointed tip, black, shiny, smooth.

Special Identifying Features

Erect, branching annual; inflorescence nodding; ocrea membrane shattering distally forming "bristles."

Toxic Properties

See comments under *Polygonum amphibium* var. *emersum*.

TOP **Seeds**
BOTTOM **Seedling**

Flowers

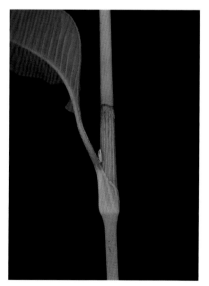

Ocrea

Pennsylvania Smartweed

Polygonum pensylvanicum L. · Polygonaceae · Buckwheat Family

Synonyms

Pinkweed; *Persicaria pensylvanica* (L.) M. Gómez

Habit, Habitat, and Origin

Erect summer annual herb; to 1.2 m tall; disturbed soils, moist areas, cultivated fields, ditches, and shorelines; native of North America east of the Rocky Mountains.

Seedling Characteristics

Hypocotyl erect or prostrate; cotyledons lanceolate, smooth, with gland-tipped trichomes covering margins.

Mature Plant Characteristics

ROOTS fibrous from taproot. **STEMS** erect, 0.3–1.2 m tall, branching, smooth with swollen nodes, reddish purple. **LEAVES** alternate, 5.0–15.0 cm long, 1.0–3.0 cm wide, simple, lanceolate, pointed at tip, glabrous or sparsely pubescent, often with purple marks in center of upper and lower surfaces; ocrea smooth, membranous, surrounding stem at nodes. **INFLORESCENCES** many flowers in a dense spike at tip of stems and branches, stalk pubescent, trichomes stalked and glanded; perianth 3.0–5.0 mm long, rose, pink, or occasionally white; sepals 5-parted, petals absent, styles 2–3. **FRUITS** includes adhering calyx containing achene. **SEEDS** achene, 2.5–3.5 mm long, 2.2–3.5 mm wide, oval or round, 2-sided, somewhat flattened with pointed tip, black, shiny, smooth, with concave indentation on one face.

Special Identifying Features

Erect summer annual; nodes swollen, smooth; ocrea surrounding stem; flower stalk pubescent, trichomes stalked and glanded; similar to ladysthumb but without hairs on top of ocrea.

Toxic Properties

See comments under *Polygonum amphibium* var. *emersum*.

TOP **Seeds**
BOTTOM **Seedling**

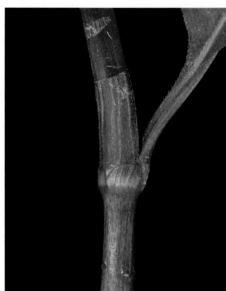

TOP **Flowers**
BOTTOM **Ocrea**

Ladysthumb

Polygonum persicaria L. · Polygonaceae · Buckwheat Family

Synonyms

Heart's ease, smartweed; *Persicaria maculosa* Gray

Habit, Habitat, and Origin

Erect or spreading summer annual; to 9.0 dm tall; disturbed damp soils, cultivated areas, fields, and shorelines; native of Europe.

Seedling Characteristics

Hypocotyl erect or prostrate; cotyledons lanceolate with rounded tips, upper and lower surfaces smooth.

Mature Plant Characteristics

ROOTS fibrous from taproot. STEMS erect or spreading, 2.0–9.0 dm tall, branched, smooth with swollen nodes, reddish. LEAVES alternate, 2.5–15.0 cm long, 5.0–18.0 mm wide, smooth, lanceolate, pointed at tip, glabrous or with short appressed pubescence, sessile or on short petiole, with watermark; ocrea membranous, surrounding nodes, fringed with hairlike bristles. INFLORESCENCES short, dense, erect, cylindrical terminal raceme less than 4.0 cm long; perianth 1.8–3.6 mm long, pink, rose purplish, green, or white, with scattered resinous glands; styles 2–3; stamen and styles not exserted. FRUITS includes adhering calyx containing achene. SEEDS achene, 2.2–2.5 mm long; oval, almost circular, lenticular, or rarely trigonous; smooth, black, shiny.

Special Identifying Features

Erect or spreading summer annual; nodes swollen; similar to Pennsylvania smartweed, but ocrea fringed by hairlike bristles.

Toxic Properties

See comments under *Polygonum amphibium* var. *emersum*.

1mm

TOP **Seeds**
BOTTOM **Seedling**

Flowers

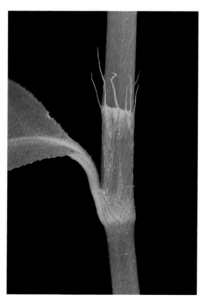

Ocrea

Red Sorrel

Rumex acetosella L. · Polygonaceae · Buckwheat Family

Synonyms

Sheep sorrel, sourgrass, Indian cane

Habit, Habitat, and Origin

Erect, slender, rhizomatous perennial herb; to 4.0 dm tall; open sandy soils, roadsides, fields, pastures, and disturbed areas; native of Eurasia.

Seedling Characteristics

Cotyledons glabrous, spatulate; leaves green, petioled.

Mature Plant Characteristics

ROOTS fibrous, slender, creeping, from long yellow taproot and rhizomelike roots with buds. STEMS erect, 2.0–4.0 dm tall, square, 1 to several from rhizomes, usually unbranched below panicle. LEAVES alternate, 5.0–20.0 mm wide, 2.0–5.0 cm long, variable in size, usually 3-lobed; terminal lobe narrowly elliptic to oblong, lateral lobes smaller, triangular, divergent; lower and median leaves petiolate, as long as blades; ocrea thin, membranous. INFLORESCENCES plants dioecious, occasionally polygamous, inflorescence sometimes half as long as shoot; pedicel jointed next to flower, outer tip lanceolate, inner tip in male florets, 1.5–2.0 mm, obovate; female florets broadly ovate, stalks about as long as calyx; calyx valve ovate, blunt apically, not membranous-winged. FRUITS includes achene and adhering calyx, 1.0–1.7 mm long, not or only slightly enlarged during maturation, reticulate-nerved. SEEDS achene, 0.9–1.5 mm long; brown, golden brown, or reddish; lustrous.

Special Identifying Features

Erect, slender, proliferating from buds on rhizomelike roots; commonly suffused with reddish purple pigment throughout; stems square; leaves arrow- or lance-shaped.

Toxic Properties

Intoxication rarely occurs from *Rumex* species but can result in depression, excessive salivation, tremors, difficulty walking, collapse, labored respiration, and renal problems.

Flowers

Mature plant

Flowering plant

1mm

TOP Seeds
BOTTOM **Seedling**

Curly Dock

Rumex crispus L. · Polygonaceae · Buckwheat Family

Synonyms

Sour dock, yellow dock

Habit, Habitat, and Origin

Erect, coarse, stout, perennial herb; to 1.5 m tall; disturbed areas, fields, cultivated areas, pastures, ditches, and waste sites; native of Eurasia.

Seedling Characteristics

Cotyledons glabrous, spatulate; first leaves glabrous, reddish, petioled.

Mature Plant Characteristics

ROOTS fibrous from fleshy taproot, yellowish orange. STEMS erect, 0.5–1.5 m tall, coarse, stout, thick, unbranched, glabrous. LEAVES alternate, 1.0–4.0 dm long, 2.0–12.0 cm wide, simple, glabrous, margins undulate and crimped, dark green, lower leaves petioled. INFLORESCENCES terminal panicle, 15.0–60.0 cm long, bisexual; petals absent; sepals 6, inner three developing wings or valves, outer three small, inconspicuous; stamens 6; styles 3. FRUITS with 3 valves or wings, 4.0–5.0 mm long, blunt apically, reticulate-nerved, brown at maturity, in clusters on panicle. SEEDS achene, 1.5–3.0 mm long, trigonous, acute tip, brown.

Special Identifying Features

Erect, coarse, stout perennial; leaf margins markedly undulate and crisped; pedicels 1 ½ times as long as valves.

Toxic Properties

See comments under *Rumex acetosella*.

Rosette

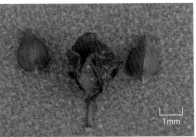

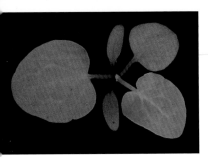

TOP Seeds
BOTTOM Seedling

Broadleaf Dock

Rumex obtusifolius L. · Polygonaceae · Buckwheat Family

Synonyms

Bitter dock, blunt-leaved dock

Habit, Habitat, and Origin

Erect, single-stemmed, perennial herb; to 1.0 m tall; marshes, ditches, and wet open woodlands; native of Europe.

Seedling Characteristics

Hypocotyl round, smooth; cotyledon ovate-deltoid, with small granules on lower surface, petiole as long as blade.

Mature Plant Characteristics

ROOTS fibrous from fleshy, yellowish taproot. STEMS erect, 0.5–1.0 m tall, simple below inflorescence, reddish-tinged, ribbed. LEAVES alternate, 15.0–30.0 cm long, 2.0–15.0 cm wide, simple, flat, margins slightly wavy, base truncate or cordate; upper leaves smaller and more pointed, lower leaves with reddish veins. INFLORESCENCES clusters of racemes, terminal, perfect; petals absent; sepals develop into valves or wings of fruit, not reflexed; valves 3.0–5.0 mm long. FRUITS sepals surrounding achene, 3.0–5.0 mm long, valves or wings with 1–3 spines. SEEDS achene, 2.2–2.5 mm long, acutely trigonous, golden to brown.

Special Identifying Features

Erect, single-stemmed perennial; taproot fleshy; fruit with 1–3 spines on valves or wings.

Toxic Properties

See comments under *Rumex acetosella*.

Mature plant

Flowers

1mm

Common Purslane

Portulaca oleracea L. · Portulacaceae · Purslane Family

Synonyms

Common portulaca, wild portulaca, purslane, pursley, pusley, pussley, wild portulac

Habit, Habitat, and Origin

Prostrate, fleshy, succulent, drought-resistant annual; stems to 5.0 dm long; cultivated areas, fields, pastures, gardens, lawns, and waste sites; native of tropical or subtropical western Asia, introduced into North America from southern Europe.

Seedling Characteristics

Cotyledons linear, glabrous, purplish red; young leaves appearing opposite with each succeeding pair oriented 90 degrees from preceding pair.

Mature Plant Characteristics

ROOTS fibrous from taproot. STEMS prostrate, 1.0–5.0 dm long, glabrous, suc-culent, fleshy, usually purplish red, root-ing at nodes. LEAVES alternate or nearly opposite, 0.4–2.8 cm long, spatulate or obovate, margins smooth, succulent. INFLORESCENCES solitary in leaf axils or several together in leaf clusters at ends of branches; petals 5, yellow, sessile; sta-mens 7–12 or 20. FRUITS capsule, 4.0–8.0 mm long, 3.0–5.0 mm wide, globular, ovoid, splitting at middle, many-seeded. SEEDS oval, 0.5 mm or less diameter, papillose.

Special Identifying Features

Prostrate, fleshy, succulent, drought-resistant annual; stems and leaves succu-lent; leaves alternate or nearly opposite; flowers in leaf axils, small, yellow.

Toxic Properties

Can cause elevated oxalate accumulation in livestock leading to metabolic and renal problems.

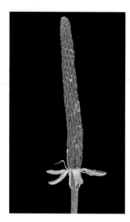

TOP **Seeds**
BOTTOM **Flower**

Flowering plant

Mousetail

Myosurus minimus L. · Ranunculaceae · Buttercup Family

Synonyms

Tiny mousetail

Habit, Habitat, and Origin

Acaulescent, prostrate to decumbent, cool-season annual; to 20.0 cm tall; moist areas of fallow and no-tillage fields, often in marshes or near open water; native of North America.

Seedling Characteristics

Basal leaves linear, very small.

Mature Plant Characteristics

ROOTS fibrous from thin taproot. **STEMS** acaulescent, 2.0–20.0 cm tall, glabrous. **LEAVES** basal, 2.0–15.0 cm long, 0.5–2.2 mm wide; linear to narrowly oblong, oblanceolate, or spatulate; margins entire; glabrous. **INFLORESCENCES** solitary at end of leafless scape, elongated spike of carpels, inconspicuous, radially symmetrical; petals 5, whitish to yellowish, quickly deciduous; sepals 5, greenish yellow; carpels 100 or more on a cylindrical to compressed, elongate receptacle 1.5–6.0 cm long, 1.5–3.0 mm wide. **FRUITS** flat, rectangular to trapezoidal, angular achene, 0.9–2.5 mm long, 1.0–1.5 mm wide, with beak 0.3–0.5 mm. **SEEDS** same as fruit.

Special Identifying Features

Prostrate to decumbent, acaulescent, glabrous, cool-season annual; leaves narrow, linear to spatulate; flowers on cylindrical to compressed spike of more than 100 carpels at top of scape, resembling a mouse's tail.

Toxic Properties

None reported.

Flowers

Flowering plant

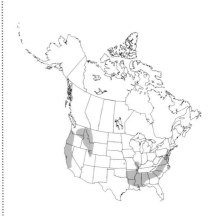

TOP Seeds
BOTTOM Seedling

Corn Buttercup

Ranunculus arvensis L. · Ranunculaceae · Buttercup Family

Synonyms

Corn crowfoot, field buttercup

Habit, Habitat, and Origin

Erect to spreading winter annual herb; to 6.0 dm tall; pastures, cultivated areas, and often covering fallow or no-till or reduced-till fields; native of Europe.

Seedling Characteristics

Cotyledons glabrous or sparsely pubescent; first leaf tricleft.

Mature Plant Characteristics

ROOTS fibrous from taproot. STEMS erect to spreading, 4.0–6.0 dm tall, glabrous or sparsely appressed hirsute pubescence. LEAVES basal rosette, alternate upward, 3.0 cm long, 5.0 mm wide, usually with 3 parts, dissected into linear lobes, margins variously sublobed or broadly toothed, long petiole. INFLORESCENCES axillary, solitary; sepals 3–5, not showy; petals usually 5, yellow, mostly obovate; stamens and carpels numerous. FRUITS aggregate of achenes, 1.5–2.5 mm long, green, turning brownish at maturity. SEEDS achene, 4.0–5.5 mm long, 3.0–4.0 mm wide, obovoid, flattened, rim and both faces coarsely spiny or papillose, slightly stipitate; beak essentially straight.

Special Identifying Features

Erect to spreading winter annual; basal and stem leaves 3-parted; fruit spiny.

Toxic Properties

None reported.

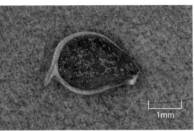

TOP Seed
MIDDLE Two-leaf seedling
BOTTOM Four-leaf seedling

Hairy Buttercup

Ranunculus sardous Crantz · Ranunculaceae · Buttercup Family

Synonyms

None

Habit, Habitat, and Origin

Erect winter annual herb; to 50.0 cm tall; fallow or no-till fields; possibly native of Europe.

Seedling Characteristics

Hypocotyl erect, pubescent; cotyledons oblong with rounded tip, sheath at base.

Mature Plant Characteristics

ROOTS fibrous from taproot. STEMS erect, 5.0–50.0 cm tall, single or branching from base, green, copiously pubescent. LEAVES basal rosette, alternate upward, 5.0–15.0 mm long, palmately 3-lobed, green, pubescent, bracteal leaves much reduced. INFLORESCENCES solitary; petals 5-parted, 5.0–9.0 mm long, 4.0–7.0 mm wide, pale to bright sulfur yellow, showy; stamens 25–50; sepals 2.0–4.0 mm long, yellowish green. FRUITS aggregate of 10–40 achenes, globose-ovoid head, 4.0–8.0 mm long, 5.0–10.0 mm wide, green, turning brownish at maturity. SEEDS achene, 3.0–4.0 mm long, 2.0–3.0 mm wide, oval, flattened, brown with green border, ring of knobby projections (tubercles) on both flat surfaces, short beak at tip.

Special Identifying Features

Erect winter annual; plant pubescent throughout; leaves palmately 3-lobed; seeds with knobby projections.

Toxic Properties

None reported.

Flowering plant

Fruit

Indian Mock-Strawberry

Duchesnea indica (Andr.) Focke · Rosaceae · Rose Family

Synonyms

Indian strawberry

Habit, Habitat, and Origin

Prostrate, low-growing perennial spreading by creeping stolons; stems to 3.0 dm long; shady locations including lawns and flowerbeds; native of Asia.

Seedling Characteristics

Rosette of trifoliate leaves resembling mature plant.

Mature Plant Characteristics

ROOTS fibrous from taproot, rooting from crowns and creeping stolons. **STEMS** prostrate, to 3.0 dm long, branched. **LEAVES** alternate, compound, leaflets 3; leaflets 2.0–4.0 cm long, elliptic to ovate, margins crenate, sparsely pubescent on lower surface. **INFLORESCENCES** borne singly on long petiole subtended by large, leafy, 3-toothed bract; petals 5, yellow, 14.0–18.0 mm wide. **FRUITS** fleshy strawberry-like berry, about 1.0 cm diameter, dry and covered with seeds, inedible. **SEEDS** kidney-shaped, 1.0 mm diameter, reddish brown, covering fleshy berry.

Special Identifying Features

Prostrate low-trailing perennial; spreading by creeping stolons; leaves trifoliate; leaflet margins crenate; flowers 5-petaled, yellow; fruit red, strawberry-like berry, inedible.

Toxic Properties

None reported.

TOP Shoots
MIDDLE Flower
BOTTOM Runners

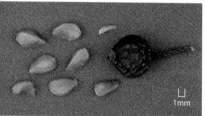

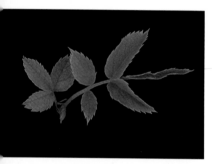

TOP Seeds and fruit
MIDDLE Sprout
BOTTOM Flowering plant

Flowers

Multiflora Rose

Rosa multiflora Thunb. · Rosaceae · Rose Family

Synonyms

Japanese rose, musk-rose

Habit, Habitat, and Origin

Erect or climbing prickly shrub; to 2.0 m long; fields, pastures, roadsides, and waste sites; native of Asia.

Seedling Characteristics

Usually not observed, leaves trifoliate and reddish, emerging from rootstock.

Mature Plant Characteristics

ROOTS fibrous from large rootstock that serves as propagating structure. STEMS erect or climbing shrub, 1.0–2.0 m long, somewhat arching, glabrous, round, reddish; prickles mostly paired, flattened, curved, broad-based. LEAVES alternate, odd-pinnately compound, 7–9 leaflets; leaflets 1.0–6.0 cm long, 0.8–3.0 cm wide, obovate to elliptic, margins sharply serrated, base cuneate or rounded, glabrous above, pubescent or rarely glabrous below, with conspicuous pectinate-serrate stipules. INFLORESCENCES pyramidal corymb, often many together; pedicels softly pubescent; petals 5, white to rarely light pink, 5.0–12.0 mm long, ovate; stamens numerous; styles exserted from hypanthium; sepals ovate-lanceolate, 12.0–15.0 mm long. FRUITS hip, 6.0–9.0 mm long, ellipsoidal, red. SEEDS achene, 2.5–3.0 mm long, densely pubescent, enclosed in ellipsoidal red hip.

Special Identifying Features

Erect or climbing prickly shrub; stems with curved, flattened, broad-based thorns; stipules conspicuously pectinate-serrate; fruit a red hip.

Toxic Properties

None reported; plants may cause mechanical injury.

Fruit

Southern Dewberry

Rubus trivialis Michx. · Rosaceae · Rose Family

Synonyms

None

Habit, Habitat, and Origin

Trailing, low-arching, thorned perennial; to several meters long; old fields and waste sites; native of North America.

Seedling Characteristics

Not seen.

Mature Plant Characteristics

ROOTS fibrous from taproot, becoming elongated and woody with age. STEMS trailing, to several meters long, slender to thickened, branching, forming thickets, pubescent, trichomes usually glandular-tipped; thorns short, curved. LEAVES usually with 5 leaflets, 2.0–10.0 cm long, 1.0–3.0 cm wide; leaflets oblong to lanceolate-ovate, acute, serrate or doubly serrate, lateral leaflets sessile to subsessile, terminal leaflet on spiny peti-

olule, persistent and turning red in late fall and winter. INFLORESCENCES solitary or rarely in 3-flowered cymes; petals obovate, white, rarely pink, 15.0–25.0 mm long, 7.0–10.0 mm wide; numerous stamens and pistils; sepals pubescent, 5.0–7.0 mm long. FRUITS aggregate of subglobose to elongate drupes, 1.5–2.0 cm long, 1.0–1.5 cm wide, green turning blackish purple at maturity, glabrous, juicy. SEEDS stone, 2.7–3.0 mm long, oblong, with irregular ridges.

Special Identifying Features

Trailing, low-arching, thorned perennial; stems with thorns and purple-tipped bristles, numerous, forming thickets; leaves with 5 leaflets, persistent and turning red in late fall and winter.

Toxic Properties

Thorns may cause mechanical injury.

TOP **Flower**
MIDDLE **Young plant**
BOTTOM **Mature plant**

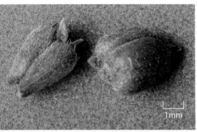

TOP Seeds
MIDDLE Seedling
BOTTOM Flower

Poorjoe

Diodia teres Walt. · Rubiaceae · Madder Family

Synonyms

Rough buttonweed, povertyweed

Habit, Habitat, and Origin

Erect to prostrate annual herb; to 8.0 dm tall; sandy or rocky soils, prairies, pastures, fields, stream valleys, and waste areas; native of North America.

Seedling Characteristics

Cotyledons ovate to oblong-ovate with round tip, glabrous; first true leaves alternate with pointed tip and prominent midvein, pubescent.

Mature Plant Characteristics

ROOTS fibrous from shallow, slender taproot. STEMS erect to prostrate, 1.5–8.0 dm long, circular to quadrangular in cross section, greenish to reddish, densely pubescent with short stiff hairs; longer pubescence retrorse, rarely unbranched. LEAVES opposite, 2.0–5.0 cm long, 2.0–6.0 mm wide, sessile, margins entire to lightly toothed, linear to elliptic-lanceolate, prominent midvein; base of leaf blade a short, membranous tube that covers node; bristly branched stipule. INFLORESCENCES petals 4-lobed, fused into tube, 4.0–6.0 mm long; corolla bluish lilac to white, axillary cluster of 1–3, sessile. FRUITS capsule, 2.5–4.0 mm long, hispid, obpyriform, green, single longitudinal furrow, pubescent, resembles swollen button, splitting into 2 nutlets. SEEDS nutlet, 3.0 mm long, oval, pubescent, light brown.

Special Identifying Features

Erect to prostrate annual; bristles stipular; flowers bluish lilac to white.

Toxic Properties

None reported.

Flowering plant

Mature plant

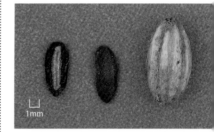

Virginia Buttonweed

Diodia virginiana L. · Rubiaceae · Madder Family

Synonyms

Large buttonweed

Habit, Habitat, and Origin

Erect, spreading, ascending, or prostrate, diffusely branched perennial herb; to 6.0 dm tall; sandy or rocky soils, lawns, turf, pastures, fields, and waste sites; native of North America.

Seedling Characteristics

Cotyledons glabrous, spatulate; leaves entire, sessile.

Mature Plant Characteristics

ROOTS fibrous from fleshy, extensively branched taproot. STEMS erect, spreading, ascending, or prostrate; 1.0–6.0 dm tall, freely branched, longitudinally ridged, with pubescence along ridges, from woody crown. LEAVES opposite, 2.0–7.0 cm long, 4.0–12.0 mm wide, sessile, elliptic to lance-shaped, entire or slightly serrulate margins, rough to touch, normally green but often found with yellow mosaic pattern; connected by interposed membranous, fringed stipule. INFLORESCENCES solitary or occasionally in groups of 2 in leaf axils, sessile; corolla white, tube filiform, 7.0–10.0 mm long, 4-lobed. FRUITS leathery, ridged, crowned by 2 prominent calyx teeth, smooth to pubescent, separating into 2 indehiscent 1-seeded carpels, resembles a swollen button. SEEDS nutlets, 6.0–7.0 mm long, remaining attached to each other in compressed oval-oblong body, splitting lengthwise, broadly rounded ribs separated by grooves, 2 per fruit.

Special Identifying Features

Erect, spreading, ascending, or prostrate perennial; leaves simple, opposite, slightly thickened, connected by fringed stipules.

Toxic Properties

None reported.

TOP Seeds and fruit
MIDDLE Seedling
BOTTOM Flower and fruit

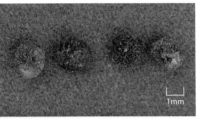

TOP Seeds
BOTTOM Flowering plant

Flowers

Catchweed Bedstraw

Galium aparine L. · Rubiaceae · Madder Family

Synonyms

Cleavers, goosegrass, goose-grass, spring-cleavers

Habit, Habitat, and Origin

Prostrate, mat-forming, reclining, shallow-rooted winter or summer annual herb; stem to 1.5 m long; fields, pastures, roadsides, fencerows, thickets, and waste sites; native of Eurasia and North America.

Seedling Characteristics

Hypocotyl and lower surface of cotyledons tinged with purple; first leaves opposite, with bristlelike hairs.

Mature Plant Characteristics

ROOTS shallow and matted, fibrous from taproot. **STEMS** prostrate, 0.2–1.5 m long, square, angles with short retrorse bristles. **LEAVES** whorls of 6–8; blade 3.0–8.0 cm long, 2.0–6.0 mm wide, linear-oblanceolate, bristle-tipped, margins retrorsely hispid. **INFLORESCENCES** clusters of 1–5 in a simple cyme, borne on peduncles; petals white, 2.0 mm diameter; sepals inconspicuously fused into hypanthium. **FRUITS** schizocarp, 2.0–4.0 mm long, 4.0–5.0 mm wide, dry; brown, gray, or black; sometimes bristly. **SEEDS** persistent with fruit segments, circular or kidney-shaped and densely covered with hooked bristles and tubercles.

Special Identifying Features

Prostrate, mat-forming, reclining winter or summer annual; leaves 6–8, whorled; stem square, covered with retrorse bristles on angles; flowers 4, small, petals white on slender peduncles.

Toxic Properties

None reported. Stems and leaves are edible.

Seedling

Florida Pusley

Richardia scabra L. · Rubiaceae · Madder Family

Synonyms

Florida purslane, Mexican clover

Habit, Habitat, and Origin

Erect or ascending annual herb; to about 8.5 dm tall; lawns, turf, flowerbeds, gardens, cultivated areas, fields, pastures, and waste sites; native of tropical America.

Seedling Characteristics

Hypocotyl smooth, green throughout or maroon-tinged in upper portion; cotyledon blades thick, smooth throughout.

Mature Plant Characteristics

ROOTS fibrous from taproot, near surface. **STEMS** erect or ascending, 1.1–8.5 dm tall, round, pilose, sparingly branched or copiously branched and diffusely spreading. **LEAVES** opposite, 2.0–8.0 cm long, pubescent, blades flat, entire, oblong or elliptic to lanceolate or ovate, acute or acuminate, wavy-margined, short petiole. **INFLORESCENCES** mostly perfect, 6-parted, in depressed terminal clusters; corolla white, 4.0–6.0 mm long, tube funnelform, usually 6-lobed; sepals lanceolate or ovate-lanceolate, 1.0–1.5 mm long, with stiff pubescence. **FRUITS** tuberculate, 3.0–4.0 mm long, leathery, separating into 4 indehiscent carpels. **SEEDS** within lobes of fruit, 2-grooved on lower face.

Special Identifying Features

Weakly erect or ascending annual; stems pubescent; flowers tubular, white, in terminal clusters.

Toxic Properties

None reported.

1mm

TOP Seeds
MIDDLE Two-leaf seedling
BOTTOM Four-leaf seedling

Flowers

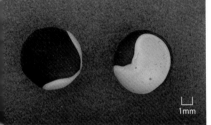

TOP Seeds
MIDDLE Seedling
BOTTOM Flower

Fruit

Balloonvine

Cardiospermum halicacabum L. · Sapindaceae · Soapberry Family

Synonyms

None

Habit, Habitat, and Origin

Climbing or trailing annual herbaceous vine; to 2.0 m long; open areas, wet ditches, field edges, roadsides, and waste sites; native of North America.

Seedling Characteristics

Cotyledons small, rectangular, soon disappearing; first leaf nearly glabrous, deeply dissected.

Mature Plant Characteristics

ROOTS fibrous from taproot. **STEMS** climbing or trailing, to 2.0 m long, round, wiry, several-ribbed, green or brownish, with axillary tendrils. **LEAVES** alternate, 2.0–4.0 cm long, 1.0–2.0 cm wide, divided or dissected, segments lanceolate, coarsely dentate or lobed, weakly pubescent. **INFLORESCENCES** imperfect, irregular; sepals 4; petals 4, white, unequal, 2.0–4.0 mm long. **FRUITS** inflated membranous capsule, 2.0–3.0 cm diameter, fleshy, drupelike, green, pubescent, subglobose or obovoid. **SEEDS** round, 5.0 mm diameter, black with white markings, very hard.

Special Identifying Features

Climbing or trailing herbaceous vine with tendrils; leaves alternate; seed with distinctive black-and-white markings, enclosed in an inflated membranous capsule.

Toxic Properties

None reported.

Infestation

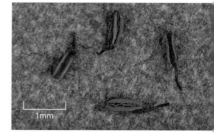

1mm

TOP Seeds
MIDDLE Sprout
BOTTOM Flowering plant

Lizardstail

Saururus cernuus L. · Saururaceae · Lizardtail Family

Synonyms

Lizard's tail, water-dragon, swamplily

Habit, Habitat, and Origin

Erect perennial herb forming colonies; to 1.2 dm tall above water level; often aquatic, in bogs, ponds, streams, and swampy areas; native of North America.

Seedling Characteristics

Emerging leaves arrowhead-shaped or cordate, alternate, petioles clasping stems.

Mature Plant Characteristics

ROOTS fibrous from taproot and slender, creeping rhizomes. STEMS erect, to 1.2 m tall above water level, jointed, appearing to zigzag, unbranched below, few ascending branches, leafy toward top of plant. LEAVES alternate, 15.0 cm long, ovate-cordate to arrowhead-shaped, margins entire, pubescent below, becoming glabrous above, petiole shorter than blade and clasping stem. INFLORESCENCES stalked, crowded raceme opposite uppermost leaves, to 30.0 cm long, white, without sepals or petals, often curving down, resembling a lizard's tail. FRUITS cluster of somewhat fleshy capsules, 3–5 carpels, depressed and rounded. SEEDS contained within indehiscent capsule when capsule separates.

Special Identifying Features

Erect perennial herb forming colonies; wetland plant; leaves arrowhead-shaped; inflorescence white and resembling a lizard's tail.

Toxic Properties

None reported.

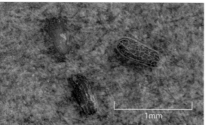

TOP Seeds
MIDDLE Seedling
BOTTOM Flowering plant

Flower

Disc Waterhyssop

Bacopa rotundifolia (Michx.) Wettst. · Scrophulariaceae · Figwort Family

Synonyms

Roundleaf bacopa

Habit, Habitat, and Origin

Prostrate, succulent, annual or perennial herb; stems to 4.0 dm long; stems floating or creeping in and around ponds, pools, lakes, marshes, and ditches; native of North America.

Seedling Characteristics

Opposite elliptical leaves; stems green to purple, pilose or hispid.

Mature Plant Characteristics

ROOTS fibrous from taproot, rooting at stem nodes. STEMS prostrate to ascending, to 4.0 dm long, round, fleshy. LEAVES opposite, 0.5–2.5 cm long, entire, sessile, glandular-punctate, palmately 5–11-veined, orbicular to obovate, submerged leaves often elliptical. INFLORESCENCES 1–3 per leaf axil, on pubescent pedicels, 5.0–15.0 mm long; corolla 8.0–10.0 mm long, white with yellowish throat, lobes shorter than tube; calyx of 5 distinct, unequal sepals 3.0–7.0 mm long; stamens 4. FRUITS capsule, 3.0–5.0 mm long, ovoid to subglobose. SEEDS cylindrical, 0.5 mm long, pale tan with darker reticular lines, minute tail on both ends.

Special Identifying Features

Prostrate to succulent floating herb, often forming mats; flowers white with yellowish throat; corolla 8.0–10.0 mm long; capsule 5.0 mm long.

Toxic Properties

None reported.

Low Falsepimpernel

Lindernia dubia (L.) Pennell · Scrophulariaceae · Figwort Family

Synonyms

Clasping false pimpernel, long-stalked false pimpernel, many-flowered false pimpernel, variable false pimpernel; *Lindernia dubia* (L.) Pennell var. *anagallidea* (Michx.) Cooper

Habit, Habitat, and Origin

Erect to ascending, small, highly branched annual herb; to 2.0 dm tall; wet ground, shallow water, and low meadows; native of North America.

Seedling Characteristics

Cotyledons petiolate, deltoid; first true leaves elliptic.

Mature Plant Characteristics

ROOTS fibrous from taproot. **STEMS** erect to ascending, 0.5–2.0 dm tall, slender, glabrous, quadrangular, highly branched from near base. **LEAVES** opposite, sessile, 0.5–2.0 cm long and 3.0–10.0 mm wide but much reduced above, ovate to elliptic, broadest at rounded or clasping base, 3–5-nerved. **INFLORESCENCES** axillary; corolla 6.0–9.0 mm long, white to lavender, all opening; lower stamens sterile; calyx lobes linear, 3.0–4.0 mm long; pedicel thin, 3.0–28.0 mm long, one to three times longer than subtending leaf. **FRUITS** capsule, 2.5–5.0 mm long, ellipsoid, usually longer than calyx lobes. **SEEDS** oblong, 1½–2 times as long as wide, brownish yellow.

Special Identifying Features

Erect to ascending, small, highly branched annual with relatively brief reproductive cycle; pedicels one to three times longer than subtending leaf; corollas all opening; leaves slightly reduced above, no longer than 2.0 cm, bases round to cordate, appear clasping, no petiole; capsule usually exceeding calyx lobes.

Toxic Properties

None reported.

Flower

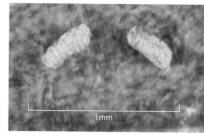

1mm

TOP Seeds
MIDDLE Seedling
BOTTOM Fruit

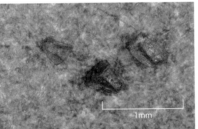

TOP Seeds
BOTTOM Seedling

Flowers

Oldfield Toadflax

Nuttallanthus canadensis (L.) D. A. Sutton · Scrophulariaceae · Figwort Family

Synonyms

Blue toadflax; *Linaria canadensis* (L.) Chaz.

Habit, Habitat, and Origin

Erect, slender, glabrous, biennial or winter annual herb; to 0.7 m tall; fallow fields, roadsides, pastures, and waste sites; native of North America.

Seedling Characteristics

Cotyledons ovate; first true leaves oblanceolate, glabrous.

Mature Plant Characteristics

ROOTS fibrous from short taproot. **STEMS** erect, 0.15–0.7 m tall, solitary; flowering stalk smooth, branched above, from basal rosette. **LEAVES** alternate, 5.0–30.0 mm long, mostly on lower portion of stem, linear; upper leaves cauline; basal leaves smaller, 3.0–12.0 mm long, opposite, subopposite, or whorled, oblanceolate or spatulate. **INFLORESCENCES** raceme; corolla spurred, 5.0–15.0 mm long including spur, strongly zygomorphic, blue to violet or purple with whitish lower lip; style 1.0 mm long; stamens 4; calyx 5-parted. **FRUITS** capsule, 2.0–3.0 mm diameter, subglobose, opening by 2 apical valves. **SEEDS** small, angled, and winged.

Special Identifying Features

Erect, slender, glabrous biennial or winter annual; basal rosette, upper leaves linear; flowers blue to violet or purple, whitish on lower lip.

Toxic Properties

None reported.

Flowering plant

Flowers

Licoriceweed

Scoparia dulcis L. · Scrophulariaceae · Figwort Family

Synonyms

Goatweed, sweet broom, sweet broom-weed, sweet broomwort

Habit, Habitat, and Origin

Erect, profusely branched annual or short-lived perennial herb; to 8.0 dm tall; open grasslands, pastures, fields, roadsides, and waste sites; native of North America.

Seedling Characteristics

Cotyledons ovate; first leaves opposite, ovate-lanceolate, glandular.

Mature Plant Characteristics

ROOTS fibrous from taproot, perennial from rootstock. STEMS erect, 3.0–8.0 dm tall, solitary or 2 or more stems from same base, pubescent especially around nodes, branched. LEAVES opposite or occasionally in whorls of 3–4, 2.0–4.0 cm long, ovate-lanceolate, short-petioled, coarsely serrate, glabrous. INFLORESCENCES in leaf axils; corolla 4-parted, white, regular, throat densely bearded; stamens 4; calyx 4-parted, 1.5–2.0 mm long. FRUITS capsule, 2.0 mm long and wide, ovoid to subglobose to widely ellipsoid, exceeding calyx, dehiscing along septum and slightly along locule. SEEDS very small, numerous, angular, somewhat shiny, brownish.

Special Identifying Features

Erect, profusely branched, annual or short-lived perennial; flowers in leaf axil, whitish, throat densely bearded.

Toxic Properties

None reported.

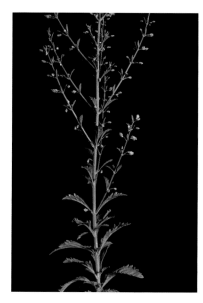

Flowering plant

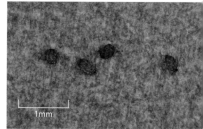

1mm

TOP Seeds
BOTTOM Seedling

Flowering plants

TOP Seeds
BOTTOM Seedling

Witchweed

Striga asiatica (L.) Kuntze · Scrophulariaceae · Figwort Family

Synonyms

Striga lutea Lour.

Habit, Habitat, and Origin

Erect, highly branched annual herb; to 4.0 dm tall; native of tropical and subtropical Africa and Asia; parasite of important grass crops; listed as a Federal Noxious Weed.

Seedling Characteristics

Underground portion of seedling stem with scales; first leaves opposite; stems round below third node.

Mature Plant Characteristics

ROOTS watery, white, attached to host plant by haustoria, no root hairs. STEMS erect, 1.5–4.0 dm tall, simple or irregularly branched, square above third node. LEAVES alternate apically and opposite basally, 1.0–5.0 cm long, 1.0–5.0 mm wide, linear or linear-elliptic, entire or remotely toothed, slightly pubescent. INFLORESCENCES axillary, solitary, showy; corolla zygomorphic, usually crimson. FRUITS 5-sided capsule, 2.0–3.0 mm long, more than 1,000 seeds per capsule. SEEDS small, 0.2 mm long, 0.1 mm wide, brown, deeply reticulated, striate.

Special Identifying Features

Erect, highly branched annual; parasite of grasses; roots attached to host plant by haustoria; stem square above third node; flowers crimson.

Toxic Properties

None reported.

Flowers

Moth Mullein

Verbascum blattaria L. · Scrophulariaceae · Figwort Family

Synonyms

None

Habit, Habitat, and Origin

Erect, sparsely branching biennial herb; to 1.2 m tall; roadsides, weedy lots, pastures, and waste sites; native of Eurasia.

Seedling Characteristics

Cotyledons spoon-shaped.

Mature Plant Characteristics

ROOTS fibrous from taproot. STEMS erect, 0.6–1.2 m tall, slender, simple, rarely with 1–3 secondary branches, glabrous below, minutely pubescent above, trichomes glandular. LEAVES basal leaves spoon-shaped, forming rosette, alternate above, 7.0–15.0 cm long, 2.0–5.0 cm wide, oblong to lanceolate or elliptic, sharply pointed, clasping, glabrous, doubly serrate, 2.0–6.0 dm diameter. INFLORESCENCES open racemes borne on pedicels, pubescent, trichomes glandular; corolla 25.0–35.0 mm wide, yellow to white, with 5 fused petals along elongated axis, 2.0–3.0 cm diameter; stamens filamentous, violet, with soft pubescence. FRUITS capsule, 5.0–10.0 mm diameter, round, 2-chambered, pubescent, trichomes glandular, many seeds per capsule. SEEDS 6-sided, 0.5–1.0 mm long, columnar, light to dark brown, covered with rows of pits.

Special Identifying Features

Erect herb from basal rosette; leaves glabrous; pedicel longer than fruit.

Toxic Properties

None reported.

TOP **Young plant**
BOTTOM **Flowering plant**

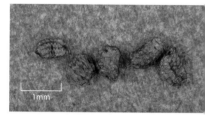

TOP **Seeds**
MIDDLE **Three-leaf seedling**

Flowers

Common Mullein

Verbascum thapsus L. · Scrophulariaceae · Figwort Family

Synonyms

Flannel-plant, great mullein, mullein, Quaker rogue, wild tobacco

Habit, Habitat, and Origin

Erect, unbranched biennial herb; to 1.8 m tall; roadsides, weedy lots, pastures, and waste sites; native of Eurasia.

Seedling Characteristics

Rosette leaves extremely fuzzy, margins slightly serrate.

Mature Plant Characteristics

ROOTS fibrous from fleshy taproot. STEMS erect, 0.9–1.8 m tall, rarely branched, stout, pubescent, woolly, trichomes branched or stellate, thick. LEAVES basal rosette, 15.0–45.0 cm long, oblong, tapering to petiole, dense woolly pubescence; stem leaves smaller and more pointed, yellow-green, dense woolly pubescence, base attached to stem continues down to next leaf. INFLORESCENCES dense, crowded, continuous terminal spike; corolla 1.5–2.5 cm diameter, 5-lobed, sulfur yellow, nearly sessile. FRUITS capsule, 6.0–10.0 mm long, ovoid. SEEDS rod-shaped, 0.8 mm wide, brown, rough, marked with ridges and grooves.

Special Identifying Features

Erect biennial; entire plant extremely pubescent with dense woolly trichomes; stems not branching; flowers fragrant.

Toxic Properties

None reported for mammals, but used as an insecticide and toxic to fish.

TOP Seeds
MIDDLE Seedling
BOTTOM Flowering plant

Purslane Speedwell

Veronica peregrina L. · Scrophulariaceae · Figwort Family

Synonyms

Neckweed

Habit, Habitat, and Origin

Erect to ascending, low-growing, cool-season annual herb; to 30.0 cm tall; cultivated areas, fields, lawns, flowerbeds, roadsides, and waste sites; native of North America.

Seedling Characteristics

Cotyledons elliptical; first true leaves opposite, oblong to linear-oblong, irregularly toothed, glabrous or pubescent, trichomes glandular.

Mature Plant Characteristics

ROOTS fibrous from short taproot. STEMS erect to ascending, 4.0–30.0 cm tall, simple or branching near base, glabrous or pubescent, trichomes glandular. LEAVES opposite below, alternate upward, 0.5–3.5 cm long, linear-oblong to oblanceolate; lower leaves petioled, with irregularly toothed crenate to serrate margins; upper leaves bracteal, sessile with slightly toothed to entire margins; glabrous or pubescent, trichomes glandular. INFLORESCENCES borne singly in bracts of a terminal raceme, solitary in upper axils; bract leaves similar to upper vegetative leaves; corolla 2.0–3.0 mm wide, inconspicuous, irregularly 4-lobed, petals white. FRUITS capsule, 2.5–4.0 mm, obcordate to obovate and notched, numerous seeds. SEEDS 0.5–1.0 mm long, plano-convex, oval, pale yellow to orange-yellow, glossy.

Special Identifying Features

Erect to ascending cool-season annual with opposite and alternate leaves; leaves oblong to oblanceolate, margins toothed to entire; flowers in raceme of inconspicuous solitary white flowers.

Toxic Properties

None reported.

1mm

TOP **Fruit**
BOTTOM **Seedling**

Flowering plant

Persian Speedwell

Veronica persica Poir. · Scrophulariaceae · Figwort Family

Synonyms

Bird's eye speedwell, common field speedwell, winter speedwell

Habit, Habitat, and Origin

Erect to ascending, low-growing, freely branched, cool-season annual herb; to 3.0 dm tall; cultivated areas, fields, lawns, flowerbeds, roadsides, pond edges, and waste sites; native of Asia.

Seedling Characteristics

Cotyledons elliptical and glabrous with smooth margins; first true leaves pubescent, serrated.

Mature Plant Characteristics

ROOTS fibrous from taproot. STEMS erect or with ascending tips, 1.0–5.0 dm long, pubescent, simple or branched, weak, prostrate, rooting at nodes. LEAVES opposite below, alternate above, 5.0–25.0 mm long, broadly ovate, 3–5 coarse teeth on each side, pubescent, short petiole. INFLORESCENCES solitary, borne on long, slender stalks in leaf axils, 15.0–30.0 mm long; corolla about 1.0 cm diameter, sky blue with dark stripes and white center; calyx lobes pointed, 3-veined, up to 8.0 mm long. FRUITS capsule, 7.0–10.0 mm wide and about two-thirds as long, heart-shaped. SEEDS numerous per capsule, 1.5–2.2 mm long, cup-shaped, rough, yellowish to pale tan.

Special Identifying Features

Erect to ascending herb; leaves pubescent with short petiole; flowers sky blue with darker stripes and white center, borne on slender stalk.

Toxic Properties

None reported.

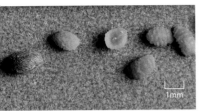

TOP **Seeds**
MIDDLE **Seedling**
BOTTOM **Flower**

Flower

Oakleaf Datura

Datura quercifolia Kunth · Solanaceae · Nightshade Family

Synonyms

Oakleaf thornapple

Habit, Habitat, and Origin

Erect, branching annual herb; to 1.5 m tall; cultivated areas, fields, pastures, and waste sites; native of North America.

Seedling Characteristics

Hypocotyl with downy pubescence; cotyledons linear.

Mature Plant Characteristics

ROOTS fibrous from thick taproot. STEMS erect, 0.5–1.5 m tall, dichotomously branched, with somewhat downy pubescence when young. LEAVES alternate, 6.0–20.0 cm long, 4.0–12.0 cm wide, oak-shaped, shallowly to deeply pinnately lobed, moderately pubescent, petioled. INFLORESCENCES solitary, tubular, pale violet to purple; anthers purple, 4.0–7.0 cm. FRUITS capsule, 3.0–4.0 cm long, erect, 4-valved, covered with large spines; spines unequal and flattened at base, more than 10.0 mm long. SEEDS oval in outline, 4.0 mm long, flattened, dark grayish to tan, with pits or shallow depressions.

Special Identifying Features

Erect, branching annual; entire plant covered by downy pubescence when young; capsule armed with stout spines more than 10.0 mm long.

Toxic Properties

Alkaloids found in *Datura* species can cause dilation of pupils, decreased digestive tract motility, decreased salivation, and mortality in livestock and humans.

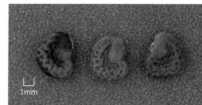

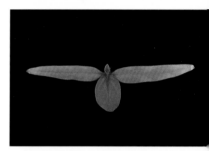

TOP **Seeds**
MIDDLE **Seedling**
BOTTOM **Fruit**

Flower

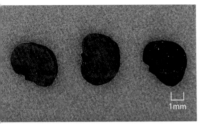

TOP Seeds
MIDDLE Seedling
BOTTOM Fruit

Jimsonweed

Datura stramonium L. · Solanaceae · Nightshade Family

Synonyms

Purple thornapple

Habit, Habitat, and Origin

Erect, stocky, many-branched herb; to 1.5 m tall; cultivated areas, fields, pastures, and waste sites; native of the Americas.

Seedling Characteristics

Hypocotyl with thickened, spindle-shaped base, smooth, often purple-tinged; cotyledon smooth or sparsely downy on upper surface.

Mature Plant Characteristics

ROOTS extensively fibrous from taproot. STEMS erect, 1.0–1.5 m tall, stout, succulent, hollow, widely branching; young stems with band of downy trichomes extending above leaf intersection, becoming smooth with age. LEAVES alternate, 8.0–20.0 cm long, 4.0–15.0 cm wide, simple, ovate, acute, margins coarsely toothed with a few large triangular teeth, nearly glabrous, long petiole, with leaf stalk extending less than halfway around stem at intersection and predominantly continuous on stem below, node appearing swollen, unpleasant odor when even slightly bruised. INFLORESCENCES solitary, in leaf axil or branch axil, pedicellate, campanulate; petals white or purple, 7.0–10.0 cm long. FRUITS capsule, 2.4–4.0 cm long, erect, round or oval in outline, usually covered with short spines. SEEDS circular or kidney-shaped in outline, 2.0–3.0 mm wide, flattened, surface rough with fine pits and ridges, 100–360 per capsule.

Special Identifying Features

Erect, stocky, many-branched; flowers large, tubular, white or purple; leaves large; bruised plants give off unpleasant odor.

Toxic Properties

See comments under *Datura quercifolia*.

Flower

Flower

Apple-of-Peru

Nicandra physalodes (L.) Gaertn. · Solanaceae · Nightshade Family

Synonyms

Chinese lantern

Habit, Habitat, and Origin

Erect, branched annual herb; to 10.0 dm tall; fields, pastures, and waste sites; native of South America, introduced from Peru.

Seedling Characteristics

Cotyledons oblong to linear, petioled; leaves simple, glabrous to sparsely pilose, ovate.

Mature Plant Characteristics

ROOTS fibrous from taproot. STEMS erect, 3.0–10.0 dm tall, succulent, stout, branched only above, ridged, glabrous to sparsely pubescent. LEAVES alternate, 5.0–20.0 cm long, 3.0–12.0 cm wide, widely lanceolate, margins lacerate-dentate or shallowly lobed, pubescent, tapering to petiole. INFLORESCENCES axillary, solitary; corolla blue or pale blue, 2.0–3.0 cm long, 3.0–4.0 cm wide, broadly campanulate, weakly lobed; calyx 1.0–2.2 cm long; sepals separate at anthesis. FRUITS berry, 1.3–2.0 cm diameter, 3–5 locules, surrounded by dry calyx. SEEDS suborbicular, 1.5–1.8 mm wide, flattened, reticulate.

Special Identifying Features

Erect annual herb; flowers blue; calyx inflated surrounding fruit; sepals distinct.

Toxic Properties

Plants contain glycoalkaloids that have caused bloating in sheep, but toxicity is not a high risk.

TOP Seeds
MIDDLE Seedling
BOTTOM Fruit

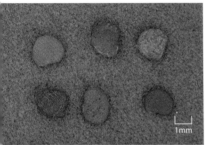

TOP **Seeds**
BOTTOM **Seedling**

Cutleaf Groundcherry

Physalis angulata L. · Solanaceae · Nightshade Family

Synonyms

Lantern plant; *Physalis lanceifolia* Nees

Habit, Habitat, and Origin

Erect, branching annual herb; to 1.0 m tall; cultivated areas, fields, pastures, open woodlands, roadsides, and waste sites; native of tropical America, possibly including the southeastern United States.

Seedling Characteristics

Hypocotyl purple-tinged, pubescent; cotyledons green, 1.0–4.0 mm long, 1.0–3.0 mm wide, glabrous or pubescent, especially along margins and veins.

Mature Plant Characteristics

ROOTS fibrous from taproot. STEMS erect, 0.4–1.0 m long, freely branching, round, green, glabrous. LEAVES alternate, 4.0–15.0 cm long, 3.0–10.0 cm wide, ovate, thin, lanceolate to widely ovate, tips acuminate; margins coarsely irregular, indented, or nearly entire; base cuneate to suborbicular. INFLORESCENCES solitary in leaf or branch axils; pedicels smooth, 1.0 cm long; corolla 6.0–10.0 mm long, slightly broader than long, dull yellowish; anthers 1.0–2.3 mm long, bluish; calyx 3.0–5.0 mm long, triangular-lobed, smooth or slightly pubescent. FRUITS berry, 8.0–11.0 mm diameter, round; surrounded by inflated, 10-ribbed, papery calyx 2.0–3.5 cm long, 1.5–2.5 cm wide, usually purple-veined, borne on peduncle 2.0–3.0 cm long. SEEDS circular to semicircular, 2.0 mm wide, flattened, light orange or straw-colored, many per berry.

Special Identifying Features

Erect, branching annual; plants smooth; berry enclosed in enlarged rounded, 10-ribbed calyx.

Toxic Properties

Physalis species contain glycoalkaloids, but toxicity has not been confirmed.

Fruit

Flower

Flower

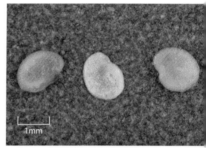

Clammy Groundcherry

Physalis heterophylla Nees · Solanaceae · Nightshade Family

Synonyms
None

Habit, Habitat, and Origin
Erect perennial herb; to 8.0 dm tall; fields, pastures, open woodlands, roadsides, and waste sites; native of North America.

Seedling Characteristics
Hypocotyl smooth or with short trichomes toward top; cotyledons green, 1.0–4.0 mm wide, 4.0–9.0 mm long, smooth or with row of trichomes on margins and midvein beneath.

Mature Plant Characteristics
ROOTS well-developed, fibrous, from thickened taproot. STEMS erect or widely spreading, 3.0–8.0 dm long, often highly branched, round, pubescent. LEAVES alternate, 3.0–11.0 cm long, 3.0–8.0 cm wide, ovate to rhombic, margins entire or shallowly indented, pubescent; trichomes widely scattered on upper surface, short and dense on margins and lower surface. INFLORESCENCES solitary in axils on pedicels; corolla yellow, 1.5–2.2 cm diameter, bell-shaped;

anthers 3.5–4.5 mm long. FRUITS berry, 1.0 cm diameter, round, 2-celled, yellow, surrounded by papery calyx 2.5–3.0 cm long. SEEDS circular, 2.0 mm wide, flattened, dull, light orange or straw-colored.

Special Identifying Features
Erect perennial; stems and leaves pubescent; anthers 3.5–4.5 mm long.

Toxic Properties
See comments under *Physalis angulata*.

Fruit

TOP Seeds
BOTTOM Sprout

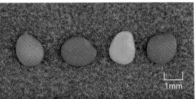

Flower

TOP Seeds
MIDDLE Seedling
BOTTOM Flower and fruit

Smooth Groundcherry

Physalis longifolia Nutt. var. *subglabrata* (Mackenz. & Bush) Cronq. ·
· Solanaceae · Nightshade Family

Synonyms

Longleaf groundcherry

Habit, Habitat, and Origin

Erect or spreading perennial herb; to 1.0
m tall; fields, pastures, open woodlands,
roadsides, and waste sites; native of
North America.

Seedling Characteristics

Hypocotyl smooth or nearly so; cotyledons 4.0–9.0 mm long, 1.0–4.0 mm
wide.

Mature Plant Characteristics

ROOTS fibrous with thickened rootstocks,
perennial. STEMS erect or spreading,
0.3–1.0 m tall, branched, green but often
purple-tinged, ridged, smooth. LEAVES
alternate, 3.0–8.0 cm long, 0.2–1.0 cm
wide, simple, ovate to ovate-lanceolate,
margins entire to coarsely toothed,
uneven at base, petioled, pubescence
slight and scattered on leaf surface.
INFLORESCENCES solitary in leaf or
branch axils on pedicel; corolla drooping, bell-shaped, yellow with dark center,
1.5–2.5 cm wide; calyx 3.0–4.0 mm long,
lobes ovate or triangular. FRUITS berry,
8.0–11.0 mm diameter, round, orange-
red to purple; surrounded by inflated,
10-ribbed, papery calyx, 5-cleft open
at mouth, 2.0–3.0 cm long, 1.5–2.0 cm
broad; peduncles 2.0–3.0 cm long. SEEDS
circular or kidney-shaped, 2.0 mm wide,
flattened, tan or straw-colored.

Special Identifying Features

Erect or spreading perennial; berry
enclosed in enlarged, 10-ribbed calyx;
thickened rootstock.

Toxic Properties

See comments under *Physalis angulata*.

American Black Nightshade

Solanum americanum P. Mill. · Solanaceae · Nightshade Family

Synonyms

None

Habit, Habitat, and Origin

Erect or spreading annual or short-lived perennial herb; to 1.0 m tall; fields, pastures, open woodlands, roadsides, and waste sites; native of North America.

Seedling Characteristics

Hypocotyl covered with small trichomes, green; cotyledons small, green on both surfaces, short trichomes on margins, midrib evident on lower surface, petiole covered with smaller trichomes.

Mature Plant Characteristics

ROOTS fibrous from shallow taproot. STEMS erect, 0.5–1.0 m tall, slender, round, ridged, with small teeth, glabrous or pubescent, becoming woody with age. LEAVES alternate, 5.0–17.0 cm long, 2.0–9.0 cm wide, simple, ovate or ovate-lanceolate, margins variable from entire to slightly crenate, petioled, pubescence highly variable and affected by environment, subglabrous to moderately pubescent. INFLORESCENCES racemiform or umbelliform clusters, star-shaped, corolla white, to 15 per cluster. FRUITS berry, round, 5.0–10.0 mm diameter, immature fruit green with white flecks nearly always present, turning black at maturity, 4–13 stone cells per berry. SEEDS round to oboval in outline, 1.4–1.8 mm wide, flattened, tan, 50–110 per berry.

Special Identifying Features

Erect or spreading annual or short-lived perennial; leaves smooth or pubescent, green; fruit with white flecks, detaches at receptacle, 4–13 stone cells per berry.

Toxic Properties

Solanum species contain varying levels of alkaloids, and many animal species are subject to intoxication from consumption of leaves, stems, and fruits. Prickles can cause mechanical injury.

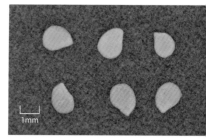

1mm

TOP Seeds
BOTTOM Seedling

Flower

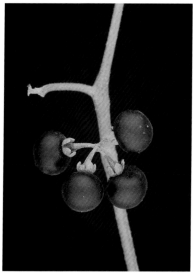

Fruit

Flowers

Red Soda Apple

Solanum capsicoides All. · Solanaceae · Nightshade Family

Synonyms

Cockroach berry, cockroach poison, devil's apple, love-apple, soda apple

Habit, Habitat, and Origin

Erect spreading perennial shrub; to 1.0 m tall; open woodlands, pastures, and fields; native of South America.

Seedling Characteristics

Hypocotyl often purple-tinged, long trichomes; cotyledons short, trichomes on both surfaces and margins; first true leaves more or less heart-shaped, pubescent on both surfaces and petiole, often purplish on lower surface.

Mature Plant Characteristics

ROOTS fibrous from taproot and rhizomes. **STEMS** erect, 0.3–1.0 m tall, spreading, green, scattered pubescence; prickles 1.0 cm long, straight, slender, broad-based, more numerous and shorter than those on leaves. **LEAVES** alternate, 6.0–18.0 cm long, 5.0–16.0 cm wide, blades ovate-triangular; sinuate-lobed, lobes subobtuse or subacute, scattered pubescence, prickles broad-based, straight, to about 1.0 cm long, scattered on petioles and on mid-sized and larger veins. **INFLORESCENCES** lateral, 2.5 cm wide, few-flowered; corolla white to cream-colored, deeply lobed, with 5 white petals; stamen yellow. **FRUITS** berry, to 5.0 cm diameter, round; pale green, yellow, and then orange-red at maturity. **SEEDS** compressed, tan to dark brown, margin winged, 80–240 per fruit.

Special Identifying Features

Erect, spreading perennial shrub; stems, petioles, and leaf veins covered with straight, broad-based prickles; mature fruit orange-red; seed with winged margin.

Toxic Properties

See comments under *S. americanum*. Fruits of *S. capsicoides* are highly toxic and in developing countries are sometimes used for rodent and insect control. Prickles can cause mechanical injury.

Flowers

Horsenettle

Solanum carolinense L. · Solanaceae · Nightshade Family

Synonyms

Bullnettle

Habit, Habitat, and Origin

Erect to spreading rhizomatous perennial herb; to 1.0 m tall; cultivated areas, fields, pastures, open woodlands, roadsides, and waste sites; native of southeastern North America.

Seedling Characteristics

Hypocotyl tough, often purple-tinged and densely covered with short, stiff, slightly retrorsely spreading trichomes; cotyledons small, glossy green above, lighter green on lower surface, upper and lower surfaces glabrous, margins with short gland-tipped trichomes.

Mature Plant Characteristics

ROOTS fibrous from taproot and creeping subterranean rhizomes. STEMS erect to spreading, 0.2–1.0 m tall, becoming thickened, zigzagged at nodes, deep purple and then woody, short stiff trichomes, prickles appearing with age.

LEAVES alternate, 7.0–12.0 cm long, 3.0–8.0 cm wide, simple, petioled, crowded, deep green above and paler beneath, elliptic-oblong to oval, margin undulate or lobed, both surfaces with sessile star-shaped trichomes, prickles on midvein and petiole, characteristic potato odor when crushed. INFLORESCENCES racemelike clusters, star-shaped, petals white or pale violet, anthers yellow. FRUITS berry, 1.0–1.5 cm diameter, round, green, turning yellow. SEEDS round or oboval in outline, 1.5–2.5 mm wide, flattened, smooth, glossy; light yellowish, orange-tan, or brown; 40–120 per berry.

Special Identifying Features

Erect to spreading perennial; stems and leaves with sharp prickles; leaves pubescent, trichomes star-shaped with 4–8 spreading rays.

Toxic Properties

See comments under *S. americanum*. Prickles can cause mechanical injury.

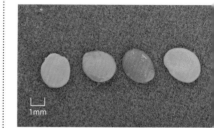

1mm

TOP **Seeds**
MIDDLE **Seedling**
BOTTOM **Fruit**

Flowers

TOP **Seeds**
MIDDLE **Seedling**
BOTTOM **Fruit**

Robust Horsenettle

Solanum dimidiatum Raf. · Solanaceae · Nightshade Family

Synonyms

Western horsenettle; *Solanum torreyi* Gray

Habit, Habitat, and Origin

Erect or spreading perennial shrub; to 1.2 m tall; cultivated areas, fields, pastures, open woodlands, roadsides, and waste sites; native of North America.

Seedling Characteristics

Hypocotyl tough, purple-tinged, smooth, stiff; cotyledons glossy green above, lighter on lower surface, smooth on both surfaces, margins with short, gland-tipped trichomes; first leaves shallowly sinuate-lobed, pubescent on both surfaces, lower surface often purplish.

Mature Plant Characteristics

ROOTS fibrous from deep taproot and lateral rhizomes. **STEMS** erect, spreading, 0.5–1.2 m long, simple, branching, rough, scattered stellate trichomes and broad-based prickles up to 12.0 mm long. **LEAVES** alternate, 8.0–20.0 cm long, 4.0–16.0 cm wide, ovate to elliptic-lanceolate, margins irregularly sinuate-lobed, surfaces and major veins with dense simple and 8–12-rayed stellate trichomes, veins and petioles with broad-based prickles to 12.0 mm. **INFLORESCENCES** few to several in terminal raceme or panicle; corolla to 5.0 cm wide, lavender to purple, shallowly lobed, star-shaped; anthers yellow. **FRUITS** berry, 1.5–3.0 cm diameter, round; green and light green mottled, yellow at maturity. **SEEDS** compressed, glossy, yellowish orange or brown, 40–120 per fruit.

Special Identifying Features

Erect or spreading perennial shrub; leaf lobes more broadly rounded than those of *S. carolinense*; stellate trichomes with 8–12 rays on lower leaf surface; corolla purple.

Toxic Properties

Intoxication from ingested fruits and leaves causes crazy-cow syndrome in cattle. Prickles can cause mechanical injury.

Silverleaf Nightshade

Solanum elaeagnifolium Cav. · Solanaceae · Nightshade Family

Synonyms

None

Habit, Habitat, and Origin

Erect to spreading perennial herb or shrub; to 1.0 m tall; cultivated areas, fields, pastures, open woodlands, roadsides, and waste sites; native of South America.

Seedling Characteristics

Hypocotyl tough, often purple-tinged, densely covered with coarse short trichomes; cotyledons small, green above, lighter on lower surface, covered with trichomes.

Mature Plant Characteristics

ROOTS fibrous from taproot and creeping rhizomes. **STEMS** erect to spreading, 0.3–1.0 m tall, branching, becoming woody with age; covered with coarse, thin, sharp prickles; silvery green, stellate trichomes. **LEAVES** alternate, 5.0–15.0 cm long, 2.0–4.0 cm wide, simple, crowded, linear to oblong or oblong-lanceolate, margins entire or merely sinuate, often deciduous basally, green above, silvery green beneath, pubescent, trichomes stellate, petioled. **INFLORESCENCES** racemelike clusters, star-shaped; corolla violet or bluish or rarely white, spreading or reflexed, lobes ovate; anthers 5, erect, bright yellow, tightly surrounding pistil. **FRUITS** berry, 1.0–2.0 cm diameter, round, green with dark streaks from top, yellow to yellowish brown at maturity, persistent, drying and wrinkled on plant. **SEEDS** round or oboval in outline, 1.5–2.5 mm wide, flattened, surface glossy; tan, yellowish orange, or dark orange; 40–120 per fruit.

Special Identifying Features

Erect to spreading perennial herb or shrub; leaves green above, silvery green beneath; flowers star-shaped, petals violet or bluish, rarely white.

Toxic Properties

See comments under *S. americanum.* Prickles can cause mechanical injury.

TOP **Flowers**
BOTTOM **Fruit**

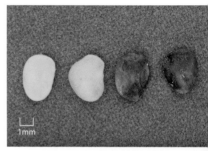

1mm

TOP **Seeds**
BOTTOM **Seedling**

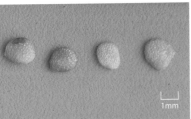

1mm

TOP **Seeds**
MIDDLE **Seedling**
BOTTOM **Flower**

Fruit

Jamaican Nightshade

Solanum jamaicense P. Mill. · Solanaceae · Nightshade Family

Synonyms

Devil's tomato

Habit, Habitat, and Origin

Erect to spreading perennial shrub; to 1.0 m tall; shaded open woodlands, pastures, and fields; native of Central America and Caribbean islands.

Seedling Characteristics

Hypocotyl short, green, with long trichomes; cotyledons small, green, with short stiff trichomes on both surfaces and margins; first true leaves more or less spatulate, in pairs, green, short stiff trichomes on both surfaces, petiole short.

Mature Plant Characteristics

ROOTS fibrous from taproot and rhizomes. **STEMS** erect to spreading, 0.5–1.0 m tall, zigzagged at nodes; prickles broad-based, retrorsely curved, to 5.0 mm long; trichomes crowded, stipitate, and stellate. **LEAVES** alternate, often paired, larger and smaller from same node, ovate-triangular, sinuate-lobed; lobes irregular, subacute; blade extending onto petiole, stipitate trichomes on both surfaces, stellate trichomes with 5–7 arms on lower surface, stellate trichomes with 6–9 arms on upper surface, prickles rarely on veins. **INFLORESCENCES** several in lateral umbels; corolla white, to 1.25 cm wide; petals 5, deeply lobed, trichomes covering outer petal surfaces. **FRUITS** berry, 8.5 mm diameter, green with darker green veining, yellowish to reddish orange at maturity, lustrous. **SEEDS** oval in outline, flattened, tan or light brownish, 20–60 per fruit.

Special Identifying Features

Erect to spreading perennial shrub; large and small leaves paired from same node; stems covered with curved prickles; flowers white; fruit bright orange-red at maturity.

Toxic Properties

See comments under *S. americanum*. Prickles can cause mechanical injury.

Nipplefruit Nightshade

Solanum mammosum L. · Solanaceae · Nightshade Family

Synonyms

Nipplefruit, dog poison

Habit, Habitat, and Origin

Erect to spreading perennial shrub; to 1.2 m tall; fields, pastures, open woodlands, roadsides, and waste sites; native of South America.

Seedling Characteristics

Hypocotyl often purple-tinged, long trichomes; cotyledons with trichomes on both surfaces, shorter trichomes on margins; first true leaves green, pubescent on both surfaces and petiole, straight broad-based prickles on main veins and petiole of second true leaf.

Mature Plant Characteristics

ROOTS fibrous from shallow, fleshy taproot and lateral roots. **STEMS** erect to spreading, 0.5–1.2 m long, green, pubescent; prickles broad-based, straight or hooked, to 2.8 cm long. **LEAVES** alternate, 10.0–30.0 cm long, 8.0–24.0 cm wide, simple, ovate-triangular, margins sinuate-lobed, lobes subobtuse or subacute, pubescent, broad-based straight prickles scattered on petiole and main veins. **INFLORESCENCES** in clusters, corolla lavender to purple, deeply lobed, 5 petals, recurved tips, to 5.0 cm wide; anthers yellow. **FRUITS** berry, 4.0–7.0 cm diameter, ovoid or ellipsoid, 1–5 basal nipples, 1 terminal nipple, pale green with dark green veins, yellow at maturity. **SEEDS** compressed, 3.0–5.5 mm wide, dark brown, lustrous, 200–460 per fruit.

Special Identifying Features

Erect to spreading perennial shrub covered with prickles; fruit large, yellow at maturity, ovoid or ellipsoid, with 1 or more nipples.

Toxic Properties

See comments under *S. americanum*. The poisonous fruit is used to kill feral dogs, rodents, and cockroaches in Central and South America; prickles can cause mechanical injury.

TOP Seeds
MIDDLE Seedling
BOTTOM Flowers

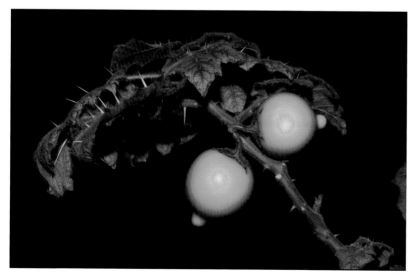

Fruit

Flowers

1mm

TOP Seeds
MIDDLE Seedling
BOTTOM Fruit

Hairy Nightshade

Solanum physalifolium Rusby · Solanaceae · Nightshade Family

Synonyms

Solanum physalifolium Rusby var. *nitidibaccatum* (Bitter) Edmonds, *Solanum sarrachoides* auct. non Sendtner

Habit, Habitat, and Origin

Erect or spreading to erect, annual or short-lived perennial herb; to 1.0 m tall; cultivated areas, fields, pastures, open woodlands, roadsides, and waste sites; native of South America.

Seedling Characteristics

Hypocotyl covered with small trichomes, green; cotyledons small, green on both surfaces, short trichomes on margins, midribs evident on lower surface; petioles covered with small trichomes.

Mature Plant Characteristics

ROOTS fibrous from shallow taproot. **STEMS** erect or spreading, 0.5–1.0 m long, slender, becoming woody, round, furrowed, ridges with small teeth, glabrous or pubescent. **LEAVES** alternate, simple, ovate or ovate-lanceolate, petioled, margins entire to slightly crenate, pubescence highly variable and affected by environment; thick, moderately long, glandular trichomes and short, stipitate-glanded trichomes, often sticky to touch. **INFLORESCENCES** racemes, star-shaped; corolla rotate, 4–5 per cluster; petals white, lobes broadly triangular; stamens 5, yellow, 2.9–4.5 mm long, tightly surrounding pistil. **FRUITS** berry, 5.0–10.0 mm diameter, round, green turning olive green to brownish green at maturity, 2–3 stone cells per berry, calyx covering half of berry at maturity. **SEEDS** round to nearly oval in outline, 1.4–1.8 mm wide, flattened, tan, 10–36 per fruit.

Special Identifying Features

Erect annual or short-lived perennial; calyx covering half of berry; 2–3 stone cells per berry.

Toxic Properties

See comments under *S. americanum*.

Flower

Jerusalem-cherry

Solanum pseudocapsicum L. · Solanaceae · Nightshade Family

Synonyms

False pepper

Habit, Habitat, and Origin

Erect, branching, leafy perennial shrub; to 19.0 dm tall; roadsides, cultivated ground, waste sites, thickets, and woodlands, but may not be hearty outside cultivation; native of southern Europe.

Seedling Characteristics

Hypocotyl green, covered with short, microscopic trichomes; cotyledons linear, green, paler underneath, glabrous except for short, gland-tipped trichomes along margins.

Mature Plant Characteristics

ROOTS fibrous from taproot. **STEMS** erect, 8.0–19.0 dm tall, branching, round, green, glabrous. **LEAVES** alternate, 3.0–10.0 cm long, simple, lanceolate to oblong or oblanceolate, margins entire or slightly undulate, margins pale, narrow at base, shining bright green, prominent veins, petiole short. **INFLORESCENCES** solitary or in small lateral clusters, corolla white, 11.0–14.0 mm wide, eciliate lobes ovate to oval; calyx lobes 2.5 mm long, lanceolate, eciliate. **FRUITS** berry, 1.0–2.0 cm diameter; green, turning yellow, then scarlet to orange at maturity. **SEEDS** compressed, 1.0–2.0 mm wide, notched, yellow-orange.

Special Identifying Features

Erect, leafy perennial shrub; leaves shiny, bright green, margins pale, tapered at base; corolla white; berry 1.0–2.0 cm diameter, green, then scarlet or orange at maturity.

Toxic Properties

See comments under *S. americanum*.

TOP **Seeds**
MIDDLE **Seedling**
BOTTOM **Fruit**

Fruit

Eastern Black Nightshade

Solanum ptycanthum Dunal · Solanaceae · Nightshade Family

Synonyms

West Indian nightshade

Habit, Habitat, and Origin

Erect, branching annual herb, or short-lived perennial herb in mild climates; to 1.0 m tall; cultivated areas, fields, pastures, open woodlands, roadsides, and waste sites; native of the Americas.

Seedling Characteristics

Hypocotyl covered with small trichomes, green; cotyledons small, green on both surfaces, short trichomes on margins, midribs evident on lower surface, petioles covered with smaller trichomes.

Mature Plant Characteristics

ROOTS fibrous from shallow taproot. STEMS erect or spreading, 0.3–1.0 m tall, slender, round, furrowed, ridges with small teeth, glabrous or pubes-cent, becoming woody with age. LEAVES alternate, simple, petioled, ovate or ovate-lanceolate, red-purple on lower surface, margins variable and affected by environment, subglabrous to moderately pubescent. INFLORESCENCES in umbel-like clusters, star-shaped, 4–5 per cluster; corolla white, often streaked or deeply tinted purple; calyx spreading. FRUITS berry, 5.0–12.0 mm diameter, round, green, purplish black or dark green at maturity, 4–15 stone cells per fruit. SEEDS compressed, 1.4–1.8 mm wide, round to oboval in outline, tan, 50–100 per fruit.

Special Identifying Features

Erect, branching annual or short-lived perennial; young leaves red-purple beneath; 4–15 stone cells per berry.

Toxic Properties

See comments under *S. americanum.*

Buffalobur

Solanum rostratum Dunal · Solanaceae · Nightshade Family

Synonyms

Beaked nightshade, Kansas thistle, prickly nightshade

Habit, Habitat, and Origin

Erect, branching annual herb; to 7.0 dm tall; fields, pastures, fencerows, roadsides, stockyards, and waste sites; native of North America.

Seedling Characteristics

Hypocotyl succulent, usually purple-tinged trichomes scattered or in vertical lines; cotyledons green, often purple-tinged beneath, 9.0–18.0 mm long, 1.5–3.0 mm wide, smooth or with clear, gland-tipped trichomes on margin and midvein underneath.

Mature Plant Characteristics

ROOTS fibrous from taproot. STEMS erect or spreading, 2.0–7.0 dm tall, zigzagged at nodes, pubescent, trichomes star-shaped; moderately to heavily armed with yellow, slender, straight prickles. LEAVES alternate, 4.0–11.0 cm long, 3.0–7.0 cm wide, elliptic or obovate, once or twice pinnately lobed; stiff straight trichomes above, especially on veins, dense star-shaped trichomes below; armed with slender straight prickles on veins, petioles 2.0–6.0 cm long. INFLORESCENCES elongate raceme, ascending; corolla bright yellow, star-shaped, 16.0–24.0 mm wide; anthers 5, yellow, one longer than others. FRUITS berry, 8.0–12.0 mm in diameter, completely enclosed by spiny, beaked calyx. SEEDS oval in outline, 1.5–2.0 mm wide, flattened, orange-brown.

Special Identifying Features

Erect or spreading herb; leaves lobed; stems, leaves, and calyx armed with stiff trichomes and slender straight prickles.

Toxic Properties

See comments under *S. americanum*. Prickles can cause mechanical injury.

Flowering plant

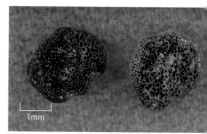

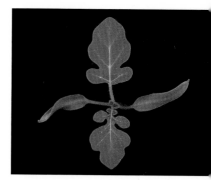

TOP Seeds
MIDDLE Two-leaf seedling
BOTTOM Young plant

Flower

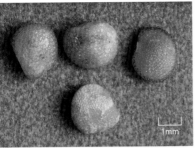

Sticky Nightshade

Solanum sisymbriifolium Lam. · Solanaceae · Nightshade Family

Synonyms

None

Habit, Habitat, and Origin

Erect or spreading perennial shrub; to 1.5 m tall; cultivated areas, fields, pastures, open woodlands, roadsides, and waste sites; native of South America.

Seedling Characteristics

Hypocotyl often purple-tinged, short trichomes; cotyledons with short trichomes on both surfaces and on margins; first true leaf sinuate, second becoming pinnatifid-lobed, scattered glandular trichomes on both surfaces and petiole, yellowish or yellowish orange prickles on main veins and petiole.

Mature Plant Characteristics

ROOTS fibrous from deep taproot and lateral rhizomes. STEMS erect or spreading, 0.4–1.5 m tall, green, glandular trichomes, prickles straight, orange to straw-colored, to 1.5 cm long. LEAVES alternate, 0.6–3.0 dm long, 0.5–1.6 dm wide, pinnatifid, lobes with mucros, acute or obtuse at apices, sinuses deeply cut, petioles broad-based; prickles straight, orange to straw-colored, to 1.5 cm on petiole and midveins; trichomes glandular. INFLORESCENCES lateral racemes; corolla white, shallowly lobed, petals 5, star-shaped, about 2.5 cm wide; trichomes sparse to dense outward. FRUITS berry, 1.0–1.5 cm diameter, green turning yellow and then bright red at maturity; calyx covered with broad-based, straight orange prickles and glandular trichomes, loosely covering and splitting as fruit matures, to 2.0 cm wide. SEEDS compressed, tan to light brownish, 8–50 per fruit.

Special Identifying Features

Erect or spreading perennial; prickles on stems, leaves, and calyx, yellow to orange; fruit 1.0–1.5 cm diameter, bright red at maturity.

Toxic Properties

See comments under *S. americanum*. Prickles can cause mechanical injury.

Fruit

Wetland Nightshade

Solanum tampicense Dunal · Solanaceae · Nightshade Family

Synonyms

Aquatic soda apple, sosumba, Tampico soda apple

Habit, Habitat, and Origin

Erect, sprawling, or clambering perennial shrub; stems to 5.0 m long; shaded wetland areas; native of Central America; listed as a Federal Noxious Weed.

Seedling Characteristics

Hypocotyl purple, pubescent; cotyledons with scattered pubescence or glabrous, short trichomes on margins; first true leaves sinuate, glabrous or with widely scattered short trichomes, main veins and petiole purple; straight, broad-based, purple prickles.

Mature Plant Characteristics

ROOTS fibrous from taproot and rhizomes. STEMS erect, sprawling, or clambering; 1.0–5.0 m long, round, green, base woody, surface granular; trichomes stellate, sparse, scattered, 5–7-armed; prickles broad-based and purplish, greenish, or straw-colored, retrorsely curved, to 4 mm long. LEAVES alternate,

6.0–20.0 cm long, 4.0–8.0 cm wide, simple, ovate-lanceolate; irregularly sinuate-lobed; prickles purplish, green, or straw-colored, broad-based, retrorsely curved, to 4.0 mm long on petioles and midveins of lower surface; upper leaf with straight prickles along midvein; surface granular; stellate trichomes sparse, scattered, 5–7-armed. INFLORESCENCES umbellate inflorescence lateral and opposite leaf, few-flowered; corolla white to creamy white; petals 5, to 1.5 cm wide, deeply lobed, not recurved. FRUITS berry, to 8.5 mm wide, green turning yellow and then orange-red at maturity, glossy. SEEDS suborbicular, flattened, light brown to light yellowish, 18–50 per fruit.

Special Identifying Features

Erect, sprawling, or clambering perennial; stems up to 5.0 m long; flowers white, in clusters; fruit green maturing to bright red.

Toxic Properties

See comments under *S. americanum.* Prickles can cause mechanical injury.

1mm

TOP **Seeds**
MIDDLE **Seedling**
BOTTOM **Flowers**

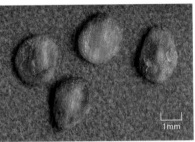

Flowers

Turkeyberry

Solanum torvum Swartz · Solanaceae · Nightshade Family

Synonyms

Bushy white solanum, susumber, Thai-eggplant

Habit, Habitat, and Origin

Perennial shrub; to 3.0 m tall; roadsides, fencerows, fields, and waste sites; native of South America, introduced into Asia and North America; listed as a Federal Noxious Weed.

Seedling Characteristics

Hypocotyl often purple-tinged, pubescent; cotyledons with short trichomes on both surfaces and on margins; first true leaves green, shallow, obtuse, unevenly lobed, pubescent; petiole green or purplish-tinged, pubescent.

Mature Plant Characteristics

ROOTS fibrous from taproot and lateral rhizomes. **STEMS** erect, 0.4–3.0 m tall, round, widely scattered broad-based prickles, retrorsely curved to straight prickles and trichomes dense, stellate, stipitate, 6–9-armed. **LEAVES** opposite, 10.0–24.0 cm long, 6.0–12.0 cm wide, simple; elliptic, oval, ovate, or oblong; entire or irregularly lobed; acute or obtuse, highly variable; broad-based prickles, downcurved to straight, very rare on midveins; woolly pubescent, stellate trichomes, stipitate, 6–9-armed. **INFLORESCENCES** lateral and forked with 2 racemes; corolla shallowly lobed, star-shaped, 1.8–2.5 cm wide, stipitate, with 6–9-armed stellate trichomes toward petal tips; petals 5, white. **FRUITS** berry, 1.0–1.5 cm diameter, round; green then yellow, orange, or brownish at maturity. **SEEDS** compressed, 2.0–3.0 mm wide, reddish or yellowish brown, 20–60 per fruit.

Special Identifying Features

Perennial shrub; leaves woolly pubescent; prickles few, broad-based; flowers white, star-shaped, in clusters; fruit green turning pale yellow or brownish.

Toxic Properties

See comments under *S. americanum*, but fruit is eaten in stir-fry dishes. Prickles can cause mechanical injury.

Tropical Soda Apple

Solanum viarum Dunal · Solanaceae · Nightshade Family

Synonyms

Pastureland nightshade, plant from hell

Habit, Habitat, and Origin

Perennial shrub; to 2.0 m tall; cultivated areas, fields, pastures, open woodlands, roadsides, turf, and waste sites; native of South America; listed as a Federal Noxious Weed.

Seedling Characteristics

Hypocotyl and cotyledons green, densely covered with trichomes; first true leaves and petioles with prickles.

Mature Plant Characteristics

ROOTS fibrous from taproot; producing shallow, long, creeping, fleshy rhizomes; to 2.0 m long. **STEMS** many-branched shrub, 0.5–2.0 m tall, green, covered with dense pubescence and broad-based straight and retrorsely hooked prickles, to 12.0 mm long. **LEAVES** opposite, 7.0–18.0 cm long, 5.0–15.0 cm wide, both surfaces and major veins covered with dense pubescence and broad-based straight prickles up to 12.0 mm long. **INFLORESCENCES** in axillary clusters, flowers white with 5 recurved petals, stamens white to cream-colored surrounding a single pistil. **FRUITS** berry, 2.1–3.5 cm diameter, smooth, round, mottled whitish to light green and dark green, yellow at maturity. **SEEDS** flattened, 2.2–2.8 mm diameter, yellowish brown, 180–420 per fruit.

Special Identifying Features

Perennial shrub; shallow root system; entire aboveground plant except fruit covered with dense pubescence and broad-based, white to yellowish prickles up to 12.0 mm long.

Toxic Properties

See comments under *S. americanum.* Prickles can cause mechanical injury.

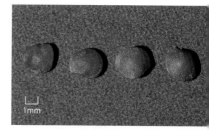

1mm

TOP **Seeds**
MIDDLE **Seedling**
BOTTOM **Fruit**

Flowers

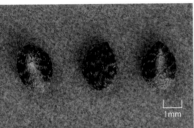

1mm

TOP Seeds
MIDDLE Two-leaf seedling
BOTTOM Young plant

Flowers

Redweed

Melochia corchorifolia L. · Sterculiaceae · Cacao or Cocoa Family

Synonyms

Chocolate weed

Habit, Habitat, and Origin

Erect, branching annual herb; to 1.5 m
tall; cultivated areas, fields, pastures,
open woodlands, roadsides, and waste
sites; native of Eurasia.

Seedling Characteristics

Sparingly pubescent, reddish, alternate
leaves.

Mature Plant Characteristics

ROOTS fibrous from taproot. STEMS erect,
0.4–1.5 m long, divergent branches from
base, reddish, pubescent, trichomes
clustered or stellate. LEAVES alternate,
2.5–7.5 cm long, 1.0–4.0 cm wide, simple,
ovate to ovate-lanceolate, margins often
slightly 3-lobed, doubly serrate, tips
acute, base truncate to obtuse or subcordate, glabrous except along principal
veins, petioles to 1.5 cm long, pubescent;
stipules 4.0–6.0 mm long, linear to
linear-lanceolate, ciliate. INFLORESCENCES
compact headlike cyme, with numerous bracts resembling stipules; petals 5,
4.0–7.0 mm long, dull purple; stamens
5; pistil 5-parted; sepals 5, united. FRUITS
capsule, 4.0–5.0 mm long, subglobose,
with 5 locules. SEEDS 1 per locule, 2.5 mm
long, brown with blackish markings,
wrinkled on back.

Special Identifying Features

Erect, branching annual; stem pubescent; leaf margins doubly serrate; flowers
in compact, headlike cymes.

Toxic Properties

None reported.

Beaked Cornsalad

Valerianella radiata (L.) Dufr. · Valerianaceae · Valerian Family

Synonyms

Cornsalad, valerian

Habit, Habitat, and Origin

Erect, dichotomously branched, succulent, annual herb; to 7.0 dm tall; fields and roadsides; native of North America.

Seedling Characteristics

Leaves alternate, sessile, soon forming a basal rosette.

Mature Plant Characteristics

ROOTS fibrous from thin taproot. STEMS erect, 1.0–7.0 dm tall, dichotomously branched, lightly pubescent to glabrous when mature, sometimes slightly angled or ridged. LEAVES lower basal leaves 2.0–6.0 cm long, 4.0–11.0 mm wide, spatulate, margins entire, obtuse tips, pubescent to glabrous; upward leaves cauline, lanceolate to lanceolate-elliptic, margins entire to dentate, sessile, attenuate bases, tips rounded to truncate. INFLORESCENCES in paired compact cymes, subtended by 2 opposite lanceolate bracts from short peduncle; corolla white, 2.5 cm diameter, actinomorphic; petals 5, united to form corolla tube; calyx absent or nearly so; stamens 3, exserted beyond corolla. FRUITS narrow ovate-tetragonal, yellowish, finely pubescent to glabrous leathery. SEEDS nutlet, 1.5–2.0 mm long, ovate, corky, smooth, with thin shallow striations on surface.

Special Identifying Features

Erect spring annual; dichotomously branched from main stem, forming vertical V-shaped side branches; inflorescences compact in square cymes.

Toxic Properties

None reported.

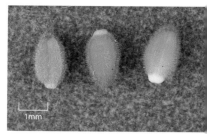

TOP **Seeds**
MIDDLE **Seedling**
BOTTOM **Flowers**

Inflorescence

1mm

TOP Seeds and fruit
MIDDLE Seedling
BOTTOM Flowers

Flowering plant

Prostrate Vervain

Verbena bracteata Lag. & Rodr. · Verbenaceae · Vervain Family

Synonyms

Bigbract verbena

Habit, Habitat, and Origin

Procumbent or decumbent annual or
short-lived perennial herb; to 5.0 dm;
fields, pastures, roadsides, and waste
sites; native of North America.

Seedling Characteristics

Pubescent, cotyledons lanceolate.

Mature Plant Characteristics

ROOTS fibrous from coarse taproot.
STEMS procumbent or decumbent,
1.0–5.0 dm long, nonstoloniferous, freely
branched from base, square, pubescent, trichomes long and stiff. LEAVES
opposite, 1.0–6.0 cm long, 0.6–3.0
cm wide, simple, lanceolate to ovate-
lanceolate; pinnately lobed, dissected,
or 3-lobed; margins toothed; pubescent,
winged petiole. INFLORESCENCES in erect
terminal spikes, usually solitary, 2.0–20.0
cm long, 1.0–2.0 cm wide; corolla bluish to lavender or purple, longer than
calyx, finely pubescent outward; calyx
2.0–3.5 mm long, lobes curved inward,
pubescent. FRUITS schizocarp, containing
4 mericarps. SEEDS nutlet, 2.0–2.2 mm
long, oblong, muricate, yellow to reddish
brown, covered with small bumps.

Special Identifying Features

Procumbent or decumbent annual or
short-lived perennial; flower spikes
erect, usually solitary, 2.0–20.0 cm long,
1.0–2.0 cm diameter.

Toxic Properties

None reported.

Flower

Field Pansy

Viola bicolor Pursh · Violaceae · Violet Family

Synonyms

Johnny-jump-up, wild pansy; *Viola rafinesquii* Greene

Habit, Habitat, and Origin

Erect, branched winter annual herb; to 10.0 cm tall; open fields, roadsides, and lawns; native of North America.

Seedling Characteristics

Basal rosette with slender stems; young leaves oval on long petioles.

Mature Plant Characteristics

ROOTS delicate, fibrous, from taproot. **STEMS** erect, 5.0–10.0 cm tall, slender, smooth, angled, often branched near base, green turning dark purple at maturity. **LEAVES** alternate, 1.0–3.0 cm long, 3.0–8.0 mm wide, obovate to linear-oblanceolate, smooth with sparsely toothed margins, petioles long; stipules large, leaflike, divided into segments. **INFLORESCENCES** stalked, irregular; petals 5, blue-violet to cream with purple or darker blue veins and yellow center, nearly twice as long as lanceolate sepals. **FRUITS** capsule, 5.0–7.0 mm long, green, smooth, shorter than mature sepals. **SEEDS** oblong, 1.0–1.3 mm long, light brown to yellowish brown.

Special Identifying Features

Small, erect annual; branched stems slender; stipules large, leaflike, divided into narrow segments at leaf petiole base; flowers stalked, blue-violet to cream-colored petals with yellow center.

Toxic Properties

None reported.

TOP Seeds
MIDDLE Seedling
BOTTOM Flowering plant

Common Blue Violet

Viola sororia Willd. · Violaceae · Violet Family

Synonyms

Dooryard violet, hooded blue violet, meadow violet, woolly blue violet; *Viola papilionacea* Pursh

Habit, Habitat, and Origin

Colony-forming perennial herb; to 16.0 dm tall; lawns, turf, gardens, flowerbeds, woodlands, and waste sites; native of North America.

Seedling Characteristics

Young leaves originating basally, blades ovate to orbicular or reniform, petiole long.

Mature Plant Characteristics

ROOTS fibrous from branching, stout, coarse rhizomes. STEMS basal crown, leaf petioles arise at or near soil level. LEAVES from basal crowns, 8.0–12.0 cm long, 6.0–10.0 cm wide, ovate to orbicular or rarely almost reniform in summer, heart-shaped, obtuse or short-tipped, crenate, cordate at base, villous especially on underside, petioles twice as long as leaves. INFLORESCENCES borne on villous peduncles, about as tall as leaves; petals typically purple or blue but range from white to dark blue or violet; lateral petals densely white or yellow-bearded, typically fading to white at base; sepals ovate-oblong, usually obtuse, finely ciliate. FRUITS capsule, 1.0–2.0 cm long, ellipsoid to cylindrical, 3-parted, green, then usually mottled brown and purple at maturity. SEEDS oblong, 2.0 mm long, dark tan or purplish brown.

Special Identifying Features

Colony-forming perennial; leaves pubescent; petioles twice as long as leaves; flowers blue to purple, borne on a long peduncle.

Toxic Properties

None reported.

TOP Seeds and fruit
MIDDLE Young plant
BOTTOM Flowering plant

Flower

Flowers

Peppervine

Ampelopsis arborea (L.) Koehne · Vitaceae · Grape Family

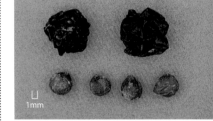

Synonyms

None

Habit, Habitat, and Origin

High-climbing, deciduous perennial vine; to several meters long; woodland edges, fencerows, thickets, field edges, and waste sites; native of North America.

Seedling Characteristics

Rarely seen, emerging sprout with bipinnately compound leaves, no tendrils, reddish.

Mature Plant Characteristics

ROOTS fibrous from large, woody rootstock; perennial. STEMS high-climbing deciduous vine, to several meters long, round, glabrous or glabrate, few or no tendrils, woody, pith white. LEAVES alternate, 8.0–10.0 cm long, 10.0–12.0 cm wide, bipinnately or tripinnately compound, leaflets to 5.0–6.0 cm long, 4.0–5.0 cm wide, glabrous, ovate, base truncate, margins coarsely toothed, petiolulate. INFLORESCENCES cyme, shorter than leaves; calyx saucer-shaped, lobes 5, usually obsolete; petals 5, separate, greenish yellow, 1.5–3.0 mm long; stamens 5. FRUITS drupe, 6.0–10.0 mm diameter, round, green turning dark bluish or purplish black at maturity, staining when crushed. SEEDS pyriform, 4.5 mm long, 2–5 per fruit.

Special Identifying Features

High-climbing woody vine; few or no tendrils; leaves bipinnately compound; fruit a drupe, dark bluish or purplish black at maturity.

Toxic Properties

None reported.

TOP **Fruit and seeds**
MIDDLE **Fruit**
BOTTOM **Young plant**

TOP **Seeds**
BOTTOM **Runner**

Virginia-creeper

Parthenocissus quinquefolia (L.) Planch. · Vitaceae · Grape Family

Synonyms

None

Habit, Habitat, and Origin

High-climbing, deciduous, perennial woody vine; several meters long; dry or rocky areas, woodlands, and fencerows; native of North America.

Seedling Characteristics

Cotyledons oblanceolate; leaves palmately compound, 3–5 leaflets, smooth.

Mature Plant Characteristics

ROOTS fibrous from taproot; developing large, woody rootstock. STEMS perennial woody vine, to several meters long, brown with white pith, 3–8-branched tendrils with adhesive disks. LEAVES alternate, palmately compound, petiolate; leaflets 3–7, usually 5, ovate-elliptic or obovate; to 15.0 cm long, 8.0 cm wide; pale beneath, smooth or occasionally pubescent, acuminate, usually coarsely serrate above middle of blade, base cuneate or oblique, petiolulate. INFLORESCENCES panicle of cymes in terminal and upper stem axils, 125–200 flowers; calyx flat, usually without lobes; petals 5, separate, yellow-green, 2.0–3.0 mm long, disk small, adnate to ovary; stamens 5, filaments short, style 0.5 mm long. FRUITS drupe, 5.0–9.0 mm diameter, black or dark blue, round. SEEDS planoconvex, 3.5–4.0 mm long, obovoid, lustrous, brown, 1–3 per fruit.

Special Identifying Features

High-climbing, deciduous, perennial woody vine; pith white; many tendrils with adhesive disks; palmately compound leaves, usually 5 leaflets.

Toxic Properties

Reported to disrupt digestion.

TOP **Inflorescence**
BOTTOM **Fruit**

Puncturevine

Tribulus terrestris L. · Zygophyllaceae · Caltrop Family

Synonyms

Caltrop, ground burnut, Mexican sandbur

Habit, Habitat, and Origin

Prostrate, mat-forming, summer annual herb; to 5.0 m diameter; sandy or well-drained soils, fields, pastures, roadsides, and railroad beds; native of Europe.

Seedling Characteristics

Cotyledons oblong, twice as wide as long, slightly pubescent.

Mature Plant Characteristics

ROOTS fibrous from taproot. STEMS prostrate, 0.3–2.4 m long, branching at base to form dense circular mats of slender branches, pubescent, trichomes stiff. LEAVES opposite, compound, short-petioled, pubescent; leaflets 3.0–4.0 mm long, oblong, 4–5 pairs, terminal pair shorter than others. INFLORESCENCES single in leaf axils, 5.0–10.0 mm wide on stalks 5.0–10.0 mm long; petals 5, yellow. FRUITS woody, star-shaped, 5.0–7.0 mm long, 5.0–6.0 mm wide; each bur with 2 long, sharp spines 4.0–7.0 mm long, each with several prickles; about 2,000 seeds per bur. SEEDS oblong, 1.5–3.0 mm long, 3.0–5.0 mm wide, yellow, remain enclosed within burs.

Special Identifying Features

Prostrate, mat-forming summer annual; branching along ground in all directions, making circular mat; leaflets small; flowers yellow with 5 petals; fruit with 5 triangular burs with sharp spines strong enough to puncture tires.

Toxic Properties

Plants contain neurotoxins and alkaloids that can cause photosensitization, depression, loss of appetite, skin lesions, and swelling; reported to cause mortality in sheep.

TOP Seeds
MIDDLE Fruit
BOTTOM Seedling

Flower

LILIOPSIDA

MONOCOTS

Flowers

Creeping Burhead

Echinodorus cordifolius (L.) Griseb. · Alismataceae · Water Plantain Family

Synonyms

Burhead

Habit, Habitat, and Origin

Relatively robust annual or short-lived perennial herb; to 5.0 dm tall; in or along edge of water; native of North America.

Seedling Characteristics

Emersed or submerged, petioles with spongy cells at base.

Mature Plant Characteristics

ROOTS fibrous. STEMS leafless scape, erect when young, becoming arched and repent with maturity, to 12.0 dm long, often producing plantlets at nodes and tip. LEAVES from base, 2.0–14.0 cm long, 0.8–11.0 cm wide, broadly ovate, with 7–9 principal veins connected by nearly straight cross-veins, petioles 5.0–50.0 cm long, grooved, spongy toward base. IN-FLORESCENCES whorls of 5–15, 12.0–20.0 mm diameter, subtended by elongated triangular bract; sepals boat-shaped with rounded apex. FRUITS achenes in heads; achene 2.0 mm long, lustrous, dorsal keel irregularly crested, with 2–4 slightly elongated glands in middle, reticulate.

Special Identifying Features

Relatively robust annual or short-lived perennial; arching and repent scape, often producing new plants at nodes; leaves broad; glands on achene.

Toxic Properties

None reported.

TOP Seeds
MIDDLE Young plant
BOTTOM Fruit

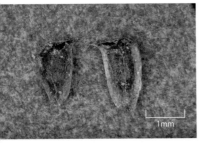

TOP Seeds
MIDDLE Seedling
BOTTOM Flowers

Flowering plant

Grassy Arrowhead

Sagittaria graminea Michx. • Alismataceae • Water Plantain Family

Synonyms

None

Habit, Habitat, and Origin

Perennial aquatic from short, stout rhizomes; to 50.0 cm tall; in or along edge of water; native of North America.

Seedling Characteristics

Juvenile, submersed, and early-formed leaves phyllodial.

Mature Plant Characteristics

ROOTS fibrous, with short, stout rhizome as overwintering structure. STEMS flowering stem, erect and unbranched from rhizome except at lowest node, to 50.0 cm tall. LEAVES emersed, petiolate, and bladed; linear to linear-lanceolate to elliptic; phyllodes of variable length and width, from a few centimeters to 50.0 cm long, 1.0–2.5 cm wide. INFLORESCENCES pistillate in lower one to two whorls and staminate above; flower stalks spreading or ascending, not recurved; petals white. FRUITS achenes in heads, to 1.0 cm across, with reflexed, persistent sepals; achene obovate to oblong, winged, dorsal wing twice as wide as ventral and with 1–2 narrow ridges; beak subulate, absent or to 0.4 mm long, below summit.

Special Identifying Features

Aquatic perennial; leaves narrow, linear; rhizomes short, stout; stalk of fruiting head not recurved.

Toxic Properties

None reported.

Common Arrowhead

Sagittaria latifolia Willd. • Alismataceae • Water Plantain Family

Synonyms

Arrowhead, duck-potato, wapato

Habit, Habitat, and Origin

Coarse, amphibious or aquatic perennial herb; to 8.0 cm tall; in or along edge of water; native of North America.

Seedling Characteristics

Juvenile and submersed leaves phyllodial, straplike, soon disappearing.

Mature Plant Characteristics

ROOTS fibrous with slender rhizomes and terminal corms as overwintering structures. STEMS flowering stem erect, to 8.0 dm, rarely branching. LEAVES from base, blades to 10.0 dm long, variable, long-petioled, spongy-inflated below, 3-lobed, orientation of basal lobes very variable. INFLORESCENCES racemes of up to 10 whorls; flower stalks ascending or spreading; sepals about 1.0 cm long; petals white, about 2.5 cm long and wide. FRUITS numerous achenes in heads, to 2.5 cm diameter, sepals persistent and reflexed; achene 2.5–3.5 mm long, 1.5–2.5 mm wide, cuneate-obovate in outline, broadly corky-winged, beak tapered.

Special Identifying Features

Coarse, amphibious or aquatic perennial from corms and rhizomes; petals white; bracts obtuse, short; plants with milky sap.

Toxic Properties

None reported.

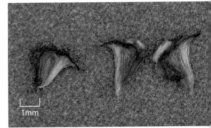

1mm

TOP Seeds
BOTTOM Seedling

TOP Flower and fruit
BOTTOM Flowering plant

California Arrowhead

Sagittaria montevidensis Cham. & Schlecht. • Alismataceae • Water Plantain Family

Synonyms

Giant arrowhead

Habit, Habitat, and Origin

Coarse, aquatic or amphibious annual herb; to 5.0 dm tall; in or along edge of water; native of South America.

Seedling Characteristics

Early stages usually submersed with phyllodial leaves, if emersed then leaf blades sagittate with spongy petioles.

Mature Plant Characteristics

ROOTS fibrous, with modified aerenchyma if in standing water. STEMS flower stem erect, to 5.0 dm tall, may be branched at lowest node, leafless. LEAVES arising from basal stem, to 2.0 dm long and wide, large, broadly ovate, sagittate, petiole stout and spongy. INFLORESCENCES pistillate or perfect, in whorls, to 12 per scape; sepals ovateorbicular, 12.0–14.0 mm long; petals longer, white with green spot; bracts to 1 cm long, ovate with acute apical tip. FRUITS numerous achenes in heads, 2.0 cm across, subtending bracts appressed; achene 2.0–3.0 mm long, 1.0 mm wide, cuneate-obovate, narrowly winged, beak horizontal or oblique.

Special Identifying Features

Coarse, aquatic or amphibious annual; leaves sagittate, petiole stout and spongy; petals white with green spot.

Toxic Properties

None reported.

TOP **Seeds**
BOTTOM **Fruit**

TOP **Flowers**
BOTTOM **Young plant**

Flower

Benghal dayflower

Commelina benghalensis L. • Commelinaceae • Spiderwort Family

Synonyms

Tropical spiderwort

Habit, Habitat, and Origin

Erect annual or perennial herb; to 3.0 dm tall; cultivated areas, fields, pastures, roadsides, and waste sites; native of tropical Africa and India; listed as a Federal Noxious Weed.

Seedling Characteristics

Cotyledon occasionally stalked, connecting seed to apical growing point; first leaf erect, ovate to ovate-elliptic, grasslike, glabrous.

Mature Plant Characteristics

ROOTS fibrous, shallow rhizomes. STEMS succulent, creeping or ascending, 1.0–3.0 dm tall, 2.0–9.0 dm wide, covered with fine pubescence, branched, climbing if supported, rooting at nodes, regenerating when broken. LEAVES ovate to ovate-elliptic, 3.0–7.0 cm long, two to three times longer than wide, venation parallel, margins entire. AERIAL INFLORESCENCES contained within green spathes, 10.0–20.0 mm long, 10.0–15.0 mm wide; petals 3, 3.0–4.0 mm long, upper two petals blue to lilac, lower petal lighter in color or white and less prominent. SUBTERRANEAN INFLORESCENCES spathes appearing as small white tubers, not opening, forced self-pollination. FRUITS pear-shaped, 2-valved capsule, splitting at maturity; aerial fruit producing 5 dimorphic seeds, 1 large and 4 small; subterranean fruit producing 1 large and 2 small seeds. SEEDS dimorphic, brownish black, rectangular to bean-shaped, 1.6–3.0 mm long, 1.3–1.8 mm wide, netted appearance, cap over embryo dislodged during germination.

Special Identifying Features

Leaf length-to-width ratio less than 3:1; trichomes red or white on apex of leaf sheath; flowers both aerial and subterranean.

Toxic Properties

None reported.

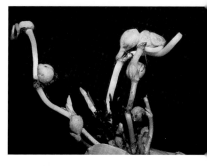

TOP **Seeds**
MIDDLE **Seedling**
BOTTOM **Underground flowers**

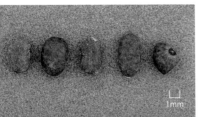

Asiatic Dayflower

Commelina communis L. • Commelinaceae • Spiderwort Family

Synonyms

Common dayflower, wandering Jew

Habit, Habitat, and Origin

Prostrate summer annual herb; to 5.0 dm; damp soil along streams or in gardens, bottomland woods, and swampy areas; native of Asia.

Seedling Characteristics

Erect, unbranched, grasslike, glabrous, becoming prostrate and profusely branched.

Mature Plant Characteristics

ROOTS fibrous. STEMS erect or repent, usually glabrous, with numerous branches rooting at nodes; stems or branches 1.5 cm long, forming large mats late in season. LEAVES lanceolate to ovate-lanceolate, 2.8–15.0 cm long, 0.7–3.6 cm wide, acuminate tips, conspicuous basal sheaths 0.7–2.5 cm long, margins ciliate. INFLORESCENCES perfect, trimerous, irregular; petals 2, lateral, partly united, blue, unequal, the smaller lower one white; anthers 6; stamens 6: 3 fertile and 3 smaller sterile ones, filaments glabrous, enclosed in keel- or boat-shaped spathe. FRUITS capsules bi- or trilocular, one locule usually abortive or empty. SEEDS brown or reddish, rugose, granular, 2.0–4.5 mm long.

Special Identifying Features

Prostrate summer annual; leaves with conspicuous basal sheaths; flowers with 2 erect blue petals and 1 inconspicuous whitish petal.

Toxic Properties

None reported.

Flower

Flower

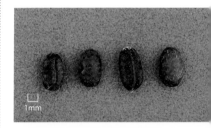

Spreading Dayflower

Commelina diffusa Burm. f. • Commelinaceae • Spiderwort Family

Synonyms

Dayflower, diffuse dayflower

Habit, Habitat, and Origin

Decumbent, diffusely branched annual herb; to 5.0 dm long; wet or seasonally wet soils near streams, marshes, floodplain woodlands, and fallow fields; native of tropical Americas.

Seedling Characteristics

Erect and unbranched initially, grasslike, glabrous, becoming prostrate and profusely branched.

Mature Plant Characteristics

ROOTS fibrous. STEMS diffusely branched, decumbent or repent, 2.0–6.0 dm long, glabrous or glabrate, rooting at nodes. LEAVES lanceolate, 2.5–8.0 cm long, 0.5–1.9 cm wide, tips acuminate or acute, sheaths 0.4–1.5 cm long, margins usually ciliate. INFLORESCENCES usually solitary spathe, acuminate, margins not fused; all petals blue; anthers 5: 2 sterile and 3 fertile, 1.7–2.5 mm long, 0.3–1.0 cm wide. FRUITS capsule, 6.0–7.0 mm long, 3.0–3.5 mm wide, bi- or trilocular, one locule usually abortive or empty. SEEDS brown or reddish, pitted-reticulate, 3.5 mm long.

Special Identifying Features

Decumbent, diffusely branched, glabrous or glabrate annual herb; margins of spathe not fused; flowers with all 3 petals blue and 5 anthers.

Toxic Properties

None reported.

TOP Seeds
MIDDLE Two-leaf seedling
BOTTOM Three-leaf seedling

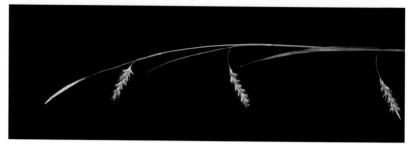

TOP Male and female inflorescences
BOTTOM Flowering plant

Rhizomes

Wolftail Sedge

Carex cherokeensis Schwein. · Cyperaceae · Sedge Family

Synonyms

Cherokee sedge

Habit, Habitat, and Origin

Erect, tufted perennial; 3.0–8.0 dm tall; open fields, pastures, partially shaded woodlands; native of North America.

Seedling Characteristics

Nascent sheaths smooth; basal leaves 3-ranked.

Mature Plant Characteristics

ROOTS brown to reddish brown, fibrous, extensively branching and creeping from short, stocky rhizomes. STEMS culm, triangular in cross section, longer than basal leaves, 3.0–8.0 dm tall. LEAVES 3-ranked, spreading, linear, 2.0–5.0 dm long, 2.5–7.0 mm wide, blades green, base brown to reddish brown. INFLO-RESCENCES racemose staminate terminal spike(s) and 4–12 lateral pistillate spikes, each subtended by a leaflike bract; spikes staminate on peduncles 0.2–2.0 cm long, flowers subtended by scales, clear to brown-tinged with green midrib, tips acute or acuminate; pistillate spikes on peduncles 0.2–10.0 cm long, perigynia 10–50, subtended by scales, clear to pale brown with green midrib, tip acuminate. FRUITS achene, 2.0–2.5 mm long, 1.2–1.7 mm wide, triangular in cross section, tan to yellowish brown; enveloped in perigynia, green to straw-colored, 5.0–6.5 mm long, 1.5–2.5 mm wide, apex tapering gradually into bidentate beak.

Special Identifying Features

Perennial with short, stocky rhizomes; tufted terminal spike staminate and lateral spikes pistillate on same culm; perigynia green to straw-colored at maturity, apex tapering into bidentate beak.

Toxic Properties

None reported.

Annual Sedge

Cyperus compressus L. · Cyperaceae · Sedge Family

Synonyms

Coco grass, common sedge

Habit, Habitat, and Origin

Erect to inclined, cespitose, annual herb; 0.5–4.0 dm tall; cultivated areas, fields, pastures, lawns, turf, flowerbeds, gardens, roadsides, open woods, and waste sites; native of Asia.

Seedling Characteristics

Glabrous nascent sheaths; 3-ranked leaves.

Mature Plant Characteristics

ROOTS reddish, fibrous, extensively branching, aromatic. STEMS culm, triangular in cross section, 0.5–4.0 dm tall. LEAVES 3-ranked, blades dark green, linear-lanceolate, usually equal to or slightly shorter than culm, 1.0–2.0 mm wide. INFLORESCENCES simple, single, nearly sessile or compound umbel-like, 1.0–7.0 cm long, with 2–6 unequal peduncles bearing spreading spikelets; bracts unequal, subtending spikelets, lowest longer than inflorescence; spikelets erect or spreading, crowded, laterally compressed, 12–14 flowers, 1.0–2.4 cm long, 2.0–3.0 mm wide; scales subtending achene, acuminate, keel-like, excurved toward end, 3.0–3.5 mm long, with pale broad hyaline margins and green midrib. FRUITS achene, 1.0–1.3 mm long, triangular, obovate, dark brown, shiny.

Special Identifying Features

Tufted annual; multiple fruiting culms from plant base; spikelets flattened, 2.0–3.0 mm wide, with toothed outline.

Toxic Properties

None reported.

TOP Seed
MIDDLE Spikelet
BOTTOM Seedling

Inflorescence

Smallflower Umbrella Sedge

Cyperus difformis L. • Cyperaceae • Sedge Family

Synonyms

Smallflower umbrella plant, variable flatsedge

Habit, Habitat, and Origin

Erect to inclined, cespitose, annual herb; to 5.0 dm tall; wet soils or shallow water, cultivated fields, roadside ditches; native of Asia.

Seedling Characteristics

Glabrous nascent sheaths; 3-ranked leaves.

Mature Plant Characteristics

ROOTS yellowish red, fibrous, extensively branching. STEMS culm, triangular, tufted with rather soft bases, equal to or taller than basal leaves, 1.0–5.0 dm tall. LEAVES 3-ranked, 2–4 per stem, linear-lanceolate, equal to or shorter than stem, 1.0–4.0 mm wide, membranous sheath enveloping stem at base.

INFLORESCENCES single or with short unequal branches bearing subglobose to lobulate glomerules, subtended by 2–3 unequal leaflike bracts; spikelets dense, spreading, crowded, linear, 5.0–8.0 mm long; tiny beadlike florets, overlapping, maturing first and shedding below as distal florets develop; scales subtending achene, obtuse tip, reddish brown or purplish with green tip. FRUITS achene, 0.5 mm long, triangular with rounded angles, obovate, pale yellowish, surface very finely granular; from seed to seed in as few as 6 weeks.

Special Identifying Features

Erect to inclined, tufted annual; multiple culms from base of plant; culms soft at base; spikelets in dense, spreading, crowded heads.

Toxic Properties

None reported.

TOP Seeds
MIDDLE Seedling
BOTTOM Bracts

Inflorescence

Deeproot Sedge

Cyperus entrerianus Boeck. • Cyperaceae • Sedge Family

Synonyms

Woodrush flatsedge; *Cyperus luzulae* (L.) Rottb. ex Retz. var. *entrerianus* (Boeck.) Barros

Habit, Habitat, and Origin

Erect, cespitose perennial; 0.3–1.0 m tall; seasonally wet soils, pastures, fields, roadside ditches, and rice levees; native of South America.

Seedling Characteristics

Smooth nascent sheaths; 3-ranked basal leaves.

Mature Plant Characteristics

ROOTS brown to reddish brown, fibrous; extensively branching from short, thick, woody rhizomes; occasional tubers. **STEMS** culm, rounded-trigonous in cross section, 0.3–1.0 m tall. **LEAVES** 3-ranked, spreading, blades dark green, linear, bases dark purple to blackish purple, shorter than culm, 1.0–7.0 dm long, 3.0–7.0 mm wide. **INFLORESCENCES** head loosely to densely round, 0.1–2.0 dm diameter; subtended by 5–10 bracts, 0.55–5.5 dm long, 1.0–7.0 mm wide; spikelets linear to broadly ellipsoid, flattened, 30–65 per inflorescence, 4.0–6.5 mm long, 1.8–3.2 mm wide; scales subtending achene, light tan or brown with green midrib, apex acute or mucronate. **FRUITS** achene, ellipsoid to narrowly ellipsoid, constricted basally, gradually tapering toward apex, 0.9–1.1 mm long, 0.3–0.4 mm wide, finely reticulate, brown, subtended by folded scale, apex acute, deciduous.

Special Identifying Features

Erect, cespitose perennial; culms rounded-trigonous; bases dark purple to blackish purple, set deep in soil; rhizomes short, thick, woody; clumps often producing 20–100 culms per year.

Toxic Properties

None reported.

Inflorescence

Mature plant

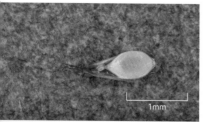

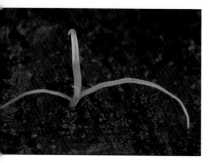

TOP Seed
MIDDLE Spikelet
BOTTOM Seedling

Inflorescence

Redroot Flatsedge

Cyperus erythrorhizos Muhl. · Cyperaceae · Sedge Family

Synonyms

Redroot sedge

Habit, Habitat, and Origin

Erect, cespitose annual; 1.0–15.0 dm tall; wet to seasonally wet soils, ditches and edges of ponds, lakes, streams, and rivers; native of North America.

Seedling Characteristics

Glabrous nascent sheaths; 3-ranked basal leaves.

Mature Plant Characteristics

ROOTS reddish to reddish purple, fibrous, extensively branching. STEMS culm, triangular, equal to or longer than basal leaves, reddish purple at base, 1.0–15.0 dm tall. LEAVES 3-ranked, blades green, linear-lanceolate, usually equal to or slightly shorter than inflorescence, up to 1.0 cm wide. INFLORESCENCES umbel-like aggregation of 4–10 unequal peduncles bearing irregular clusters of nearly sessile spikes with elongate axes of 15–70 spreading spikelets; bracts 4–10, usually much longer than inflorescence, 2.0–8.0 mm wide; spikelets crowded, divergent-linear, 6–34 flowers, reddish brown, 0.3–2.0 cm long, 1.0 mm wide; scales subtending achene, keeled distally, short acute tip, 1.5 mm long, green, tawny brown or reddish with age. FRUITS achene, 0.8 mm long, triangular, elliptic, whitish, tan at maturity and shiny.

Special Identifying Features

Erect, cespitose annual; multiple fruiting stems from reddish purple plant base; rhizomes and tubers absent.

Toxic Properties

None reported.

Yellow Nutsedge

Cyperus esculentus L. · Cyperaceae · Sedge Family

Synonyms

Chufa, yellow nutgrass

Habit, Habitat, and Origin

Erect, persistent, colonial perennial; 3.0–8.0 dm tall; cultivated areas, fields, pastures, lawns, turf, roadsides, wetlands, and disturbed areas; native of North America.

Seedling Characteristics

Small, inconspicuous, rarely encountered.

Mature Plant Characteristics

ROOTS fibrous, extensively branched from tubers, rhizomes, or basal bulbs. **STEMS** culm, triangular, borne individually from tuber or basal bulb, as long as or shorter than basal leaves. **LEAVES** 3-ranked, mostly basal; blade green, linear, 5.0–6.0 mm wide, prominent midvein, flat or slightly corrugated, usually length of culm or longer, with long attenuated tip. **INFLORESCENCES** umbel-like, composed of several unequally stalked spikes subtended by unequal leaflike bracts usually as long as inflorescence; spikelets linear, yellowish brown or straw-colored, 1.0–3.0 cm long with several flowers, flattened, 2-ranked, stamens 3, style 3-cleft; scales subtending achene, yellowish, with acute tip. **FRUITS** achene, 1.5 mm long, triangular, narrowing gradually from square-shouldered apex toward base, granular, brownish gray to brown, production and viability variable. **TUBERS** spherical, smooth, solitary, terminal from basal bulb or rhizomes, sweet to taste, buds positioned near apical end, can produce 10 or more rhizomes from tubers or basal bulbs.

Special Identifying Features

Erect, persistent, colonial perennial; inflorescence yellow; leaves gradually tapering to sharp point; tubers not in chains, smooth.

Toxic Properties

None reported. Tubers are food for humans, livestock, and wildlife.

TOP Seedling
BOTTOM Inflorescence

1mm

TOP Spikelet
BOTTOM Rhizome and tubers

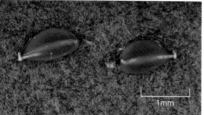

TOP **Seeds**
BOTTOM **Spikelet**

Inflorescence

Seedling

Rice Flatsedge

Cyperus iria L. • Cyperaceae • Sedge Family

Synonyms

Annual sedge, umbrella sedge

Habit, Habitat, and Origin

Erect to inclined, cespitose annual; 2.0–5.0 dm tall; wet to seasonally wet soils in open fields; native of Eurasia.

Seedling Characteristics

Glabrous nascent sheaths, 3-ranked basal leaves.

Mature Plant Characteristics

ROOTS yellowish red, fibrous, extensively branching. STEMS culm, triangular, tufted, usually larger than basal leaves, 2.0–6.0 dm tall. LEAVES 3-ranked, blades green, linear-lanceloate, usually shorter than culm, 3.0–6.0 (to 8.0) mm wide; margins rough, especially toward apex, membranous sheath enveloping culm at base. INFLORESCENCES simple or compound, usually open, up to 20.0 cm long, with elongate rather dense spikes; subtended by 3–7 unequal leaflike bracts, the lowest longer than inflorescence; spikelets erect and spreading, crowded, 6–24-flowered, 5.0–13.0 mm long, 1.5–2.0 mm wide, rachilla wingless; scales subtending achene, broadly obovate, retuse, mucronate, 1.2–1.7 mm long, with golden brown sides and green midvein. FRUITS achene, 1.0–1.5 mm long, triangular, obovate, brown.

Special Identifying Features

Erect to inclined, cespitose annual; no tubers; multiple fruiting stems from plant base.

Toxic Properties

None reported.

Flatsedge

Cyperus odoratus L. • Cyperaceae • Sedge Family

Synonyms

Rusty flatsedge

Habit, Habitat, and Origin

Erect, inclined, cespitose annual; 1.0–8.0 dm tall; shallow ditches, damp to seasonally wet soils around lakes, ponds, streams, and rivers; native of North America.

Seedling Characteristics

Glabrous nascent sheaths; 3-ranked leaves.

Mature Plant Characteristics

ROOTS brownish, fibrous, extensively branching. STEMS culm, triangular, equal to or longer than basal leaves, 1.0–8.0 dm tall. LEAVES 3-ranked; blade green, linear-lanceolate, usually equal to or shorter than inflorescence, up to 1.0 cm wide. INFLORESCENCES variable from compact, 5.0–6.0 cm long and wide, to spreading with sessile spikes and unequal primary and secondary peduncles, up to 1.5 dm long and 3.0 dm wide; subtending bracts 3–10, longer than inflorescence, up to 3.0 dm long; spikelets variable, 4–30, at right angles to axis, not flattened; scales subtending achene, closely overlapping, hard, rounded on back, oblong, obtuse or acute tip, 1.5–3.0 mm long, midrib green, margins chestnut brown, nerved, shiny. FRUITS achene, 1.0–2.0 mm long, triangular, oblong or narrowly obovate, often asymmetrical, brown.

Special Identifying Features

Erect, inclined, cespitose annual; multiple fruiting stems from base; rhizomes and tubers absent.

Toxic Properties

None reported.

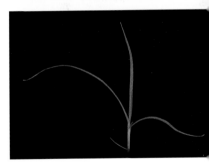

TOP Seed
MIDDLE Spikelet
BOTTOM Seedling

Inflorescence

TOP **Spikelet**
MIDDLE **Rhizome
and tubers**
BOTTOM **Seedling**

Purple Nutsedge

Cyperus rotundus L. • Cyperaceae • Sedge Family

Synonyms

Coco, purple nutgrass

Habit, Habitat, and Origin

Erect, persistent, colonial perennial; to 9.0 dm tall; cultivated areas, fields, pastures, lawns, turf, and disturbed areas; native of Eurasia.

Seedling Characteristics

Inconspicuous, very rarely seen.

Mature Plant Characteristics

ROOTS fibrous, extensively branched, from tubers and bulbs, slender rhizomes, white and fleshy, covered with scale leaves when young, becoming fibrous. **STEMS** culm, 1.0–9.0 dm tall, triangular, solitary from tuber or basal bulb, longer than basal leaves. **LEAVES** 3-ranked, mostly basal; blade green, linear, 2.0–6.0 mm wide, prominent midvein, flat or slightly corrugated, usually shorter than culm, abruptly tapering at tip. **INFLO-RESCENCES** umbel-like, several sessile and unequally stalked spikes, subtended by unequal leaflike bracts shorter than inflorescence; spikelets linear, dark, reddish purple or reddish brown, 3 per spike, loosely disposed, flattened, stamens 3, style 3-cleft, scales subtending achene, purplish. **FRUITS** achene, 1.5 mm long, triangular, base and apex obtuse, granular, dull, olive gray to brown or blackish covered with network of gray lines; production and viability variable. **TUBERS** oblong, irregularly shaped, rough, in chains connected by rhizomes, bitter to taste, buds scattered over surface, forming 1–3 rhizomes from tubers or basal bulbs.

Special Identifying Features

Erect, persistent, colonial perennial; inflorescence purple; tubers connected in chains, rough, bitter to taste; leaves abruptly tapering to acute tip.

Toxic Properties

None reported; tubers are edible but have a bitter or peppery taste.

Inflorescence

Inflorescence

TOP Seed
MIDDLE Spikelets
BOTTOM Seedling

False Nutsedge

Cyperus strigosus L. · Cyperaceae · Sedge Family

Synonyms

Cyperus hansenii Britt.

Habit, Habitat, and Origin

Erect to inclined, cespitose, annual or perennial; 30–100 cm tall; damp to seasonally wet soils in ditches, roadsides, fields, and pastures, and along edges of ponds, lakes, streams, and rivers; native of North America.

Seedling Characteristics

Glabrous nascent sheaths; 3-ranked basal leaves.

Mature Plant Characteristics

ROOTS brownish, fibrous, extensively branching. STEMS culm, triangular, equal to or longer than basal leaves, 30–100 cm tall, arising from a reddish purple bulblike base. LEAVES 3-ranked; blade green, firm, linear-lanceolate, equal to or shorter than inflorescence, 2.0–8.0 mm wide. INFLORESCENCES umbel-like, composed of several unequally stalked spikes each bearing 20–40 spreading spikelets, subtended by 3–10 bracts, longest ones longer than inflorescence, 1.0–8.0 mm wide; spikelets lanceolate to linear, pointed, 3.0–30.0 mm long; scales subtending achene, flattened, oblong to linear, midvein green, margins golden yellow to tawny brown. FRUITS achene, 1.5–2.0 mm long, triangular, linear-oblong, often curved, brown to gray, surface with very fine bumps.

Special Identifying Features

Erect to inclined, cespitose annual or perennial; multiple fruiting stems arising from plant base; base slightly enlarged, reddish purple; rhizomes and tubers absent.

Toxic Properties

None reported.

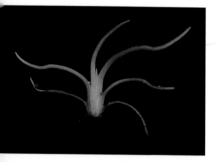

TOP Seed
MIDDLE Seedling
BOTTOM Spikelets

Inflorescence

Annual Fringerush

Fimbristylis annua (All.) Roem. & Schult. • Cyperaceae • Sedge Family

Synonyms

Fimbristylis baldwiniana (Schult.) Torr. & Gray

Habit, Habitat, and Origin

Erect, cespitose annual; to 50.0 cm tall; seasonally wet soils, cultivated fields, pastures, old fields, and edges of ponds, lakes, streams, and rivers; native of North America.

Seedling Characteristics

Pubescent nascent sheaths; 3-ranked basal leaves.

Mature Plant Characteristics

ROOTS brown to reddish brown, fibrous, extensively branching. STEMS culms in clump, compressed or triangular, equal to or longer than basal leaves, 5.0–50.0 cm tall. LEAVES 3-ranked, narrow, spreading; blade green, linear, taller leaves one-half to equal length of culm, 1.0–1.5 (to 2.0) mm wide; sheath with bristly trichomes, back with rough trichomes or smooth, ligule present. INFLORESCENCES compound, open, spreading, ascending branches of unequal lengths, taller than wide; subtended by 3–4 bracts, longest one longer than inflorescence; spikelets ovoid to lance-ovoid, 3.0–8.0 mm long, 1.5–3.0 mm wide, tan, brown, or reddish brown; scales broadly oblong to ovate, 2.0 mm long, tan, brown, or reddish brown, acute tips. FRUITS achene, 1.0 mm long, lenticular or obovoid; rows of rectangular pits with random warts or rarely smooth, white to brown.

Special Identifying Features

Erect, cespitose annual; multiple fruiting stems from plant base; inflorescence spreading; spikelets ovoid to lance-ovoid; rhizomes and tubers absent.

Toxic Properties

None reported.

Globe Fringerush

Fimbristylis miliacea (L.) Vahl • Cyperaceae • Sedge Family

Synonyms

Fimbristylis littoralis Gaud.

Habit, Habitat, and Origin

Erect, cespitose annual; to 4.5 dm tall; wet, sandy, often-disturbed soils, cultivated areas, wet meadows, coastal plains, and edges of ponds, lakes, streams, and rivers; native of Asia.

Seedling Characteristics

Small, glabrous, nascent sheaths; 3-ranked leaves.

Mature Plant Characteristics

ROOTS fibrous, light brown to tan. STEMS culms, tufted, 1.4–4.5 dm tall, strongly angled, bases soft. LEAVES smooth, up to 20.0 cm long, 1.5–2.0 mm wide; sheaths smooth, flattened, distichous, in fans, no ligule. INFLORESCENCES bisexual, decompound cyme, rounded spikelets; spikelets rounded to subglobose, 1.5–4.0 mm long, 1.5–2.5 mm wide, dark reddish brown; scales subtending achene, ovate to orbiculate. FRUITS achene, obovoid, 0.5–0.6 mm long, 0.5 mm wide, longitudinally ribbed, transversely lined, tubercate, straw-colored.

Special Identifying Features

Erect, cespitose annual; leaf without ligule; achene obovoid with wartlike tubercules; spikelets rounded; leaves flattened in fan-shaped subclumps within tufted plant.

Toxic Properties

None reported.

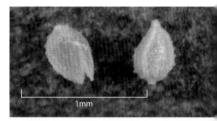

TOP Seeds
MIDDLE Fruit
BOTTOM Seedling

Inflorescence

TOP Seeds
MIDDLE Spikelets
BOTTOM Young plants

Inflorescence

Green Kyllinga

Kyllinga brevifolia Rottb. · Cyperaceae · Sedge Family

Synonyms

Kyllinga, perennial greenhead sedge; *Cyperus brevifolius* (Rottb.) Endl. ex Hassk.

Habit, Habitat, and Origin

Erect, rhizomatous perennial; rarely to 50.0 cm tall; damp to seasonally wet soils, lawns, turf, fields, roadsides, disturbed areas, cemeteries, ditches, and waste sites; native of Asia.

Seedling Characteristics

Glabrous nascent sheaths; 3-ranked basal leaves.

Mature Plant Characteristics

ROOTS brown, fibrous, extensively branching from rhizomes. **STEMS** culm, triangular, longer than basal leaves, 1.0–20.0 (rarely to 50.0) cm tall. **LEAVES** basal, blade flat, green, linear, shorter than culm, 2.0–21.0 cm long, 1.5–3.5 mm wide, ligule absent. **INFLORESCENCES** solitary, simple, round or cylindrical, 20–100 spikelets, 2.0–8.0 cm long; subtended by 2–4 leaflike bracts, flat, longer than inflorescence; spikelets pale greenish or reddish brown, ovate; floral scales 2, ovate to elliptic, one above and one below, enclosing achene. **FRUITS** achene, 1.0–1.3 mm long, 0.6–0.8 mm wide, 2-sided, broadly ovate to oblong, tan to reddish brown.

Special Identifying Features

Erect, rhizomatous perennial; inflorescence round, oval, or dome-shaped; spikelets laterally compressed; seed surrounded by 2 scales.

Toxic Properties

None reported.

Hardstem Bulrush

Schoenoplectus acutus (Muhl. ex Bigelow) A. & D. Löve ·
Cyperaceae · Sedge Family

Synonyms

Bull tule, bulrush, great bulrush, hard-stemmed bulrush, hear-stem bulrush, heterochetous bulrush, pointed bulrush, tule, viscid great bulrush; *Scirpus acutus* Muhl. ex Bigelow

Habit, Habitat, and Origin

Erect, rhizomatous perennial; to 4.5 m tall; marshy areas, cultivated fields, roadside ditches, and edges of ponds, lakes, streams, and rivers; native of North America.

Seedling Characteristics

Inconspicuous, hairlike, 3-ranked nascent stems.

Mature Plant Characteristics

ROOTS fibrous, extensively branched from a stout, fleshy rhizome. STEMS culm, thick, roundish or obscurely triangular, pithy, to 2.5 cm thick at base, tapering to tip. LEAVES few, poorly developed, to 7.5 cm long, well-developed sheath enclosing stem. INFLORESCENCES 10–30 spikelets on drooping peduncles of unequal length, each peduncle arising from one point, subtending bract appearing as continuation of stem; spikelets lance-ovoid, 20–50 flowers, 8.0–15.0 mm long; scales oblong-ovate, 5.0 mm long, pale brown, red-dotted, with short awn and scabrous midvein. FRUITS achene, 1.8–3.0 mm long, obovate, tan or light brown; 6 slender, downwardly barbed bristles at base.

Special Identifying Features

Erect, rhizomatous perennial; inflorescence borne on a thick, nearly round, pithy culm arising from a thick, fleshy rhizome; forming extensive colonies; leaves poorly developed.

Toxic Properties

None reported.

TOP Inflorescence
BOTTOM Infestation

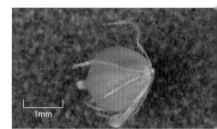

1mm

TOP Seed
BOTTOM Seedling

Brazilian Egeria

Egeria densa Planch. • Hydrocharitaceae • Frog's-bit Family

Synonyms

Brazilian elodea, Brazilian waterweed, anacharis, giant elodea; *Anacharis densa* (Planch.) Vict., *Elodea densa* (Planch.) Casp.

Habit, Habitat, and Origin

Perennial, dioecious, submersed herb, sometimes rooted in mud or floating; up to 20 m long; native of South America, introduced as an aquarium plant and escaped.

Seedling Characteristics

In the United States, fragmented stems capable of regeneration.

Mature Plant Characteristics

ROOTS fibrous. STEMS elongate, 3.0–20.0 m long, 2.0–3.0 mm thick, sparingly branched. LEAVES opposite below, otherwise usually in whorls of 4 or 5, short internodes, nearly linear, 1.0–5.0 cm long, 2.0–5.0 mm wide, finely serrate margins, sessile, bright green, flaccid. INFLORESCENCES dioecious, pedicels up to 2.0 cm above water surface, 2–4 staminate flowers per spathe, no pistillate flowers produced within North America; spathes in upper leaf axils; 3 white petals, 7.0–11.0 mm long; 3 green sepals, 2.0–4.0 mm long. FRUITS none produced within North America. SEEDS none produced within North America; reproduction is by fragmentation because no female plants have been found in North America.

Special Identifying Features

Perennial, dioecious, submersed herb; leaves in whorls of 4 or 5; principal leaves 1.2–4.0 cm long; short internodes make plants appear very leafy; 2–4 staminate flowers per spathe.

Toxic Properties

None reported.

Stems

Flowers

Flower | Flowering bud

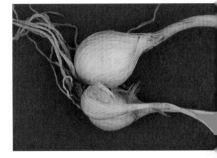

Multiple bulbs

Wild Onion

Allium canadense L. • Liliaceae • Lily Family

Synonyms

Canadian onion

Habit, Habitat, and Origin

Erect perennial herb from underground bulb; to 4.0 dm tall; fields, prairies, meadows, lawns, cultivated areas, and waste sites; native of North America.

Seedling Characteristics

Single flat, slightly keeled leaf.

Mature Plant Characteristics

ROOTS fibrous, attached to bulb with brown, fibrous outer coat, white and shiny underneath. STEMS erect, smooth, to 5.0 dm, terminated by inflorescence. LEAVES mostly basal, flat, but noticeably keeled, narrowly linear, soft, flaccid.

INFLORESCENCES roundish cluster of pink flowers atop scape. AERIAL BULBLETS from within flower head or 3-valved seed capsules containing 3–6 seeds. SEEDS roundish, glossy black, with wing-like ridge.

Special Identifying Features

Herbaceous perennial from underground bulb with reticulate outer coat; leaves flat and noticeably keeled, not hollow; distinctive onion odor.

Toxic Properties

When plants are ingested in high quantities toxic effects include hemolytic anemia and Heinz bodies, especially in cats, dogs, and horses.

TOP Underground bulb
BOTTOM Young plant

Flowers

Aerial bulblets

Wild Garlic

Allium vineale L. • Liliaceae • Lily Family

Synonyms

Crow garlic, field garlic, ramp, scallions

Habit, Habitat, and Origin

Erect perennial herb; to 9.0 dm tall; fields, pastures, roadsides, and waste sites; native of Europe.

Seedling Characteristics

Rarely observed from seed, shoots from bulbs identical with leaves.

Mature Plant Characteristics

ROOTS fibrous, attached to bulb. STEMS slender, smooth, waxy, 3.0–9.0 dm tall, leafy to near middle. LEAVES slender, hollow, nearly round, attached to lower half of stem, mostly 10.0–20.0 cm tall. INFLORESCENCES greenish white, small, on short stems above aerial bulblets, wholly or partially replaced by bulblets.

AERIAL BULBLETS wholly or partially replacing flower in umbel; ovoid, usually tipped by long, fragile, rudimentary leaf. SEEDS black, flat on one side, about 3.0 mm long, produced in spring and germinating the following fall, formed only occasionally. Bulbs round-ovoid, 1.2 cm, enclosed within thin membranous sheath.

Special Identifying Features

Erect perennial herb reproducing from seeds, aerial bulblets, and underground bulblets; leaves round, hollow, with a distinctive odor from allylsulfide that imparts a disagreeable flavor and odor to agricultural products such as milk and wheat.

Toxic Properties

See comments under *Allium canadense*.

Flowering plant

Star-of-Bethlehem

Ornithogalum umbellatum L. • Liliaceae • Lily Family

Synonyms

Nap-at-noon, sleepy dick, summer snowflake

Habit, Habitat, and Origin

Cool-season, succulent perennial herb; to 4.0 dm tall; roadsides, fields, turf, and gardens; native of Europe, escaped from cultivation.

Seedling Characteristics

Single narrow leaf, rarely seen.

Mature Plant Characteristics

ROOTS fibrous. **STEMS** erect, smooth scape, 1.0–4.0 dm tall. **LEAVES** narrow, smooth, channeled, with conspicuous pale green to whitish stripe near midrib on upper surface, 2.0–6.0 mm wide, arising from base of plant. **INFLORES-CENCES** corymbous with 3–20 flowers becoming racemose, flowers perfect, with 6 distinct white tepals and a bright green stripe down center of back, oblong or elliptic-lanceolate, 10.0–25.0 mm long, 3.0–6.0 mm wide; stamens 6, with flattened filaments tapering to tip; superior ovary; bracts to 4.0 cm long, with pedicels 2.0–8.0 cm long; flowers open in sunlight at noon, close at night and on cloudy days. **FRUITS** trilocular, oblong-ovoid, yellow-green capsule with several seeds per locule. **SEEDS** globose to ovoid. **BULBS** ovoid, with membranous coating, white to pale brown, producing offsetting bulblets.

Special Identifying Features

Cool-season, succulent perennial from bulbs; leaves conspicuously pale green, whitish stripe near midrib on upper surface; scape topped by white flowers with broad green midstripe on lower surface, often confused with *Allium* species.

Toxic Properties

Plants contain cardiotoxins and glycosides toxic to humans and livestock; highest concentrations are in flowers and bulbs.

TOP **Bulb and shoots**
BOTTOM **Flower**

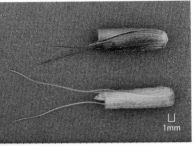

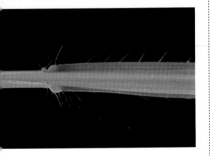

TOP Seeds
BOTTOM Collar

Seedling

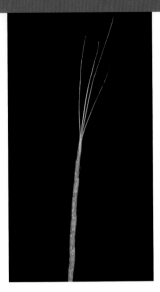

Inflorescence

Jointed Goatgrass

Aegilops cylindrica Host · Poaceae · Grass Family

Synonyms

Jointgrass

Habit, Habitat, and Origin

Tufted, erect winter annual; to 60.0 cm tall; roadsides, railroad beds, fields, and waste sites; native of Europe.

Seedling Characteristics

Sparsely pubescent leaf blade, pubescent sheath margin, membranous ligule.

Mature Plant Characteristics

ROOTS fibrous. STEMS erect, bent at lower nodes, to 60.0 cm tall, nodes glabrous. LEAVES 3.0–12.0 cm long, 2.0–3.0 mm wide, pubescent on upper surface and often on lower surface; sheaths pubescent especially on margins, short auricles at mouth; ligule 0.6 mm long, membrane with rough margin. INFLORESCENCES narrow spike, 5.0–11.0 cm long excluding awns, 2.0–5.0 mm wide, composed of 5–10 spikelets; spikelets 2–5-flowered, 8.0–12.0 mm long, sessile; glumes 2, similar, thick, asymmetrical, scabrous with 1 awn, 6.0–9.5 mm long; lemma thin on back but with thick margins, slightly larger than glumes, lemma of uppermost spikelet with awn 3.5–8.0 cm long; palea about as long as lemma, pubescent on nerves. FRUITS caryopsis, usually 2 or more, falling as a joint of spike or spikelet.

Special Identifying Features

Stiff, thick-walled trichomes on leaves; distinctive rounded, awned inflorescence.

Toxic Properties

None reported.

Creeping Bentgrass

Agrostis stolonifera L. • Poaceae • Grass Family

Synonyms

Carpet bent, carpet bentgrass, redtop, redtop bent, seaside bentgrass; *Agrostis alba* L. var. *palustris* (Huds.) Pers., *A. palustris* Huds.

Habit, Habitat, and Origin

Erect or decumbent, spreading, often mat-forming cool-season perennial from rhizomes or stolons; 2.0–15.0 dm tall; fields, pastures, turf, roadsides, railroad beds, and waste sites; native of Europe.

Seedling Characteristics

Leaf blades rolled in bud, lower leaves flat, glabrous to scabrous, sheath glabrous, membranous; ligule margin jagged, 2.0–7.0 mm long, rounded at tip.

Mature Plant Characteristics

ROOTS fibrous from stolons, rhizomes, and lowermost stem nodes. STEMS 2.0–15.0 dm tall, decumbent or erect, hollow, stout or weak. LEAVES 2.0–20.0 cm long, 2.0–6.0 mm wide, linear, glabrous to scabrous, rolled or folded in bud, mature leaf blades flat or folded; sheaths rounded to slightly keeled, glabrous, opening nearly to base; ligule 2.0–7.0 mm long, membranous, rounded, margin jagged. INFLORESCENCES dense panicle, open when young and contracted at maturity, 4.0–40.0 cm long; spikelets compressed to round, 1 floret, green to reddish purple; glumes lanceolate to elliptical, 1.5–3.5 mm long, smooth, keels scabrous; lemma lanceolate, narrowed to pointed tip, glabrous, shorter than glumes, 1.2–2.8 mm long, awnless or rarely with slender straight awn; palea shorter than lemma, 2-nerved. FRUITS caryopsis, 1.0–1.4 mm long, reddish brown.

Special Identifying Features

Rapid-spreading cool-season perennial from rhizomes and stolons; ligule membranous, sheath smooth; inflorescence dense panicle, contracted to somewhat open, often reddish to purple.

Toxic Properties

Plants may be a source of cyanide toxicity but are not considered a serious risk.

Inflorescence

TOP **Seeds**
MIDDLE **Stolon**
BOTTOM **Collar**

Carolina Foxtail

Alopecurus carolinianus Walt. • Poaceae • Grass Family

Synonyms

None

Habit, Habitat, and Origin

Tufted, erect or ascending winter annual; to 0.5 m tall; cultivated areas, fields, pastures, roadsides, and waste sites; native of North America.

Seedling Characteristics

Glabrous leaf sheath, scabrous leaf blade, membranous ligule.

Mature Plant Characteristics

ROOTS fibrous. STEMS erect or ascending, occasionally bent or rarely spreading at base, to 0.5 m tall, sometimes rooting at lower nodes. LEAVES 7.0–15.0 cm long, 0.6–3.2 mm wide, upper and lower surfaces scabrous; sheaths smooth, slightly inflated; ligule membranous, 1.5–5.0 mm long, pointed, often sharp. INFLORESCENCES erect, cylindrical panicle, 2.0–5.5 cm long; spikelets 1-flowered, 2.2–2.4 mm long; first glume 2.1–2.3 mm long, pubescent, second glume 2.1–2.3 mm long, pubescent; lemma 1.9–2.1 mm long, glabrous, awn from lower back 3.2–4.2 mm long, straight or bent; palea absent. FRUITS caryopsis, 1.0–1.5 mm long, enclosed in glumes and lemma.

Special Identifying Features

Narrow, dense, cylindrical, fuzzy seedhead; large, pointed membranous ligule.

Toxic Properties

None reported.

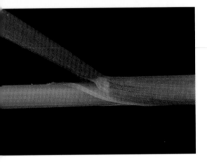

TOP Seeds
BOTTOM Collar

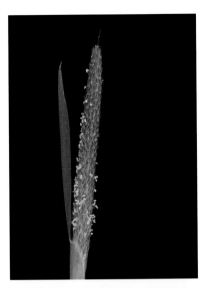

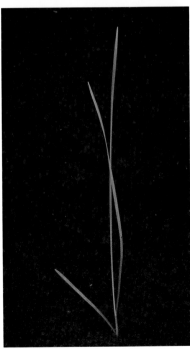

TOP Inflorescence
BOTTOM Seedling

Mature plant

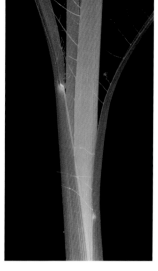

Collar

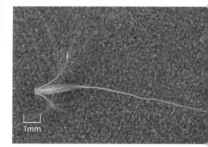

1mm

TOP **Seed**
BOTTOM **Seedling**

Broomsedge

Andropogon virginicus L. · Poaceae · Grass Family

Synonyms

Broom grass

Habit, Habitat, and Origin

Summer perennial bunchgrass; 0.5–1.5 m tall, usually in small tufts; common in abandoned pastures and waste sites; native of North America.

Seedling Characteristics

Glabrous to hispid leaf blades.

Mature Plant Characteristics

ROOTS fibrous, densely but often shallowly rooted. STEMS 0.5–1.5 m tall, upper two-thirds freely branched. LEAVES 0.3 m long, 2.0–3.0 mm wide, pilose on upper surface near base; sheaths glabrous to villous; ligule strongly ciliate. INFLORESCENCES elongate, simple to much-branched raceme, 2.0–3.0 cm long; spikelets 3.0–4.5 mm long; delicate, straight awn 1.0–2.0 cm long, pedicel long-villous; glumes 2, flat to slightly concave or convex, coriaceous, narrow, the second sometimes slightly keeled; lemmas 2, fertile and sterile, sterile lemma shorter than glumes, hyaline; fertile lemma awned; palea hyaline, beard usually longer than spikelet. FRUITS caryopsis purplish or yellowish, linear-ellipsoid, 2.0–3.0 mm long.

Special Identifying Features

Tufted grass with a branched, racemose inflorescence with conspicuous feathery white tufts.

Toxic Properties

Andropogon species are reported to be cyanogenic and to host endophytic fungi.

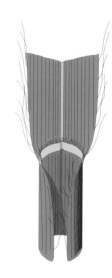

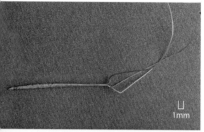

TOP **Seed**
BOTTOM **Seedling**

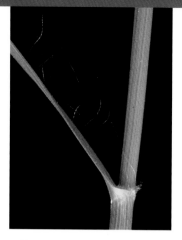

Collar

Inflorescence

Prairie Threeawn

Aristida oligantha Michx. • Poaceae • Grass Family

Synonyms

Few-flowered aristida, oldfield threeawn, plains three-awn grass

Habit, Habitat, and Origin

Erect, slender, branched, tufted warm-season annual; to 7.0 dm tall; open woodlands, prairies, fields, pastures, roadsides; native of North America.

Seedling Characteristics

Leaves glabrous below, scabrous above; ligule a short membrane, inconspicuous, pubescent.

Mature Plant Characteristics

ROOTS fibrous. **STEMS** erect, 1.5–7.0 dm long, hollow, wiry, branched from base and nodes, glabrous to scabrous. **LEAVES** linear, narrow, flat to loosely involute, tapering to filiform point, 1.0–2.0 mm wide, 10.0–30.0 cm long, glabrous below, glabrous or scabrous above, sometimes pilose near ligule, glabrous on underside; sheaths glabrous to slightly scabrous, open most of length;

ligule membranous with inconspicuous fringe of trichomes, 0.1–0.5 mm long. **INFLORESCENCES** open panicle, lax, 5.0–33.0 cm long, lower spikelets paired, upper solitary; lateral panicles dense with short internodes and few flowers; spikelets large, 1 fertile, short pedicel, disarticulating above glumes; lower glume 12.0–33.0 mm long, tip pointed or awned, 1.0–4.0 mm long; upper glume 15.0–37.0 mm long, glabrous, tip divided, awn erect, 7.0–17.0 mm long; lemma pilose, 10.0–28.0 mm long, central awn 2.5–7.0 cm long, lateral awns shorter, awns spirally curved or bent at base; palea shorter than lemma, lacking awns. **FRUITS** caryopsis, linear, shed inside lemma.

Special Identifying Features

Slender, branching, wiry summer annual; narrow leaves; ligule with inconspicuous fringe of trichomes on basal membrane; lemma 3-awned.

Toxic Properties

The long awns can injure grazing animals.

Giant Reed

Arundo donax L. • Poaceae • Grass Family

Synonyms

None

Habit, Habitat, and Origin

Erect, stout, warm-season perennial; 2.0–8.0 m tall; moist soils in open areas, damp soils, fields, pastures, wetland areas, and old house sites; native of southern Asia, Europe, and northern Africa.

Seedling Characteristics

Seldom produced; leaves and sheaths glabrous, ligule a membrane fringed with minute trichomes.

Mature Plant Characteristics

ROOTS fibrous from knotty rhizomes. STEMS erect, stout, glabrous, round, hollow, 2.0–8.0 m tall, unbranched or with few branches; forming large, dense colonies. LEAVES linear, flat or folded, glabrous, rough margins, thick midvein below, 40.0–100.0 cm long, 20.0–70.0 mm wide; sheaths glabrous; ligule membranous with minute fringe of trichomes. INFLORESCENCES large, dense panicle with extensive ascending branches, 30.0–60.0 cm tall; usually silvery, whitish, or sometimes purplish; spikelets obovate to obtriangular, somewhat flattened, 10.0–15.0 mm long, with 2–5 florets, rachilla glabrous; glumes glabrous, lanceolate, 3–5-nerved, 8.0–13.0 mm long, brownish to purplish; lemma lanceolate, tapering to 2-toothed tip or short awn, covered with long, dense, silky trichomes, 5.0–12.0 mm long, 3–5-nerved; palea shorter than lemma, pubescent at base, 3.0–5.0 mm long. FRUITS caryopsis, narrowly oblong, seldom produced.

Special Identifying Features

Tall, stout, warm-season perennial forming dense colonies from knotty rhizomes; inflorescence in dense, branched, usually silvery panicle with smooth spikelet stalks and pubescent lemmas.

Toxic Properties

None reported.

TOP Rhizomes
MIDDLE Collar
BOTTOM Leaves

Inflorescence

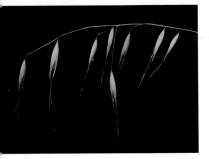

TOP Seed
BOTTOM Inflorescence
branch

Inflorescence

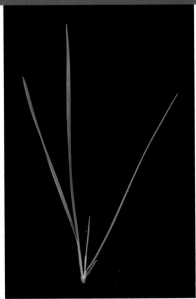

Seedling

Wild Oat

Avena fatua L. · Poaceae · Grass Family

Synonyms

None

Habit, Habitat, and Origin

Erect, tufted annual; to 1.2 m tall; cultivated areas, fields, roadsides, railroad beds, and waste sites; native of Europe.

Seedling Characteristics

Sheath glabrous or pubescent, blades usually glabrous, ligule a distinct membrane.

Mature Plant Characteristics

ROOTS fibrous. STEMS erect, stout, sometimes bent at base, to 1.2 m tall. LEAVES 7.0–34.0 cm long, 2.0–11.0 (to 18.0) mm wide, usually glabrous, occasionally pubescent; sheaths glabrous or pubescent; ligule a membrane to 5.0 mm long. INFLORESCENCES open panicle with 2–4-flowered dangling spikelets, 15.0–31.0 cm long, 7.0–20.0 cm wide; spikelets 2–4-flowered, 2.0–3.2 cm long; glumes glabrous, strongly nerved, 1.9–3.1 cm long; lemma 1.3–2.1 mm long, stiffly pubescent, awned from near midpoint of back; awn bent, twisted, 2.0–5.0 cm long; palea 9.6–13.1 mm long, 2-keeled. FRUITS caryopsis, lemma and palea attached.

Special Identifying Features

Stout stem, membranous ligule to 5.0 mm long; inflorescences with spreading branches and dangling spikelets with 2 or 3 bent awns.

Toxic Properties

The long awns can injure grazing animals.

Rescuegrass

Bromus catharticus Vahl • Poaceae • Grass Family

Synonyms

Prairiegrass, rescue brome; *Bromus uniloides* Kunth, *Bromus willdenowii* Kunth

Habit, Habitat, and Origin

Tufted, erect winter annual or biennial; to 1.0 m tall; cultivated areas, fields, pastures, roadsides, railroad beds, and waste sites; native of South America.

Seedling Characteristics

Soft and succulent, laterally flattened and usually somewhat keeled; sheaths nearly glabrous to densely pubescent with fine, straight, spreading trichomes, membranous ligule; leaf blades pubescent.

Mature Plant Characteristics

ROOTS fibrous. STEMS erect, to 1.0 m tall, glabrous. LEAVES to 35.0 cm long, 5.0–12.0 mm wide, may be pubescent on upper side only; sheaths closed, may be pubescent on lower leaves; ligule a 4.3 mm membrane with frayed margin, obtuse, lanceolate to erose. INFLORESCENCES open, drooping panicle 6.0–27.0 cm long, composed of paired branches up to 15.0 cm long; spikelets 4–9-flowered, 2.2–3.7 cm long, 4.5–10.4 mm wide, pediceled; first glume 6.1–12.6 mm long, second glume longer than first; lemma laterally flattened, sharply keeled, glabrous or pubescent, awnless or with short awn (1.0–3.5 mm); palea similar to lemma, slightly shorter. FRUITS caryopsis, essentially glabrous and smooth, lemma and palea attached.

Special Identifying Features

Spikelets laterally compressed, lemma keeled, no auricles.

Toxic Properties

The long awns of various species of *Bromus* can injure grazing animals, with tetanus a possible result; plants may be infected with fungi that cause ergot.

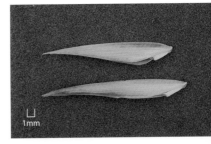

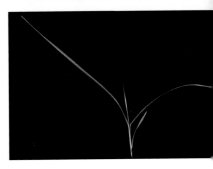

TOP Seeds
MIDDLE Collar
BOTTOM Seedling

Inflorescence

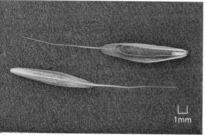

Seeds

Hairy Chess

Bromus commutatus Schrad. • Poaceae • Grass Family

Synonyms

Bald brome

Habit, Habitat, and Origin

Erect, coarse-stemmed annual; to 1.2
m tall; cultivated areas, fields, pastures,
roadsides, railroad beds, and waste sites;
native of Mediterranean region and
Eurasia.

Seedling Characteristics

Leaf sheath and blade pubescent, ligule
membranous.

Mature Plant Characteristics

ROOTS fibrous. STEMS erect, to 1.2 m tall,
nodes pubescent. LEAVES 5.0–25.0 cm
long, most 2.0–8.0 mm wide, pubescent
on both surfaces, sheaths pubescent
with retrorsely pointing trichomes or
glabrous; ligules membranous, glabrous
or pilose, 1.0–2.5 mm long. INFLORES-
CENCES panicle with ascending branches,
12.0–25.0 cm long; spikelets 8–11-flow-
ered, 1.8–3.3 cm long including awns,
pediceled; first glume 5.2–5.7 mm long
and scabrous, second glume 6.5–7.0 mm
long and scabrous; lemma 7.7–8.0 mm
long, awned from between 2 teeth, awn
8.0–11.0 mm long; palea within 0.1 mm
of length of lemma. FRUITS caryopsis,
lemma and palea attached.

Special Identifying Features

Spikelets on long pedicels, multiflow-
ered, awned; leaf sheath pubescent with
retrorsely pointing trichomes; leaf blade
and nodes pubescent.

Toxic Properties

See comments under *Bromus catharticus*.

TOP **Inflorescence**
BOTTOM LEFT **Collar**
BOTTOM RIGHT **Seedling**

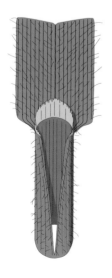

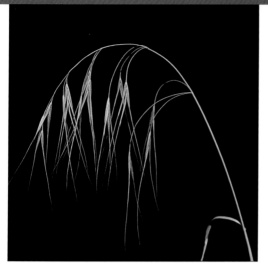

Inflorescence

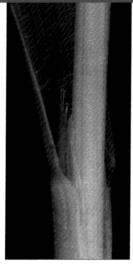

Collar

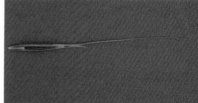

TOP Seed
BOTTOM Seedling

Ripgut Brome

Bromus diandrus Roth · Poaceae · Grass Family

Synonyms

Ripgutgrass; *Bromus rigidus* Roth

Habit, Habitat, and Origin

Thick-stemmed, erect, pubescent annual; to 70.0 cm tall; cultivated areas, fields, pastures, roadsides, railroad beds, and waste sites; native of Mediterranean region and southwestern Europe.

Seedling Characteristics

Blades and sheath pubescent, ligule membranous.

Mature Plant Characteristics

ROOTS fibrous. STEMS erect to leaning, usually thick, up to 70.0 cm tall, sometimes branching at lower nodes. LEAVES 4.0–22.0 cm long, 1.0–9.8 cm wide, pubescent on both surfaces; sheath pubescent, partially closed; ligule 3.0–5.5 mm long, membranous with jagged margins. INFLORESCENCES 8.0–19.0 cm long with wide-spreading branches; spikelets 7.0–9.1 cm long including awns, 1.5–2.1 cm wide; first glume 1.5–2.1 cm long, second glume 2.0–2.8 cm long, both with a few scattered trichomes; lemma 2.0–2.4 cm long, few scattered longer trichomes, awned from between 2 teeth, awns 3.5–5.0 cm long; palea 7.0–17.0 mm long. FRUITS caryopsis; lemma, palea, and caryopsis fall together but are not attached.

Special Identifying Features

Pubescent leaves; pubescent, partially closed sheath; membranous ligule; very long, thin awns on lemma.

Toxic Properties

See comments under *Bromus catharticus*.

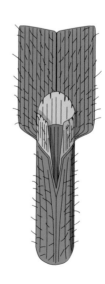

Soft Brome

Bromus hordeaceus L. · Poaceae · Grass Family

Synonyms

Soft chess; *Bromus mollis* auct. non L.

Habit, Habitat, and Origin

Cool-season annual; to 4.0 dm tall; prefers undisturbed or protected areas but found in cultivated areas, fields, roadsides, railroad beds, and waste sites; native of Europe.

Seedling Characteristics

Leaf surface and sheath pubescent, leaves twisted clockwise.

Mature Plant Characteristics

ROOTS fibrous. STEMS weak, usually geniculate, typically 2.0–4.0 dm tall. LEAVES soft, flat or folded, glabrous or sparsely hirsute, most 2.0–6.0 mm wide; at least lower sheaths densely covered with soft, spreading trichomes; ligule with prominent membrane. INFLORESCENCES contracted, densely flowered panicle or raceme, 3.0–10.0 cm long, pedicels and branches typically shorter than spikelets; spikelets 1.5–2.0 cm long, with 5–9 closely overlapping florets; glumes large, broad, soft, and hirsute, first glume 3–5-nerved, second glume 5–7-nerved; lemma hirsute, similar to glumes in texture, usually 7.0–10.0 mm long with awn 5.0–9.0 mm long, apical teeth less than 3.0 mm long; palea 1.5–2.0 mm shorter than lemma, covered with soft, short trichomes. FRUITS caryopsis, 6.0–6.5 mm long, 1.5–2.0 mm wide, tuft of trichomes at apex, palea attached.

Special Identifying Features

Annual, decumbent, hirsute plant; glumes and lemmas with soft pubescence.

Toxic Properties

See comments under *Bromus catharticus*.

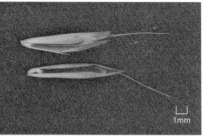

TOP **Seeds**
MIDDLE **Collar**
BOTTOM **Seedling**

Inflorescence

Smooth Brome

Bromus inermis Leyss. • Poaceae • Grass Family

Synonyms

Arctic brome, Hungarian brome, northern awnless brome, pumpelly brome

Habit, Habitat, and Origin

Erect or ascending, cool-season perennial from creeping rhizomes; to 1.3 m tall; cultivated areas, fields, pastures, roadsides, railroad beds, and waste sites, occasionally sod-forming; native of Europe.

Seedling Characteristics

Leaves glabrous, ligule membranous, auricles absent to sometimes vestigial.

Mature Plant Characteristics

ROOTS fibrous from rhizomes, with dark brown scales. STEMS erect or sometimes decumbent at base from extensively creeping rhizomes, 0.4–1.3 m tall. LEAVES linear, flat, glabrous, margins rough, 9.0–40.0 cm long, 4.0–15.0 mm wide, with conspicuous M- or W-shaped constriction; sheaths fused at margins, glabrous or with long, soft trichomes; ligule membranous, short, rounded with margin slightly jagged, 0.5–2.0 mm long; auricles absent or inconspicuous. INFLORESCENCES initially an open, 7.0–24.0 cm panicle with many spikelets, contracted at maturity by ascending branches; spikelets stalked, 15.0–40.0 mm long, cylindrical, with 4–13 florets; glumes lanceolate and glabrous, first (lower) glume 4.0–8.0 mm long and 1-veined, second (upper) glume 6.0–11.0 mm and 3-veined; lemma lanceolate, glabrous, 9.0–14.0 mm long, 3–5-veined, awnless or with awn 0.5–2.5 mm long; palea shorter than lemma, 2-nerved. FRUITS caryopsis, somewhat flattened, with groove on one side.

Special Identifying Features

Cool-season perennial growing in loose colonies from sod-forming rhizomes; ligule membranous; inflorescence a panicle with ascending branches.

Toxic Properties

See comments under *Bromus catharticus*.

Inflorescence

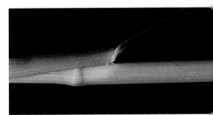

TOP Seeds
MIDDLE Collar
BOTTOM Flower

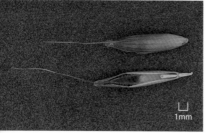

Seeds

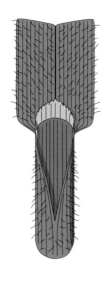

Japanese Brome

Bromus japonicus Thunb. ex Murr. • Poaceae • Grass Family

Synonyms

None

Habit, Habitat, and Origin

Erect annual; to 1.0 m tall; cultivated areas, fields, pastures, roadsides, railroad beds, and waste sites; native of Europe.

Seedling Characteristics

Pubescent sheath and blades, membranous ligule.

Mature Plant Characteristics

ROOTS fibrous. STEMS erect, often bending and rooting at lower nodes, to 1.0 m tall. LEAVES 4.0–24.0 (to 30.0) cm long, 2.0–7.0 mm wide, pubescent on both surfaces; sheaths pubescent; ligule membranous, to 1.7 (rarely to 3.5) mm long. INFLORESCENCES open panicle, 17.0–30.0 cm long, 6.0–13.0 cm wide, lower branches drooping at maturity; spikelets 5–10 (to 13)-flowered, 2.0–3.2 cm long including awns, green at maturity; glumes 4.0–7.6 mm long, minutely scabrous, second longer than first; lemma 7.2–9.1 mm long, scabrous, tip bent-awned from between 2 teeth; palea about 2.0 mm shorter than lemma, with 2 ciliate keels. FRUITS caryopsis, lemma and palea attached.

Special Identifying Features

Leaf sheath and blade pubescent; ligule membranous; inflorescence with thin branches; spikelets bent-awned, not flat, often drooping, green at maturity.

Toxic Properties

See comments under *Bromus catharticus*.

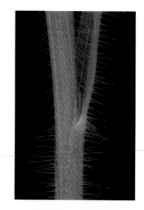

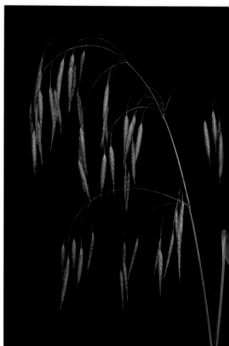

TOP LEFT Collar
TOP RIGHT Seedling
BOTTOM Inflorescence

Cheat

Bromus secalinus L. • Poaceae • Grass Family

Synonyms

Chess, rye bromegrass

Habit, Habitat, and Origin

Tufted, erect winter annual; to 1.2 m tall; cultivated areas, fields, pastures, roadsides, railroad beds, and waste sites; native of Europe.

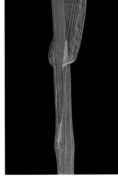

Seedling Characteristics

Glabrous leaf sheath and blade, membranous ligule.

Mature Plant Characteristics

ROOTS fibrous. STEMS erect, to 1.2 m tall or occasionally bent at lower nodes, nodes pubescent. LEAVES 10.0–21.0 cm long, 4.5 mm wide, upper sides sometimes pubescent, pubescent below; sheaths closed, glabrous; ligule membranous, toothed, 2.0–3.0 mm long, prominent. INFLORESCENCES erect, loose or contracted panicle, 5.0–23.0 cm long; spikelets 4–7-flowered, 17.0–21.0 mm long, pediceled; first glume 4.0–6.0 mm long, 3–5-veined, second glume 6.0–8.0 mm long; lemma 6.0–9.0 mm long, with apical teeth, awn 1.5–9.0 mm long or may be rudimentary, inrolled at maturity; palea pubescent on keels, slightly shorter to slightly longer than lemma. FRUITS caryopsis, lemma and palea attached, all nearly equal in length.

Special Identifying Features

Tufted winter annual, sheaths glabrous or lower sheaths loosely pubescent; inflorescence a loose panicle, glumes glabrous or scabrous.

Toxic Properties

See comments under *Bromus catharticus*.

TOP **Inflorescence**
BOTTOM LEFT **Seedling**
BOTTOM RIGHT **Collar**

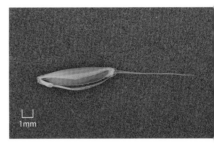

1mm

Seed

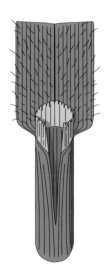

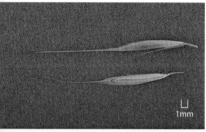

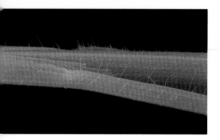

Downy Brome

Bromus tectorum L. • Poaceae • Grass Family

Synonyms

Drooping bromegrass

Habit, Habitat, and Origin

Tufted, erect winter annual; to 60.0 dm tall; cultivated areas, fields, pastures, roadsides, railroad beds, and waste sites; native of southern Europe.

Seedling Characteristics

Softly pubescent leaves and sheath; prominent membranous ligule; seedling leaves usually twisted, long, thin.

Mature Plant Characteristics

ROOTS fibrous. STEMS erect, to 7.0 dm tall, internodes glabrous or strigillose. LEAVES 3.0–21.0 cm long, 2.0–4.5 mm wide, flat, pubescent on both surfaces; sheath closed on lower leaves, pubescent; ligule a prominent membrane with frayed margin, to 2.0 mm long. INFLORESCENCES loose panicle, 4.0–18.0 cm long, flexuous and drooping; spikelets 4–8-flowered, 1.9–2.4 cm long including awns; first glume 4.0–9.0 mm long, 1-veined, second glume 7.0–13.0 mm long, 3–5-veined; lemma 9.0–12.0 mm long, with awn 10.0–18.0 mm long, pubescent; palea 1.2–2.1 mm shorter than lemma, ciliate on nerves. FRUITS caryopsis, lemma and palea attached.

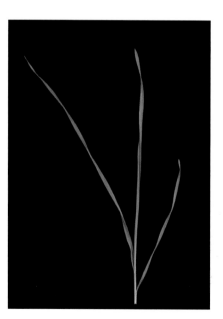

Special Identifying Features

Flexuous, drooping panicle branches with long (25.0 mm) awns; sheath pubescent; nodes glabrous, internodes glabrous or strigillose; glumes glabrous or hairy.

Toxic Properties

See comments under *Bromus catharticus*.

Seedling

Inflorescence

Female Inflorescence

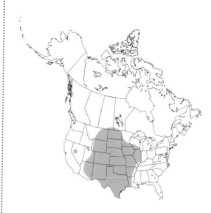

Buffalograss

Buchloë dactyloides (Nutt.) Engelm. • Poaceae • Grass Family

Synonyms

None

Habit, Habitat, and Origin

Stoloniferous, sod-forming, warm-season perennial grass with stems; to 25.0 cm tall; open areas, prairies, fields, and pastures; native of North America.

Seedling Characteristics

Leaves glabrous to hispid.

Mature Plant Characteristics

ROOTS shallow, fibrous. **STEMS** solid, 8.0–25.0 cm tall, nodes smooth; perennial from stolons. **LEAVES** flat, rolled in bud, hispid, 2.0–12.0 cm long, 1.0–2.5 mm wide, ligule a ciliate membrane 0.5 mm long; sheaths open, rounded on back, glabrous except for marginal pubescence. **INFLORESCENCES** dioecious; male with 2–3 unilateral spicate branches, female flowers in burlike clusters enclosed in leaf sheath; spikelets: male 2-flowered, both anther-bearing; female 1-flowered; glumes: male broad, unequal, shorter than lemma, 1–2-nerved; female glabrous, unequal, first glume usually reduced; lemma: male thin, 3-nerved, glabrous; female thin but firm, glabrous, 3-nerved; palea: both male and female similar to lemma. **FRUITS** caryopsis, ovate or oblong, brown, 2.0–2.5 mm long, burlike.

Special Identifying Features

Female plants with burlike clusters of seeds; male plants with spiked, flaglike inflorescence.

Toxic Properties

None reported.

TOP Seeds
MIDDLE Male Inflorescence
BOTTOM Seedling

Southern Sandbur

Cenchrus echinatus L. • Poaceae • Grass Family

Synonyms

Hedgehoggrass

Habit, Habitat, and Origin

Tufted annual ascending from bent base; to 1.0 m long; fields, pastures, roadsides, railroad beds, and waste sites; native of American tropics.

Seedling Characteristics

Leaf surface like sandpaper and often reddish.

Mature Plant Characteristics

ROOTS fibrous. STEMS round, 0.2–1.0 m long, geniculate or trailing, erect tips 14.0–40.0 cm long. LEAVES to 25.0 cm long, 2.0–10.0 mm wide, upper surface scaberulous and sparsely papillose-pilose, lower surface glabrous, margins pubescent; sheaths compressed with protruding keel, pubescent on

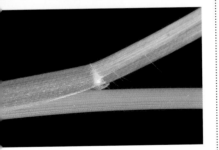

TOP **Seed**

MIDDLE **Seedling**

BOTTOM **Collar**

margins; ligule a ciliate, pubescent ring, 0.6–2.0 mm long. INFLORESCENCES cylindrical raceme, 2.5–12.0 cm long, 1.4–2.1 cm wide, crowded with burs, each 5.0–10.0 mm long; spikelets 2–4 per bur, 5.0–6.0 mm long, sessile; first glume 1.0–3.0 mm long, 1–3-veined, second glume 4.0–5.0 mm long, 3–7-veined; lemma 5.0–6.0 mm long; palea 5.0–6.0 mm long. FRUITS caryopsis, 2.0 mm long, within bur.

Special Identifying Features

Spiny bur with 1 whorl of united flattened spines with 1 to several whorls of shorter bristles below; pubescent leaf sheath; flowering late spring through autumn.

Toxic Properties

Spiny bur can cause mechanical injury.

Inflorescence

Longspine Sandbur

Cenchrus longispinus (Hack.) Fern. · Poaceae · Grass Family

Synonyms

Field sandbur; *Cenchurus incertus* M. A. Curtis

Habit, Habitat, and Origin

Occasionally erect but usually bending and spreading, annual grass; to 55.0 cm tall when erect and to 1.0 m long when spreading; fields, pastures, roadsides, railroad beds, and waste sites; native of North America.

Seedling Characteristics

Very narrow, glabrous leaf sheaths and blades; ligule a tiny fringed membrane.

Mature Plant Characteristics

ROOTS fibrous. **STEMS** occasionally erect but usually bending and spreading, to 55.0 cm tall when erect and to 1.0 m long when spreading, sometimes rooting at lower nodes. **LEAVES** 6.0–17.0 cm long, 3.0–7.0 mm wide, scabrous on upper surface or occasionally with a few scattered trichomes, glabrous on lower surface; sheaths glabrous except for a few trichomes at top on margin; ligule a membrane to 0.2 mm long fringed with trichomes to 1.2 mm long. **INFLORESCENCES** raceme of burs, 4.0–10.0 cm long, 1.2–1.8 cm wide; spikelets 2-flowered, 1–3 per bur, 5.1–6.2 mm long; first glume 2.2–2.4 mm long, glabrous, second glume as long as spikelet, glabrous; sterile lemma as long as spikelet, glabrous, fertile lemma firm, as long as spikelet, scabrous near tip; palea fertile, firm, slightly shorter than lemma. **FRUITS** caryopsis, ovoid, within intact spikelet or bur.

Special Identifying Features

Burs pubescent, with more than 45 spines; leaf blades scabrous on upper surface.

Toxic Properties

See comments under *Cenchrus echinatus*.

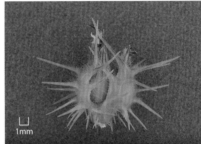

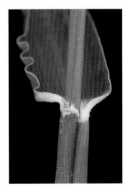

TOP Seed
MIDDLE Seedling
BOTTOM Collar

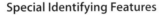

Inflorescence

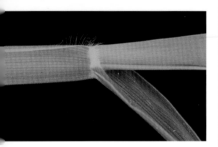

TOP **Seeds**
BOTTOM **Collar**

Inflorescence

Seedling

Field Sandbur

Cenchrus spinifex Cav. • Poaceae • Grass Family

Synonyms

Cenchrus incertus M. A. Curtis, *Cenchrus pauciflorus* Benth.

Habit, Habitat, and Origin

Tufted annual or short-lived perennial from basal stolons; to 80.0 cm tall; fields, pastures, roadsides, railroad beds, and waste sites; native of and widely distributed in the Americas.

Seedling Characteristics

Leaf sheath and leaf blade glabrous except for trichomes at mouth and on sheath margin; narrow-fringed membranous ligule.

Mature Plant Characteristics

ROOTS fibrous, with stolons. **STEMS** 25.0–80.0 cm tall, erect to ascending, branching and rooting at nodes. **LEAVES** 10.0–25.0 cm long, 3.0–7.0 mm wide, glabrous; sheaths compressed, occasionally pubescent on lower nodes and/or sheath margin; ligule a narrow membrane fringed with trichomes, to 1.3 mm long. **INFLORESCENCES** raceme, 3.0–8.0 cm long, burs not crowded on raceme, each 3.0–8.0 mm wide excluding spines; spikelets 1–3 in each bur, 4.0–6.5 mm long, 2.2 mm wide; first glume 1.8 mm long, 1.5 mm wide, second glume 4.0 mm long, 2.0 mm wide; lemma similar to second glume; palea 5.0 mm long. **FRUITS** caryopsis, ovoid, within intact spikelet or bur.

Special Identifying Features

Bur with 8–40 spines; leaf surface rough.

Toxic Properties

See comments under *Cenchrus echinatus*.

Feather Fingergrass

Chloris virgata Sw. • Poaceae • Grass Family

Synonyms

None

Habit, Habitat, and Origin

Tufted annual; 9.0 dm tall; fields, pastures, roadsides, railroad beds, and waste sites; native of tropical Americas.

Seedling Characteristics

Slightly rough, upright, with compressed leaf sheaths.

Mature Plant Characteristics

ROOTS fibrous. **STEMS** ascending to spreading from decumbent base, 4.0–9.0 dm tall, branching and rooting at lower nodes. **LEAVES** flat, rough above, 2.0–7.0 mm wide, with sparse long trichomes; sheaths compressed, keeled, shorter than internodes; ligule membranous, 1.0 mm long. **INFLORESCENCES** 7–16 spikes atop stem, arranged fingerlike, erect to spreading, 2.0–8.0 cm long, pale green or tawny, feathery or silky; spikelets arranged in 2 rows along one side of a continuous rachis, each having a perfect flower and several sterile ones; first glume 1-nerved, slightly rough on keel, narrow-pointed on both ends, 1.5–2.0 mm long, second glume as first, 2.0–2.5 mm long; lemma rough, 3.5 mm long, 3–5-nerved, slightly humpbacked on keel, pubescent, awn 5.0–10.0 mm long and protruding from midnerve; palea 2-keeled, as lemma. **FRUITS** caryopsis, within lemma and palea, pubescent, feathery.

Special Identifying Features

Inflorescence consisting of many spikes originating at a stem terminus, like a feather duster.

Toxic Properties

None reported.

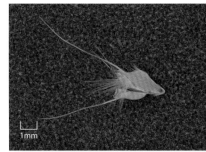

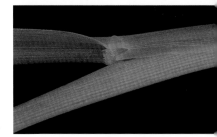

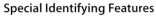

Inflorescence

TOP Seeds
MIDDLE Collar
BOTTOM Seedling

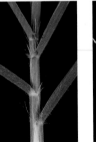

TOP **Seeds**
BOTTOM LEFT **Collar**
BOTTOM RIGHT **Inflorescence**

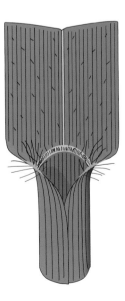

Stolon

Bermudagrass

Cynodon dactylon (L.) Pers. • Poaceae • Grass Family

Synonyms

Couchgrass, wiregrass

Habit, Habitat, and Origin

Rhizomatous and stoloniferous, warm-weather, spreading perennial; to 3.0 dm tall; cultivated areas, fields, pastures, gardens, roadsides, railroad beds, and waste sites; possibly native of Eurasia.

Seedling Characteristics

Leaf sheath smooth, collar smooth in young seedlings; ligule fringed, membranous.

Mature Plant Characteristics

ROOTS fibrous, arising from nodes of stolons and rhizomes. STEMS arising from stolons and rhizomes, erect, rooting at nodes, internode length variable, terminating in an inflorescence, usually slightly flattened. LEAVES linear, 5.0–16.0 cm long, 2.0–5.0 mm wide, upper surface sometimes pubescent; sheaths pubescent on lower surface, with tuft of erect trichomes on margin at collar; ligule an inconspicuous membrane with ciliate rim, to 0.3 mm long. INFLORESCENCES erect on stem, to 0.5 m tall, composed of 3–9 finger-like spikes 3.0–10.0 cm long; spikelets 2.0–3.2 mm long with 1 fertile floret, sessile; both glumes similar, 2.7 times as long as spikelet; lemma keeled, acute, with fringe of trichomes on keel, 1.9–3.1 mm long, awnless; palea narrow, slightly shorter than lemma. FRUITS caryopsis, light brown, microscopically striate, hulled or without glumes.

Special Identifying Features

Forms a turf of fine leaves; tuft of erect trichomes on sheath margins at collar.

Toxic Properties

Grass may cause nitrate and/or nitrite intoxication, is cyanogenic, and may host endophytic fungi that cause tremorgenic syndromes.

Orchardgrass

Dactylis glomerata L. • Poaceae • Grass Family

Synonyms

Cocksfoot

Habit, Habitat, and Origin

Tufted, erect, cool-season perennial; to 12.0 dm tall; fields, pastures, roadsides, railroad beds, and waste sites; native of Eurasia.

Seedling Characteristics

Flattened, keeled, glabrous or slightly scabrous culm; leaves with rough margins; ligule membranous.

Mature Plant Characteristics

ROOTS fibrous, rarely with short rhizomes. **STEMS** erect, 4.0–12.0 dm tall, hollow, flattened, glabrous to slightly scabrous. **LEAVES** blades flat or folded, 3.5–43.0 cm long, glabrous to scabrous, with scabrous margins, midvein prominent and scabrous; sheaths closed toward base, margins overlapping and fused, glabrous and fused, keeled and flattened; ligule membranous, lacerate, (obtuse) rounded, 2.0–13.0 mm long. **INFLORESCENCES** panicle, stiffly erect to spreading, branches few, 3.0–20.0 cm long; spikelets 2–5-flowered, 5.0–9.0 mm long, clustered in one-sided groups on each branch; glumes unequal, 2.5–7.0 mm long, keeled, first glume glabrous, second glume often ciliate on keel, often short-awned; lemma acute to short-awned, scabrous to pubescent, keeled, margins ciliate, lowest 4.5–8.0 mm long, awn up to 1.8 mm long; palea acute and shorter than lemma, 2-forked at tip. **FRUITS** caryopsis, lemma and palea attached.

Special Identifying Features

Tufted cool-season perennial; flattened, smooth to scabrous stem; blade margins rough, midrib rough and prominent; spikelets on only one side of each branch of panicle.

Toxic Properties

Associated with tetany syndrome in cattle, less often in sheep.

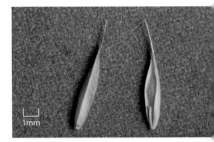

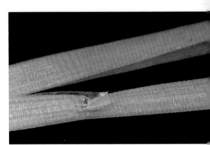

TOP **Seeds**
BOTTOM **Collar**

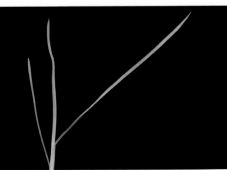

TOP **Inflorescence**
BOTTOM **Seedling**

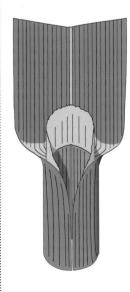

TOP **Seeds**
MIDDLE **Seedling**
BOTTOM **Collar**

Crowfootgrass

Dactyloctenium aegyptium (L.) Willd. • Poaceae • Grass Family

Synonyms

Egypt grass

Habit, Habitat, and Origin

Annual grass, bending and rooting at lower nodes, tips ascending; to 70.0 cm tall; fields, lawns, pastures, roadsides, railroad beds, and waste sites; native of Eurasian tropics.

Seedling Characteristics

Sheath glabrous; blade usually glabrous, margin with long, stiff trichomes; ligule a fringed membrane.

Mature Plant Characteristics

ROOTS fibrous. STEMS bending and rooting at lower nodes, tips ascending, to 70.0 cm tall. LEAVES blade 2.0–30.0 cm long, 2.0–9.0 mm wide, glabrous or with long stiff trichomes, margins ciliate; sheaths glabrous; ligule a membrane to 1.0 mm long, fringed with trichomes to 0.8 mm long. INFLORESCENCES 1–7 spikes at tip of stem, spikes 1.0–6.2 cm long, 3.0–7.0 mm wide, rachis projecting as a point beyond spikelets; spikelets 3–5-flowered, 2.6–4.0 mm long; glumes nearly equal, 1.5–2.2 mm long, first glume awnless, second glume with hooked awn from back of tip; lemma 2.2–3.0 mm long, with sharply scabrous keel, awnlike tip; palea slightly shorter than lemma, 2-keeled, scabrous on keels. FRUITS caryopsis, lemma and palea attached.

Inflorescence

Special Identifying Features

Plant glabrous, blade margins ciliate, inflorescence like crow's foot, spikelets on lower side of rachis.

Toxic Properties

Reported to be cyanogenic.

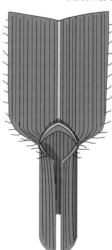

Tropical Crabgrass

Digitaria bicornis (Lam.) Roem. & Schult. ex Loud. • Poaceae • Grass Family

Synonyms

Asian crabgrass

Habit, Habitat, and Origin

Annual with stems either erect or bending and rooting at lower nodes; to 1.0 m

Inflorescence

tall; fields, pastures, roadsides, railroad beds, and waste sites; native of Eurasia.

Seedling Characteristics

Leaf blade usually with scattered stiff trichomes on both surfaces; leaf sheaths pubescent; ligule membranous.

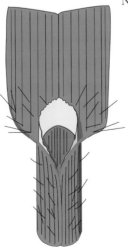

Mature Plant Characteristics

ROOTS fibrous. STEMS 0.3–1.1 m tall, erect or ascending, often decumbent. LEAVES 2.0–12.0 cm long, 3.0–6.0 mm wide, flat, smooth or usually with scattered stiff trichomes on both surfaces near base; sheaths pubescent; ligule membranous, 1.0–2.7 mm long. INFLORESCENCES 2–9 spreading, digitate branches, 3.0–10.0 cm long; spikelets 2-flowered, the lower male or sterile, 2.5–3.1 mm long; first glume less than 0.3 mm long, rounded at tip, second glume 1.8–2.3 mm long; lemma glabrous or appressed, pubescence becoming widely divergent at maturity on upper spikelets; palea 2.5–3.0 mm long, nerveless, smooth. FRUITS caryopsis, entire spikelet attached.

Special Identifying Features

Flat leaf blade with stiff trichomes on both surfaces, leaf sheath pubescent; seed head with all branches at same point (digitate); spikelets 2.5–3.1 mm long; second glume 1.5 mm long or longer.

Toxic Properties

None reported.

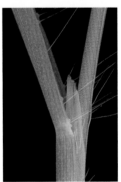

TOP **Seeds**
MIDDLE **Seedling**
BOTTOM **Collar**

Seeds

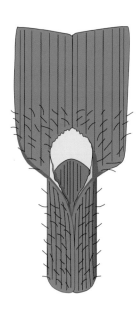

Southern Crabgrass

Digitaria ciliaris (Retz.) Koel. • Poaceae • Grass Family

Synonyms

Summergrass; *Digitaria adscendens* (Kunth) Henr.

Habit, Habitat, and Origin

Tufted, prostrate to spreading summer annual; to 1.0 m tall or broad; cultivated areas, fields, pastures, lawns, turf, roadsides, railroad beds, and waste sites; introduced, presumably native of Eurasia.

Seedling Characteristics

Pubescent leaves, leaf sheaths, and collars; membranous ligule.

Mature Plant Characteristics

ROOTS fibrous, with roots from stem nodes. **STEMS** 25.0–100.0 cm long and tall, decumbent to spreading, branched, rooting at nodes, nodes pubescent. **LEAVES** 5.0–18.0 cm long, 3.0–10.0 mm wide, upper leaves occasionally pubescent on upper surface only; sheaths densely pubescent including margins and mouth; ligule a membrane 1.0–3.0 mm long with truncate frayed margin. **INFLORESCENCES** composed of 2–9 fingerlike branches, each 3.0–21.0 cm long; spikelets 2-flowered, 2.5–3.5 mm long, 0.7–0.8 mm wide; first glume minute, second glume two-thirds length of spikelet, 1.5 mm long, pubescent; lemma equaling spikelet length, pubescent; palea as lemma. **FRUITS** caryopsis, entire spikelet including glumes attached.

Special Identifying Features

Distinguished from large crabgrass by length of second glume (two-thirds length of floret), seedlings not distinguishable from large crabgrass.

Toxic Properties

None reported.

TOP LEFT **Collar**
TOP RIGHT **Seedling**
BOTTOM **Inflorescence**

Smooth Crabgrass

Digitaria ischaemum (Schreb. ex Schweig.) Schreb. ex Muhl. •
Poaceae • Grass Family

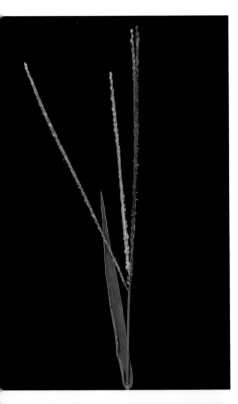

Synonyms

Fingergrass, smooth summergrass

Habit, Habitat, and Origin

Tufted, decumbent summer annual; to
0.6 m; cultivated areas, fields, pastures,
lawns, turf, roadsides, railroad beds, and
waste sites; native of Europe.

Seedling Characteristics

Leaf blades and sheaths glabrous, few
trichomes at mouth only, ligule membranous.

Mature Plant Characteristics

ROOTS fibrous. STEMS prostrate to decumbent, up to 0.6 m tall, branching at
lower nodes but not rooting, nodes glabrous. LEAVES 5.0–14.0 cm long, 2.0–7.0
mm wide, glabrous on both surfaces;
sheath glabrous, closed, trichomes at
mouth only; ligule a membrane, 1.0–2.0
mm long, with even margins. INFLORES-
CENCES 2–6 fingerlike branches, each
10.0 cm long; spikelets 1.8–2.1 mm long,
0.8–0.9 mm wide; first glume hyaline,
obscure, or absent; second glume equaling spikelet length, with short mush-
roomlike trichomes; lemma equaling
spikelet length, with short trichomes;
palea as lemma. FRUITS caryopsis, entire
spikelet including glumes attached.

Special Identifying Features

Glabrous sheath and leaves, mushroom-
like trichomes on spikelet.

Toxic Properties

None reported.

Seeds

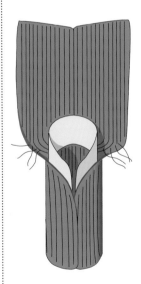

TOP **Inflorescence**
BOTTOM LEFT **Seedling**
BOTTOM RGHT **Collar**

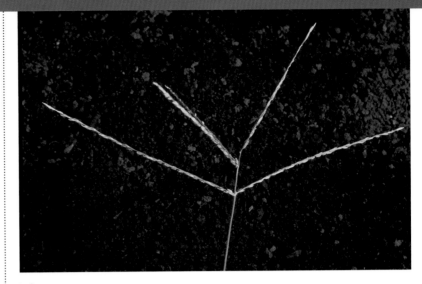

Inflorescence

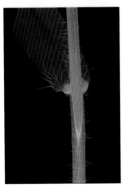

TOP Seeds
MIDDLE Seedling
BOTTOM Collar

Large Crabgrass

Digitaria sanguinalis (L.) Scop. • Poaceae • Grass Family

Synonyms

Hairy fingergrass

Habit, Habitat, and Origin

Tufted, prostrate to spreading summer annual; to 0.7 m tall or broad; cultivated areas, fields, pastures, lawns, turf, roadsides, railroad beds, and waste sites; native of Europe.

Seedling Characteristics

Leaf blades and sheaths usually densely pubescent, ligule membranous.

Mature Plant Characteristics

ROOTS fibrous. STEMS prostrate, spreading, branched at older nodes, rooting at nodes to 0.7 m. LEAVES 3.0–20.0 cm long, 3.0–10.0 mm wide, usually pubescent on both surfaces; sheaths pubescent, closed; ligule a 1.0–2.0 mm membrane, truncate with uneven teeth on margin. INFLORESCENCES 4–6 branches, each 4.0–18 cm long; spikelets 2-flowered, 2.5–3.3 mm long, 0.9 mm wide, subsessile to pediceled; first glume minute, second glume twice length of spikelet, pubescent; lemma equaling spikelet in length; palea as lemma. FRUITS caryopsis, entire spikelet detaches and falls.

Special Identifying Features

Second glume twice length of spikelet, shorter than second glume of southern crabgrass.

Toxic Properties

None reported.

Saltgrass

Distichlis spicata (L.) Greene · Poaceae · Grass Family

Synonyms

Desert saltgrass, island saltgrass, seashore saltgrass; *Distichlis stricta* (Torr.) Rydb.

Habit, Habitat, and Origin

Erect to ascending, mat-forming perennial; to 50.0 cm tall; cultivated areas, fields, pastures, lawns, turf, roadsides, railroad beds, and waste sites; native of North America.

Seedling Characteristics

Leaves folded or C-shaped in bud, scabrous or trichomes fine above, glabrous below; sheath glabrous or trichomes fine; ligule membrane fringed with trichomes, flanked by long trichomes.

Mature Plant Characteristics

ROOTS fibrous from creeping, scaly rhizomes or stolons. STEMS erect, decumbent base, 8.0–50.0 cm tall, glabrous, solid to hollow, internodes short; female plants shorter than males. LEAVES linear, 2.0–12.0 cm long, involute to flat, stiffly ascending, scalelike near base, 2-ranked upward, scabrous or finely pubescent above, glabrous below; sheath margins overlapping, glabrous or fine trichomes; ligule a thin membrane, 0.2–0.6 mm long, trichome fringe often flanked by longer trichomes. INFLORESCENCES contracted panicle, 2.0–8.0 cm tall, male extending above leaves, female equal to or shorter than leaves; spikelets flattened, 3–20 florets, light green or purple-tinged, 6.0–25.0 mm long; glumes ovate to lanceolate, glabrous, keeled, first glume 1.5–7.8 mm long, second glume 2.5–8.5 mm long; lemma ovate to lanceolate, glabrous, sharply pointed, male 3.0–7.0 mm long, female 3.0–10.0 mm long; palea equal to lemma, trichomes short on female. FRUITS caryopsis, ovate, straw-colored, few produced.

Special Identifying Features

Warm-season perennial from creeping rhizomes; ligule membranous, fringed with trichomes; inflorescence a short contracted panicle; dioecious.

Toxic Properties

None reported.

TOP Seeds
MIDDLE Seedling
BOTTOM Collar

Inflorescence

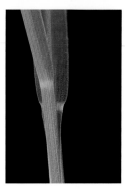

TOP **Seeds**
BOTTOM **Collar**

Inflorescence

Seedling

Junglerice

Echinochloa colona (L.) Link • Poaceae • Grass Family

Synonyms

Awnless barnyardgrass, watergrass

Habit, Habitat, and Origin

Decumbent to erect, tufted summer annual with several to many stems from base; to 1.0 m tall; cultivated areas, fields, pastures, wet ditches, roadsides, railroad beds, and waste sites; native of Europe.

Seedling Characteristics

Glabrous leaves and leaf sheath, no ligule.

Mature Plant Characteristics

ROOTS fibrous. **STEMS** spreading to erect-ascending, bent at nodes, diverging, glabrous, to 1.0 m tall, rooting at nodes. **LEAVES** 10.0–40.0 cm long, 3.0–10.0 cm wide, sometimes with purple bands, glabrous or sparsely pubescent above; sheaths glabrous; ligules absent. **INFLORESCENCES** panicle 5.0–20.0 cm long with 8–10 branches, each 1.0–4.0 cm long; spikelets 2.3–2.8 mm long, 1.0–1.5 mm wide, subsessile; first glume as long as spikelet, second glume equaling spikelet; lemma similar to second glume; palea equaling spikelet, shiny. **FRUITS** caryopsis, hulled or spikelet including glumes attached.

Special Identifying Features

Simple, awnless spikelets; purple leaf brand not present on young plants; no ligule.

Toxic Properties

None reported.

Inflorescence

Seedling

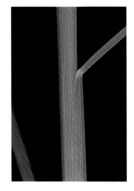

TOP Seeds
BOTTOM Collar

Barnyardgrass

Echinochloa crus-galli (L.) Beauv. • Poaceae • Grass Family

Synonyms

Billiondollargrass, watergrass

Habit, Habitat, and Origin

Tufted, erect summer annual; up 2.0 m tall; cultivated areas, fields, pastures, wet ditches, roadsides, railroad beds, and waste sites; native of Eurasia.

Seedling Characteristics

Glabrous leaves and leaf sheath, no ligule.

Mature Plant Characteristics

ROOTS fibrous. STEMS erect 1.5–2.0 m tall, often bent and branched at lower nodes, glabrous. LEAVES 10.0–50.0 cm long, 5.0–30.0 mm wide, glabrous but rough on both surfaces; sheaths glabrous; ligules absent. INFLORESCENCES terminal, nodding panicle 10.0–40.0 cm long with numerous appressed or spreading branches, the lower, larger ones rebranching; spikelets 2.6–4.4 mm long, 1.0–1.8 mm wide, subsessile or short-pediceled; first glume one-third to one-half as long as spikelet, hispid on nerves, second glume equaling spikelet; lemma awned or awnless, hardened, apex obtuse or broadly acute; palea large, well-developed. FRUITS caryopsis, hulled or spikelet attached.

Special Identifying Features

No ligule; crowded, rebranched spikelets; may be awned or awnless.

Toxic Properties

Consumption may cause photosensitization.

> **Japanese millet (*Echinochloa frumentacea* Link)**, widely planted as waterfowl food, was considered a variety of barnyardgrass as *Echinochloa crus-galli* (L.) Beauv. var. *frumentacea* (Link) W. F. Wight until recently. Lemmas of Japanese millet lack awns, upper lemmas are wider and longer than upper glumes, and panicles do not disarticulate at maturity. Barnyardgrass lemmas are often awned, upper lemmas are not longer than glumes, and spikelets disarticulate at maturity.

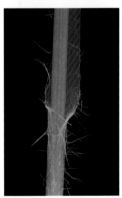

TOP Seed
MIDDLE Seedling
BOTTOM Collar

Goosegrass

Eleusine indica (L.) Gaertn. • Poaceae • Grass Family

Synonyms

Bullgrass, crowfootgrass, white crab-grass, wiregrass

Habit, Habitat, and Origin

Tufted, warm-season annual; to 8.5 m tall; cultivated areas, fields, pastures, lawns, turf, roadsides, railroad beds, and waste sites; native of Eurasia.

Seedling Characteristics

Leaf sheath margins broad, whitish to translucent, distinctly flattened, smooth throughout; ligule fringed, membranous.

Mature Plant Characteristics

ROOTS fibrous. STEMS erect to spreading, to 8.5 m tall, branched, nodes and internodes glabrous. LEAVES 5.0–35.0 cm long, 3.0–8.0 mm wide, glabrous or occasionally sparsely pubescent; sheaths compressed, margin and mouth pubescent; ligule a fringed, uneven membrane, 0.9 mm long. INFLORESCENCES 1–13 fingerlike spikes, each 4.0–15.0 cm long, 3.0–7.0 mm wide, frequently 1 spike below terminal clusters; spikelets 3–6 fertile flowers, 3.0–7.5 mm long, sessile; first glume 2.4–3.0 mm long, winged on keel, second glume 2.7–4.2 mm long, acute; lemma glabrous, laterally compressed, equaling second glume in length; palea shorter than lemma. FRUITS caryopsis, dark brown, usually hulled, sometimes with whitish pericarp persisting.

Inflorescence

Special Identifying Features

Stems flattened, whitish green, in young plants almost parallel to ground.

Toxic Properties

Reported to be cyanogenic.

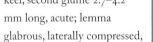

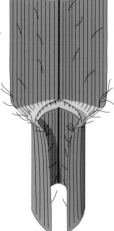

Quackgrass

Elymus repens (L.) Gould · Poaceae · Grass Family

Synonyms

Couch grass, quack, quitch-grass, twitch-grass, twitch, witch-grass, quick-grass, scotch; *Agropyron repens* (L.) Beauv., *Elytrigia repens* (L.) Desv. ex B. D. Jackson

Habit, Habitat, and Origin

Erect, warm-season perennial from rhizomes, often bending out and up at base; to 1.1 m tall; fields, pastures, roadsides, railroad beds, woodland edges, and waste sites; native of Mediterranean region.

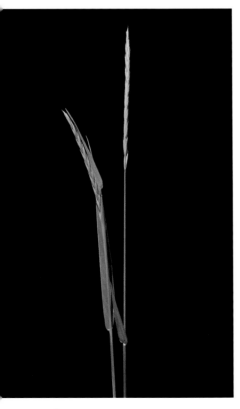

Inflorescence

Seedling Characteristics

Auricles present, sheaths pubescent to glabrous; blade pubescent to glabrous on upper surface, glabrous on lower surface; ligule to 0.4 mm long, membranous.

Mature Plant Characteristics

ROOTS fibrous, rhizomes present. STEMS erect, often bent out and upward at base, to 1.1 m tall. LEAVES blades 4.0–30.0 cm long, 3.0–8.0 (to 16.0) mm wide, upper surface pubescent to glabrous, lower surface glabrous; sheaths glabrous or pilose; ligule membranous, truncate. INFLORESCENCES spike of many several-flowered spikelets, 6.0–23.0 cm long, 0.7–1.7 cm wide; spikelets 4–8-flowered, 11.0–18.0 mm long including awns; glumes strongly nerved, 8.0–13.0 (to 15.0) mm long including awns; lemma 8.5–15.4 mm long including awn, firm; palea 6.8–8.4 mm long, 2-keeled, keels scabrous, tip rounded. FRUITS caryopsis, lemma and palea attached.

Special Identifying Features

Rhizomes present; auricles present; inflorescence a spike of many several-flowered spikelets.

Toxic Properties

Elymus species are reported to cause tetany and ergot, and to be cyanogenic.

1mm

TOP Seed
BOTTOM RIGHT Seedling
BOTTOM LEFT Collar

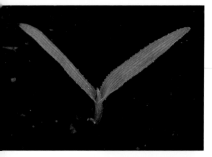

TOP **Seeds**
MIDDLE **Seedling**
BOTTOM **Collar**

Stinkgrass

Eragrostis cilianensis (All.) Vignolo ex Janch. • Poaceae • Grass Family

Synonyms

Skunkgrass, snakegrass; *Eragrostis megastachya* (Koeler) Link

Habit, Habitat, and Origin

Tufted, spreading or ascending, warm-season annual; to 6.0 dm tall; cultivated areas, fields, pastures, lawns, turf, roadsides, railroad beds, and waste sites; native of Europe.

Seedling Characteristics

Leaf sheaths glabrous with scattered glands on veins, sheath overlapping stem; ligule a fringe of short, dense trichomes, 1.0 mm or less long, surrounded by longer trichomes at collar region 1.0–3.0 mm long.

Mature Plant Characteristics

ROOTS fibrous. **STEMS** ascending, sharply bent below, tufted, branched at base and above, 1.0–6.0 dm tall, glabrous, ring of glands below nodes. **LEAVES** flat to somewhat involute, 5.0–15.0 cm long, 2.0–8.0 mm wide, conspicuous nerves, with glands on nerves and margins, rarely pubescent; sheaths overlapping, glabrous except for pilose throat; ligule ciliate, 0.4–0.8 mm long. **INFLORESCENCES** dense panicle, 5.0–20.0 cm long, 2.0–8.0 cm wide, pyramidal, lower panicle branches not verticillate; spikelets compressed, pedicellate, 10–40-flowered, 3.0–20.0 mm long, 2.0–4.0 mm wide, ovate-lanceolate; glumes membranous, unequal, first glume 1.2–2.2 mm long, second glume slightly longer, boat-

Inflorescence

shaped, 1–3-nerved; lemma membranous, broadly elliptic to ovate, 2.0–2.8 mm long, glandular on keel, 3-nerved, lateral nerves prominent; palea hyaline, about two-thirds as long as lemma. **FRUITS** caryopsis, ovoid, 0.5–0.8 mm long, separating from lemma and palea at maturity.

Special Identifying Features

Glabrous, ascending to spreading culms; panicles dense, pyramidal; scattered glands, disagreeable odor when crushed.

Toxic Properties

None reported.

India Lovegrass

Eragrostis pilosa (L.) Beauv. • Poaceae • Grass Family

Synonyms

Southern speargrass

Habit, Habitat, and Origin

Tufted annual ascending from decumbent base, occurring in clumps; up to 60 cm tall; cultivated areas, fields, pastures, lawns, turf, roadsides, railroad beds, and waste sites; native of North America.

Seedling Characteristics

Leaf sheath and margins broad, veins prominent on leaf blades.

Mature Plant Characteristics

ROOTS fibrous. **STEMS** 10.0–60.0 cm tall, frequently bent, with glabrous nodes

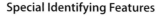

and internodes, branched at base and above. **LEAVES** basal and low on stem, up to 10.0 cm long, 0.5–4.0 mm wide, upper surface scaberulous, lower surface glabrous; sheaths glabrous, pilose at throat of sheath and under collar region; ligule ciliate, 0.1–0.3 long. **INFLORESCENCES** flowering June–August, open panicle, 4.0–20.0 cm long, 2.0–10.0 cm wide, spreading branches with spikelets on distal one-third, branches at a lower node whorled; spikelets pediceled, 3–17 flowers, 2.0–6.0 mm long, usually less than 1.0 mm wide, narrowly ovate, grayish green with reddish tips; first glume 1-nerved, keeled, 0.8–1.0 mm long, second glume 1.2–1.6 mm long, ovate-acute; lemma keeled, 1.2–1.8 mm long, membranous with hyaline tip, lateral-nerved; palea shorter than lemma, hyaline, persistent, ciliate on keels. **FRUITS** caryopsis, reddish, short-cylindrical, 0.5–0.9 mm long.

Special Identifying Features

Tufted annual ascending from decumbent base; plants without scattered glands.

Toxic Properties

None reported.

Inflorescence

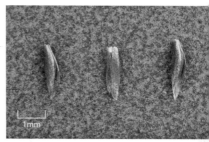

TOP Seeds
BOTTOM Seedling

Seeds

Southwestern Cupgrass

Eriochloa acuminata (J. Presl) Kunth · Poaceae · Grass Family

Synonyms

Eriochloa gracilis (Fourn.) A. S. Hitchc.

Habit, Habitat, and Origin

Erect or ascending, tufted warm-season annual with 1 to several stems from base, sometimes bending and rooting at lower nodes; to 1.0 m tall; roadsides, railroad beds, and waste sites; native of North America.

Seedling Characteristics

Usually glabrous, ligule a fringe of trichomes to 1.3 mm long.

Mature Plant Characteristics

ROOTS fibrous. STEMS erect or ascending, bending and sometimes rooting at lower nodes, 1 to several from base, to 1.0 m tall. LEAVES blade 3.0–24.0 cm long, 3.0–11.0 mm wide, flat, usually glabrous; sheaths usually glabrous; ligule a fringe of trichomes to 1.3 mm long.

INFLORESCENCES panicle 5.0–20.0 cm long, 0.5–2.5 cm wide, with upwardly appressed branches having spikelets in 2 rows on lower side of branches; spikelets 2-flowered, the second fertile, 3.8–5.2 mm long, sharply pointed; first glume forming a cup, second glume long-pubescent, sharply pointed, as long as spikelet; lemma 2.4–2.8 mm long, glabrous, awn 0.1–0.4 mm long; palea about as long as lemma, tip rounded with no awn. FRUITS caryopsis, glumes, lemmas, and palea fall attached.

Special Identifying Features

Plants usually glabrous; ligule with fringe of trichomes to 1.3 mm long; inflorescence upwardly appressed branches, spikelets with cup at base and long trichomes in 2 rows on lower side of branch.

Toxic Properties

None reported.

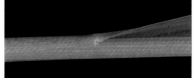

TOP LEFT **Seedling**
BOTTOM LEFT **Collar**
RIGHT **Inflorescence**

Prairie Cupgrass

Eriochloa contracta Hitchc. • Poaceae • Grass Family

Synonyms

None

Habit, Habitat, and Origin

Erect, tufted annual sometimes bending at lower nodes; to 0.9 m tall; roadsides, railroad beds, and waste sites; native of the Americas.

Seedling Characteristics

All parts finely pubescent, ligule a tiny membrane with longer trichomes on upper margin.

Mature Plant Characteristics

ROOTS fibrous. STEMS erect, many from base, often bent at lower nodes, nodes pubescent, to 0.9 m tall. LEAVES 3.0–15.0 cm long, 2.0–8.0 mm wide, pubescent on both surfaces; sheaths pubescent; ligule a minute membrane to 0.2 mm long, fringed with trichomes to 0.9 mm long. INFLORESCENCES panicle 6.0–14.0 cm long, 0.6–1.7 cm wide, with upwardly appressed branches having spikelets in 2 rows on lower side of branches; spikelets 2-flowered, the second fertile, 3.3–4.0 mm long; first glume forming a cup, second glume pubescent, pointed, as long as spikelet; lemma 2.2–2.5 mm long; palea about as long as lemma, tip round and awnless. FRUITS caryopsis, falls with glumes, lemmas, and palea attached.

Special Identifying Features

Entire plant including nodes pubescent; panicle branches still upwardly appressed against stem; spikelets in 2 rows on lower side of branch with cup at base.

Toxic Properties

None reported.

Seeds

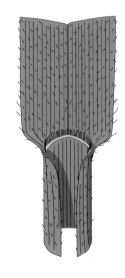

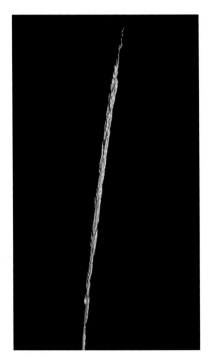

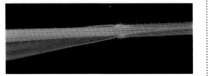

LEFT Inflorescence
TOP RIGHT Seedling
BOTTOM RIGHT Collar

Common Velvetgrass

Holcus lanatus L. • Poaceae • Grass Family

Synonyms

Meadow softgrass

Habit, Habitat, and Origin

Erect or prostrate, tufted or clump-forming warm-season perennial; to 2.0 m tall; fields, pastures, roadsides, railroad beds, turf, and waste sites; native of Europe.

Seedling Characteristics

Leaves densely soft-pubescent, velvety, light green; membranous ligule with minutely pubescent, jagged margin.

Mature Plant Characteristics

ROOTS fibrous. STEMS erect or spreading, 0.3–2.0 m tall, many tillers, lower internodes soft-villous and velvety, upper internode glabrous. LEAVES linear, flat, 4.0–45.0 cm long, 4.0–12.0 mm wide, pale green, covered with dense, soft trichomes; sheaths soft-villous, rounded on back, slightly compressed, margins open; ligule membranous, 1.0–4.5 mm long, minutely pubescent, margin jagged. INFLORESCENCES congested, dense, erect panicles, 5.0–20.0 cm long, branches ascending, soft-pubescent, grayish with purplish tinge; spikelets compressed, breaking below glumes, 2-flowered, 3.5–5.5 mm long; glumes thin, keeled, with a soft-villous surface and long trichomes on keel, purplish, concealing lemmas; first glume lanceolate, 3.0–6.0 mm long, 1-veined, perfect; second glume ovate, 3-veined, staminate, awned; lemma ovate to oblong-elliptic, glabrous, shiny, 1.7–4.0 mm long, uppermost floret with 1.0–2.0 mm awn with hooked tip; palea slightly shorter than lemma, membranous, 2-nerved. FRUITS caryopsis, elliptic, yellow, 1.5–2.0 mm long.

Special Identifying Features

Tufted perennial; plants light green to gray-green; trichomes soft, dense, velvety; ligule membranous with fringe of minute trichomes.

Toxic Properties

Reported to be cyanogenic, but health risk is considered low.

TOP Seeds
MIDDLE Seedling
BOTTOM Collar

Inflorescence

Inflorescence

Seedling

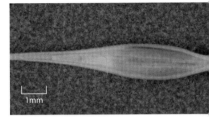
1mm

Foxtail Barley

Hordeum jubatum L. • Poaceae • Grass Family

Synonyms

Flickertail, skunktail, squirreltail barley, squirreltail grass

Habit, Habitat, and Origin

Coarse, tufted annual or short-lived perennial with ascending stems often decumbent at base; to 80.0 cm tall; fields, pastures, roadsides, railroad beds, and waste sites; native of North America.

Seedling Characteristics

Bluish green, thin; blades and sheath pubescent or smooth; ligule membranous.

Mature Plant Characteristics

ROOTS fibrous. STEMS 25.0–80.0 cm tall, tufted, ascending from a usually decumbent base. LEAVES 3.0–22.0 cm long, 2.0–5.0 mm wide, sandpapery, pubescence present or absent on both surfaces; sheaths glabrous or pubescent; ligule a membrane, flat at top, 0.2–1.1 mm long. INFLORESCENCES foxtail-like, 6.0–20.0 cm long

including awn, 4.0–8.0 cm wide including awn; spikelets in threes, 1 floret per spikelet; middle spikelet fertile, 5.0–8.0 mm long, sessile; lateral spikelets sterile or male, 5.0–8.0 mm long, on short stalks (pedicels), awns 4.0–8.0 cm long; glumes 0.2–6.6 cm long and very narrow with short, stiff trichomes, awns 5.0–8.0 cm long; lemma 3.6–8.0 mm long with short, stiff trichomes, awns 2.5–6.2 cm long; palea 3.6 mm long, flat, pointed, veins with fine trichomes. FRUITS caryopsis, cluster of 3 spikelets falls attached.

Special Identifying Features

Tufted, many-stemmed plant; membranous ligule; seedhead with long-awned spikelets attached in threes, central spikelet fertile.

Toxic Properties

The long awns can injure grazing animals, and plants may cause nitrate and/or nitrite intoxication.

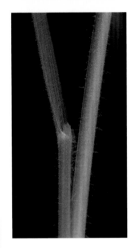
TOP **Seed**
BOTTOM **Collar**

Seeds

Little Barley

Hordeum pusillum Nutt. • Poaceae • Grass Family

Synonyms

None

Habit, Habitat, and Origin

Erect, tufted annual; to 0.5 m tall; cultivated areas, fields, pastures, lawns, turf, roadsides, railroad beds, and waste sites; native of the Americas.

Seedling Characteristics

Sheath pubescent or glabrous, blades usually pubescent, ligule membranous.

Mature Plant Characteristics

ROOTS fibrous. STEMS erect, 1 to several from base, to 0.5 m tall. LEAVES 3.0–12.0 cm long, 2.0–4.0 mm wide, glabrous or pubescent and often scabrous, auricles small; sheaths pubescent or glabrous; ligules 0.5–0.8 mm long, membranous. INFLORESCENCES dense, spikelike raceme, 2.0–7.0 cm long, 4.0–7.0 mm wide; spikelets 1-flowered, awned; central spikelet sterile, 11.0–17.0 mm long including awns, lateral 2 spikelets sterile, 10.0–14.0 mm long including awns, stalked; glumes scabrous, awned; first glumes on central spikelet expanded, 11.0–17.0 mm long including awns, first glumes on lateral spikelets expanded, 11.0–14.0 mm long including awns; lemmas scabrous, awned; lemmas on central spikelet 10.0–13.5 mm long including awns, lemmas on lateral spikelet 3.7–7.5 mm long including awns; palea thin, not awned, pubescent at tip. FRUITS caryopsis, fertile and sterile spikelets fall attached.

Special Identifying Features

Thin, erect stem; pubescent or glabrous leaf sheaths and blades; narrow, dense terminal inflorescence densely covered with awns.

Toxic Properties

The long awns can injure grazing animals.

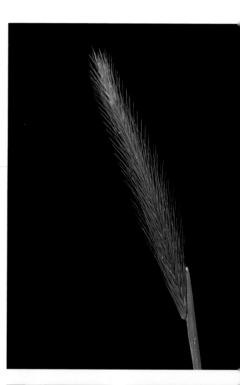

TOP Inflorescence
BOTTOM RIGHT Collar
BOTTOM LEFT Seedling

Inflorescences

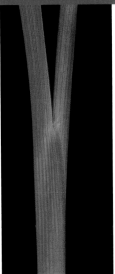

Collar

TOP **Seed**
BOTTOM **Seedling**

Cogongrass

Imperata cylindrica (L.) Beauv. • Poaceae • Grass Family

Synonyms

Speargrass

Habit, Habitat, and Origin

Erect, dense, spreading perennial from scaly rhizomes, leaves tufted at base; to 1.2 m tall; open fields and roadsides; native of Asia and the Indian subcontinent; listed as a Federal Noxious Weed.

Seedling Characteristics

Sheath glabrous or pubescent; blade glabrous except for tuft of trichomes at base on upper surface; ligule a small, fringed membrane.

Mature Plant Characteristics

ROOTS fibrous with scaly rhizomes.
STEMS erect to spreading, to 1.2 m tall.
LEAVES 8.0–133.0 cm long, 2.0–8.0 (to 18.0) mm wide, glabrous except for tuft of trichomes on upper surface at base, basal portion of some blades narrowed and resembling a petiole, midvein off center; sheaths glabrous to pubescent; ligule a fringed membrane to 1.1 mm long. **INFLORESCENCES** terminal, silky panicle 4.0–21.0 cm long, 1.0–3.5 cm wide, plumelike; spikelets 2-flowered, 3.5–4.3 mm long, silky trichomes at base, occurring in pairs, one pediceled and one sessile; glumes nearly equal, pubescent, as long as spikelet; lemma slightly shorter than glumes, thin, papery; palea nearly equal to lemma, thin, papery. **FRUITS** caryopsis; entire spikelet falls attached.

Special Identifying Features

Scaly rhizome; tuft of trichomes at base of upper surface of blade; midvein off center; flowering immediately after emerging in spring or just prior to frost.

Toxic Properties

None reported, but sharp young shoots injure feet or hooves, creating openings for infection.

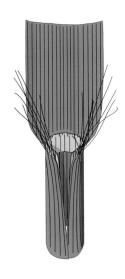

1mm

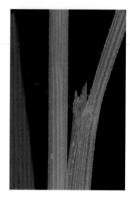

TOP **Seeds**
BOTTOM **Collar**

Seedling

Inflorescence

Green Sprangletop

Leptochloa dubia (Kunth) Nees · Poaceae · Grass Family

Synonyms

None

Habit, Habitat, and Origin

Erect, warm-season, tufted perennial; to 1.2 m tall; cultivated areas, fields, roadsides, railroad beds, wet ditches, and waste sites; native of southern North America and South America.

Seedling Characteristics

Lower leaf sheaths sparsely pubescent, leaves glabrous to sparsely pubescent, ligule a thin membrane with pubescent margin.

Mature Plant Characteristics

ROOTS fibrous. STEMS erect, 0.2–1.2 m tall, 4.0 mm wide, wiry, growing in tufts. LEAVES flat or folded, upper leaves loosely involute, 5.0–30.0 cm long, 2.0–10.0 mm wide but usually less than 5.0 mm, lower side glabrous, upper side scabrous; sheaths round and glabrous, lowermost keeled and pilose; ligule an inconspicuous membrane with a dense ciliate margin. INFLORESCENCES elongate panicle, 10.0–45.0 cm long, with 7–14 racemose ascending branches 3.0–15.0 cm long; spikelets flattened, 5.0–10.0 mm long with 3–12 florets; glumes lanceolate, 2.5–5.0 mm long with pointed tips; lemma glabrous, oblong-ovate to obovate, 3.3–5.0 mm long, bluntly pointed to truncate tip, midnerve sometimes extending to short point, or sometimes tip minutely notched; palea absent. FRUITS caryopsis, elliptic to obovate, 1.9–2.3 mm long, slightly flattened, often with shallow longitudinal groove.

Special Identifying Features

Summer perennial growing in dense tufts; ligule an inconspicuous membrane with fringed margin; elongated panicle with 7–14 racemose ascending branches.

Toxic Properties

Reported to be cyanogenic.

Inflorescence   Seedling

Bearded Sprangletop

Leptochloa fusca (L.) Kunth var. *fascicularis* (Lam.) N. Snow •
Poaceae • Grass Family

Synonyms

Leptochloa fascicularis (Lam.) Gray

Habit, Habitat, and Origin

Erect or spreading, tufted annual; to 1.0
m tall; cultivated areas, fields, pastures,
wet ditches, roadsides, and waste sites;
native of North America.

Seedling Characteristics

Sheath scabrous, blade scabrous on both
surfaces, ligule membranous.

Mature Plant Characteristics

ROOTS fibrous. STEMS erect or
spreading, branched, to 1.0 m
tall. LEAVES 5.0–57.0 cm long,
1.0–5.0 mm wide, scabrous on
both surfaces, often tightly rolled
when dry; sheaths scabrous; ligule
membrane to 6.5 mm long. INFLO-
RESCENCES panicle, 8.0–26.0 cm

long, 3.0–11.0 cm wide, with 6–36 stiff
erect branches; spikelets 5–12-flowered,
4.8–10.6 mm long, uppermost flowers
reduced, in 2 rows on one side of branch;
first glume 2.1–2.8 mm long, second
glume 3.0–5.0 mm long and sometimes
short-awned; lemma lowermost, 2.5–5.2
mm long including short awn from
between 2 teeth, lateral nerves ending in
small teeth, pubescent on lower central
nerve and margins; palea 1.7–3.8
mm long, often pubescent on
nerves. FRUITS caryopsis, lemma
and palea attached.

Special Identifying Features

Uppermost leaf sheathing and
often longer than inflorescence.

Toxic Properties

None reported.

TOP Seeds
BOTTOM Collar

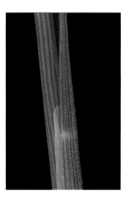

Red Sprangletop

Leptochloa panicea (Retz) Ohwi • Poaceae • Grass Family

Synonyms

Leptochloa filiformis (Lam.) Beauv.

Habit, Habitat, and Origin

Erect, tufted, summer annual with bent, often branching stems; to 1.2 m tall; cultivated areas, fields, roadsides, wet ditches, and waste sites; apparently native of the Americas.

Seedling Characteristics

Sheath often red, pubescent on lower leaves; blades glabrous; ligule fringed, membranous.

Mature Plant Characteristics

ROOTS fibrous. **STEMS** erect, bent at nodes and branches, to 1.2 m tall, often dwarfed to 10.0 cm tall. **LEAVES** 15.0–30.0 cm long, 3.0–10.0 mm wide, usually glabrous; sheaths pubescent, often red; ligule a 1.0–2.0 mm fringed membrane. **INFLORESCENCES** panicle 10.0–40.0 cm long with few to numerous slender racemes, 5.0–15.0 cm long; spikelets 2–4-flowered, 1.5–2.5 mm long; first glume acute, narrow, 1.4–2.2 mm long, second glume broader than first, slightly shorter; lemma 1.0–1.6 mm long, usually pubescent on nerves; palea similar to lemma. **FRUITS** caryopsis, lemma and palea attached or not.

Special Identifying Features

Filiform, branching panicle; branches red or green.

Toxic Properties

None reported.

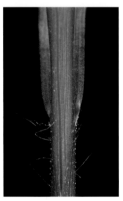

TOP Seeds
MIDDLE Seedling
BOTTOM Collar

Inflorescence

Amazon Sprangletop

Leptochloa panicoides (J. Presl) A. S. Hitchc. • Poaceae • Grass Family

Synonyms

Tighthead sprangletop.

Habit, Habitat, and Origin

Tufted, erect summer annual; to 1.0 m tall; cultivated areas, fields, roadsides, wet ditches, marshes, and waste sites; native of Brazil.

Seedling Characteristics

Leaf sheath and blade glabrous or scabrous.

Mature Plant Characteristics

ROOTS fibrous. STEMS coarse, stiffly erect, to 1.0 m tall, branched. LEAVES flat, usually smooth, occasion-

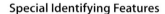

ally scabrous on margins, 5.0–12.0 mm wide; sheaths glabrous or scabrous, tightly compressed at base, lowermost sheaths keeled; ligule a truncate, erose, glabrous membrane 1.0–4.0 mm long. INFLORESCENCES erect, spreading panicle, 12.0–30.0 cm long, 4.0–8.0 cm wide with 40–90 short, crowded, erect, spreading branches 3.0–6.0 cm long; spikelets 4–7-flowered, 4.0–5.0 mm long; first glume narrow, acute, 1.6–2.0 mm long, second glume 1.6–2.0 mm long, much broader than first; lemma 2.0–2.8 mm long, pubescent on margins near base, apex widely acute, minutely lobed, with short, small, abrupt tip; palea broad, almost as long as lemma, ciliate, similar to lemma. FRUITS caryopsis; lemma and palea attached, but not glumes.

Special Identifying Features

Characteristic long, narrow panicle; keeled sheath; long ligule; green midrib.

Toxic Properties

None reported.

Inflorescence

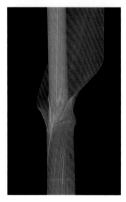

TOP Seeds
MIDDLE Seedling
BOTTOM Collar

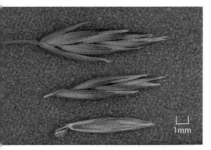

TOP **Spikelets**
BOTTOM **Collar**

Tall Fescue

Lolium arundinaceum (Schreb.) S. J. Darbyshire • Poaceae • Grass Family

Synonyms

Meadow fescue; *Festuca arundinacea* Schreb.

Habit, Habitat, and Origin

Erect, tufted perennial; to 1.4 m tall; cultivated areas, fields, pastures, lawns, turf, road-sides, railroad beds, and waste sites; native of Europe.

Seedling Characteristics

Smooth to scabrous, prominent ciliated auricles, membranous ligule.

Mature Plant Characteristics

ROOTS fibrous, without rhizomes. STEMS erect, glabrous, 0.7–1.4 m tall. LEAVES flat, 5.0–45.0 cm long, most 3.0–8.0 mm wide, smooth to scabrous; sheaths smooth to scabrous with promi-nent, ciliated auricles; ligule a collarlike membrane, 0.2–0.6 mm long, auricles prominent, ciliated. INFLORESCENCES panicle, 10.0–30.0 cm long, contracted and narrow; spikelets short-pediceled, appressed, 10.0–15.0 mm long, 5–7-flowered; glumes glabrous, lanceolate, acute, membranous on margins; first glume 1-nerved, 4.0–6.0 mm long, second glume 3-nerved, longer than first; lemma rounded dorsally, 6.0–9.0 mm long, 5-nerved, smooth or minutely rugose, awnless or with short awn 1.0–4.0 mm; palea simi-lar to lemma. FRUITS caryopsis, elements separating individually.

Special Identifying Features

Prominent, ciliated auricles.

Toxic Properties

Festuca species are reported to be cyanogenic, but risk of adverse effects is considered low.

Seedling

Inflorescence

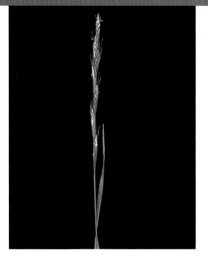

Inflorescence

Seedling

1mm

Seeds

Perennial Ryegrass

Lolium perenne L. · Poaceae · Grass Family

Synonyms

Common ryegrass, English ryegrass, English raygrass, ryegrass; *Lolium perenne* L. var. *perenne*

Habit, Habitat, and Origin

Erect (sometimes decumbent at base), cool-season, short-lived perennial; 0.3–1.2 m tall; cultivated areas, fields, pastures, lawns, turf, roadsides, railroad beds, and waste sites; native of Europe.

Seedling Characteristics

Leaves rolled or rolled in bud, leaves glabrous to scabrous, membranous ligule, conspicuous auricles.

Mature Plant Characteristics

ROOTS fibrous. STEMS slender hollow culms, erect or sometimes decumbent at base, basal portion commonly reddish, growing in clumps from short stolons. LEAVES linear, glabrous, sometimes upper surface scabrous, rolled to folded in bud, mature leaves flat to slightly involute, 4.0–30.0 cm long,

2.0–7.0 mm wide; sheaths rounded to slightly keeled, glabrous, open nearly to base, auricle usually present and conspicuous, 0.5–3.0 mm long; ligule membranous, short, rounded to truncate, 0.5–2.5 mm long. INFLORESCENCES compressed spike, 5.0–25.0 cm long; rachis margins rough; spikelets 2-ranked, 5.0–22.0 mm long with 2–10 florets; glumes linear-lanceolate, soft, 4.0–12.0 mm long, one-half to two-thirds as long as spikelet; lemma lanceolate, 5.0–7.0 mm long, awnless or short-awned; palea short-pointed, short trichomes on keel. FRUITS caryopsis, oblong or lanceolate, somewhat flattened, slightly narrowing at base.

Special Identifying Features

Cool-season perennial growing in clumps from short stolons; ligule membranous; conspicuous auricles; inflorescence a spike with 2-ranked spikelets.

Toxic Properties

See comments under *Lolium multiflorum*.

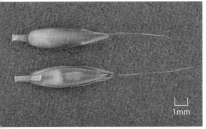

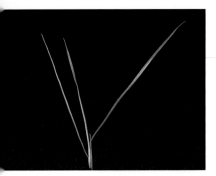

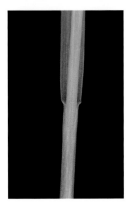

TOP Seeds
MIDDLE Seedling
BOTTOM Collar

Inflorescence

Italian Ryegrass

Lolium perenne L. ssp. *multiflorum* (Lam.) Husnot · Poaceae · Grass Family

Synonyms

Lolium multiflorum Lam.

Habit, Habitat, and Origin

Tufted winter annual, or persistent as biennial or occasionally perennial; to 1.3 m tall; cultivated areas, fields, pastures, lawns, turf, roadsides, railroad beds, and waste sites; native of Europe.

Seedling Characteristics

Leaf sheath and blade glabrous, ligule membranous, auricles usually present but may be absent on young seedlings.

Mature Plant Characteristics

ROOTS fibrous. STEMS erect, branching at base to 1.3 m tall, glabrous. LEAVES 6.0–36.0 cm long, 4.0–10.0 mm wide; sheaths open, culminating in prominent auricles, glabrous; ligule a membrane to 2.4 mm long, with frayed margin. INFLORESCENCES slender spike, 10.0–40.0 cm long, 5–38 spikelets; spikelets 4–17-flowered, 8.0–34.0 mm long excluding awns, sessile; first glume absent except in terminal spikelet, second glume 4.0–15.0 mm long; lemma 3.0–8.0 mm long, awnless or with awns to 11.0 mm long; palea equaling lemma, with scabrous or short trichomes on nerves. FRUITS caryopsis, lemma and palea attached.

Special Identifying Features

Auricles and lack of trichomes distinguish this plant from *Bromus* species.

Toxic Properties

Consumption of *Lolium* fruits may cause disorientation (called tares), staggers in grazing animals, and photosensitization, all thought to be caused by endophytic fungal toxins.

Wirestem Muhly

Muhlenbergia frondosa (Poir.) Fern. • Poaceae • Grass Family

Synonyms

None

Habit, Habitat, and Origin

Erect or decumbent perennial, prostrate or sprawling with age; to 1.0 m tall; moist soils, woodland edges, open woodlands, and disturbed areas; native of North America.

Seedling Characteristics

Leaf blades flat, scabrous, short; ligule erose, membranous.

Mature Plant Characteristics

ROOTS fibrous with creeping, scaly rhizomes. **STEMS** relatively stout, glabrous below nodes, freely branching, finally decumbent and rooting at lower nodes, to 1.0 m tall. **LEAVES** flat, scabrous, 10.0–15.0 cm long, 3.0–7.0 mm wide; sheaths glabrous, rounded, most shorter than culm internodes; ligule membranous, erose-ciliate, 0.8–1.4 mm long. **INFLORESCENCES** terminal or axillary, partly enclosed in sheath, dense panicle, up to 10.0 cm long with densely flowered branches; spikelets 3.0 mm long excluding awn; first glume acute, acuminate, or short-awned, 2.0–4.0 mm long, second glume as first; lemma 2.9–3.6 mm long, awnless or with awn 1.0–2.0 mm long, pubescent on callus; palea as lemma. **FRUITS** caryopsis; lemma and palea, but not glumes, attached.

Special Identifying Features

Plants perennial, rhizomatous; mature leaf blades flat.

Toxic Properties

None reported.

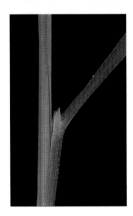

TOP **Seeds**
BOTTOM **Collar**

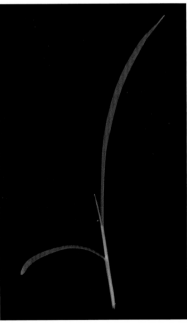

LEFT **Inflorescence**
RIGHT **Seedling**

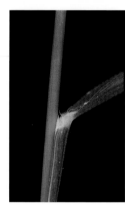

TOP **Seed**
BOTTOM **Collar**

Inflorescence

Seedling

Nimblewill

Muhlenbergia schreberi J. F. Gmelin • Poaceae • Grass Family

Synonyms

None

Habit, Habitat, and Origin

Perennial spreading herb; to 60.0 cm tall; moist soils, shaded habitats, lawns, gardens, open woodlands, and wetlands; native of North America.

Seedling Characteristics

Glabrous, linear blade, short, erose, membranous ligule.

Mature Plant Characteristics

ROOTS fibrous, with stolons. **STEMS** to 60.0 cm tall, geniculate, branching and rooting at nodes, glabrous. **LEAVES** 3.5–6.0 cm long, 1.0–3.0 mm wide, pubescent at mouth only; sheaths shorter than internode, glabrous; ligule minute, erose or lacerate, ciliolate, 0.5 mm long. **INFLORESCENCES** terminal and axillary contracted panicles, 5.0–15.0 cm long; spikelets with 1 fertile flower, 2.0 mm long, excluding awns, pediceled; first glume minute or absent, second glume 0.1–0.3 mm long; lemma 2.0 mm long, with 2.0–5.0 mm awn from tip; palea as lemma. **FRUITS** caryopsis, lemma and palea attached.

Special Identifying Features

Minute glumes, no rhizomes, narrow panicles, leaf blades resembling bermudagrass.

Toxic Properties

None reported.

Red Rice

Oryza sativa L. • Poaceae • Grass Family

Synonyms

None

Habit, Habitat, and Origin

Erect, tufted summer annual; to 80.0 cm tall; cultivated areas, fields, roadsides, wet ditches, and waste sites; native of Asia; conspecific with cultivated rice.

Seedling Characteristics

Leaf sheath and blade smooth, glabrous; ligule a triangular

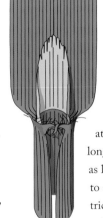

membrane, 1.5 cm long; trichomes at mouth, only on older seedlings.

Mature Plant Characteristics

ROOTS fibrous. **STEMS** erect, tufted, to 80.0 cm, glabrous at nodes. **LEAVES** 30.0–60.0 cm long, 7.0–20.0 mm wide, 30 times as long as broad, older leaves hispid to scabrous; sheaths glabrous, with trichomes at mouth along with well-developed auricles; ligule a long, triangular membrane, 10.0–15.0 mm. **INFLORESCENCES** loose, erect panicle with or without awns, to 20.0 cm long; spikelets 1-flowered, pediceled, 7.0–10.0 mm long, 2.0–2.5 mm wide; glumes rudimentary; lemma firm, minutely reticulate, keel-shaped and awnless or with stout awn, 3–5-nerved; palea equal to lemma but not awned, 2-nerved. **FRUITS** caryopsis, pericarp red on surface only or red throughout, straw-colored, or black; lemma and palea usually attached.

Special Identifying Features

Red pericarp; scabrous to hispid leaves and seeds.

Toxic Properties

Reported to be cyanogenic, but risk is considered low.

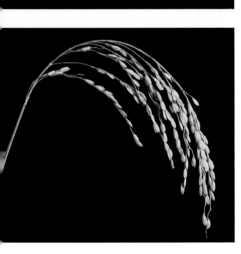

TOP **Seedling**
BOTTOM **Inflorescence**

TOP **Seed**
BOTTOM **Collar**

Witchgrass

Panicum capillare L. • Poaceae • Grass Family

Synonyms

Ticklegrass

Habit, Habitat, and Origin

Tufted, erect or decumbent, freely branching annual; to 0.8 m tall; cultivated areas, fields, pastures, lawns, turf, gardens, roadsides, and waste sites; native of North America.

Seedling Characteristics

Densely pubescent.

Mature Plant Characteristics

ROOTS fibrous. STEMS erect or decumbent at base, to 0.8 m. LEAVES flat and wide, folded in bud, pubescent on both surfaces, 10.0–25.0 cm long; sheaths papillose-hispid with spreading trichomes; ligule with short, stiff trichomes, more or less connate and membranous below, 0.7–2.0 mm long. INFLORESCENCES large, diffuse panicle, to 40.0 cm long, with spikelets widely spaced; spikelets glabrous, 2.0–3.5 mm long; first glume half as long as spikelet, 5–7-nerved, second glume equaling spikelet in length; lemma equaling spikelet; palea 1.5 mm long, smooth, shiny. FRUITS caryopsis, falling with lemma and palea attached.

Special Identifying Features

Plant pubescent; open, spreading panicle.

Toxic Properties

None reported.

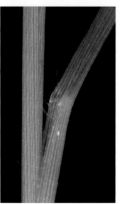

TOP Seeds
MIDDLE Seedling
BOTTOM Collar

Inflorescence

Fall Panicum

Panicum dichotomiflorum Michx. • Poaceae • Grass Family

Synonyms

Smooth witchgrass

Habit, Habitat, and Origin

Decumbent to erect summer annual with bent and branched nodes; to 2.2 m tall; cultivated areas, fields, pastures, lawns, turf, roadsides, railroad beds, and waste sites; native of eastern United States and West Indies.

Seedling Characteristics

Leaf sheaths and blades glabrous; ligule fringed, membranous.

Mature Plant Characteristics

ROOTS fibrous. STEMS erect, bent at

nodes and branched, to 2.2 mm tall. LEAVES 10.0–50.0 cm long, 8.0–20.0 cm wide, upper surface may be pubescent; sheaths compressed, glabrous; ligule a membrane fringed with trichomes, to 3.0 mm long. INFLORESCENCES panicle, 10.0–75.0 cm long, with numerous open, flexuous branches; spikelets 2.0–3.2 mm long, 0.7–1.0 mm wide, pediceled; first glume 0.4–0.8 long, second glume equaling spikelet; lemma similar to second glume; palea 1.8–2.2 mm long. FRUITS caryopsis, smooth and glossy, flattened, oblong, entire spikelet attached or glumeless.

Special Identifying Features

Thick, compressed sheath; bent stems; distinctive spikelet.

Toxic Properties

Toxicological problems may occur when used as forage, but the risk is considered low.

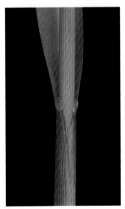

TOP Seeds
MIDDLE Seedling
BOTTOM Collar

Inflorescence

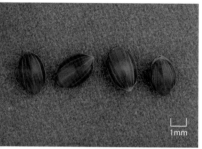

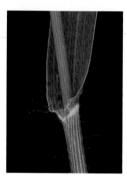

Seedling

Inflorescence

Wild-proso Millet

Panicum miliaceum L. • Poaceae • Grass Family

Synonyms

Broomcorn millet, hog millet, proso millet

Habit, Habitat, and Origin

Erect summer annual, branched at base; to 2.0 m tall; cultivated areas, fields, pastures, lawns, turf, roadsides, railroad beds, and waste sites; native of temperate regions of Asia.

Seedling Characteristics

Leaves and stems pubescent, leaf blades wide, ligule a fringe of dense trichomes fused at base.

Mature Plant Characteristics

ROOTS fibrous. **STEMS** erect, 0.3–2.0 m tall, branching at base, glabrous to hirsute, hollow. **LEAVES** flat, glabrous to pubescent, 5.0–40.0 cm long, 4.0–18.0 mm wide; sheaths open, rounded to slightly keeled, nearly glabrous to hispid (commonly with long, spreading trichomes); ligule 1.0–3.0 mm long with fringe of trichomes from a prominent basal membrane. **INFLORESCENCES** spreading panicle, 11.0–40.0 cm long, slightly nodding; spikelets with 2 florets, 4.0–5.4 mm long, ovoid, tapering to pointed tip; lower floret sterile; first glume 2.5–3.6 mm long, 7-nerved, second glume 3.8–4.8 mm long, 9–11-nerved; lower lemma similar to second glume in size but 9–15-nerved, upper lemma smooth, plump, 2.7–3.7 mm long; palea 0.9–1.9 mm long, margins inrolled. **FRUITS** caryopsis, smooth, shiny, olive brown to black, glumeless.

Special Identifying Features

Ligule a fringe of dense trichomes from basal membrane, olive brown to black seed that often adheres to roots of seedling; blooming summer until frost.

Toxic Properties

Toxicological problems reported when used in forage, but the risk is considered low.

Torpedograss

Panicum repens L. • Poaceae • Grass Family

Synonyms

None

Habit, Habitat, and Origin

Erect, warm-season perennial from creeping rhizomes; to 8.0 dm tall; wetland areas, fields, roadsides, lawns, turf, and waste sites; native of South America.

Seedling Characteristics

Leaves glabrous or sparsely pubescent, sheaths covered with soft trichomes, ligule a thin membrane fringed with trichomes, seed seldom produced.

Mature Plant Characteristics

ROOTS fibrous, from creeping rhizomes. STEMS erect or bent at base from extensively creeping rhizomes, 3.0–8.0 dm tall, base enclosed by bladeless sheaths. LEAVES linear, flat or folded, glabrous or with scattered soft trichomes, 5.0–20.0 cm long, 2.0–5.0 mm wide; sheaths covered with scattered soft trichomes, lower sheaths without blades, auricles absent; ligule a short membrane fringed with trichomes, 1.0 mm long. INFLORESCENCES open panicle with ascending branches, 7.0–12.0 cm tall; spikelets compressed, ovate, pale, 2.2–3.0 mm long, solitary, 2 florets; glumes truncate, first glume 0.5–1.0 mm long, obscurely nerved, second glume as long as spikelet, 5–7-nerved; lemma smooth and shiny, 2.0–3.0 mm long; palea hardened. FRUITS caryopsis, not produced by plants found in the United States.

Special Identifying Features

Warm-season rhizomatous perennial; glabrous or with sparsely scattered soft trichomes; base of stem enclosed by bladeless sheaths; inflorescence an open panicle, rarely producing viable seed in the United States.

Toxic Properties

None reported.

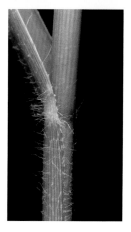

TOP **Rhizome**
MIDDLE **Shoot**
BOTTOM **Collar**
LEFT **Inflorescence**

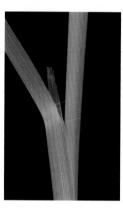

Dallisgrass

Paspalum dilatatum Poir. • Poaceae • Grass Family

Synonyms

Paspalum

Habit, Habitat, and Origin

Warm-season, subrhizomatous perennial; to 1.2 m; cultivated areas, fields, pastures, lawns, turf, roadsides, railroad beds, and waste sites; native of South America, introduced circa 1875.

Seedling Characteristics

Leaf sheaths pubescent near base of plant, otherwise glabrous; ligule membranous, acute to obtuse.

Mature Plant Characteristics

ROOTS fibrous with short rhizomes. STEMS 0.4–1.5 m tall, tufted, erect to slightly spreading, nodes and internodes glabrous. LEAVES to 36.0 cm long and 12.0 mm wide, glabrous except near base, which has long trichomes; sheaths pubescent near base of plant, glabrous toward top; ligules membranous, 2.5 mm long, acute to obtuse. INFLORESCENCES 3–7 erect branches (racemes) 4.0–10.0 cm long and not paired on stem; spikelets 2 florets, 2.8–4.0 mm long, 1.0–1.5 mm wide, ovate, tip pointed; first glume absent, second glume as long as spikelet and pubescent; sterile lemma similar to second glume, fertile lemma nerveless, rounded; palea same as fertile lemma. FRUITS caryopsis, 3.0–4.0 mm long, 2.0–2.5 mm wide, falling with glumes and sterile lemma attached.

Special Identifying Features

Ligule membranous; trichomes at base of leaf; 3–7 racemes, not paired, with long trichomes in axils.

Toxic Properties

Used as forage, *Paspalum* is associated with tremorgenic syndromes caused by fungal endophytes.

Seedling

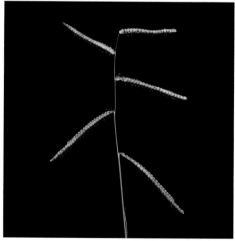

Inflorescence

Knotgrass

Paspalum distichum L. • Poaceae • Grass Family

Synonyms

Paspalum paspaloides (Michx.) Scribn.

Habit, Habitat, and Origin

Erect or ascending, mat-forming, warm-season perennial with slender stolons; 0.8–1.2 m tall; cultivated areas, fields, pastures, lawns, turf, roadsides, railroad beds, and waste sites; native of North America.

Seedling Characteristics

Leaves flat or folded, glabrous, short trichomes on margins near base, sheath glabrous, trichomes near tip, ligule membranous.

Mature Plant Characteristics

ROOTS fibrous from spreading stolons. **STEMS** erect or decumbent to ascending, solid, slightly compressed, forming tufts, 0.8–1.2 m tall. **LEAVES** flat or folded, mostly glabrous with sparse trichomes on margins, often long trichomes near collar, 2.0–22.0 cm long; sheaths slightly keeled, conspicuously swollen, glabrous or short trichomes on margins, long trichomes on surface; ligules membranous, 0.6–2.5 mm long. **INFLORESCENCES** panicle of paired or closely spaced, spikelike, 1-sided, branched racemes, ascending, 1.5–13.0 cm long; spikelets elliptic to ovate, tip acute, 2 florets, 2.5–4.0 mm long, pale or purple-tinged; glumes ovate, scabrous along midnerve; first glume absent or present in variable form, up to 2.5 mm long, second glume elliptic to narrowly ovate, acute tip, minute trichomes, 2.3–3.0 mm, awnless; sterile lemma essentially flat, barely exceeding fertile lemma, acute tip, glabrous or minute trichomes; fertile lemma elliptic, leathery, compressed, pale, glossy, with fine striations, 2.3–2.8 mm long; palea glabrous, leathery, thickened, hardened. **FRUITS** caryopsis, broadly oblong-elliptic.

Special Identifying Features

Erect or decumbent, mat-forming perennial from slender stolons; conspicuous pubescent nodes; ligule membranous; panicle of paired or closely spaced, spikelike, 1-sided branched racemes.

Toxic Properties

See comments under *Paspalum dilatatum.*

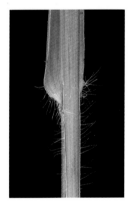

TOP Seeds
BOTTOM Collar

Inflorescence

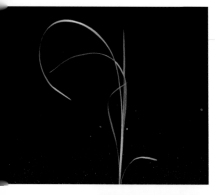

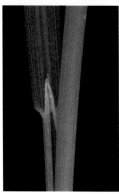

Longtom

Paspalum lividum Trin. • Poaceae • Grass Family

Synonyms

None

Habit, Habitat, and Origin

Perennial herb growing in tufts from decumbent or creeping base; 0.5–1.0 m tall; fields, pastures, lawns, turf, roadsides, railroad beds, and waste sites; native of North America.

Seedling Characteristics

Leaf and sheath glabrous, a few long trichomes at mouth; ligule a membrane with a ragged margin.

Mature Plant Characteristics

ROOTS fibrous, arising from nodes of creeping stolons. STEMS solitary or few in a tuft, glabrous, decumbent to erect, rooting at lower nodes, stem nodes darker than internodes. LEAVES linear, 15.0–25.0 cm long, 3.0–6.0 mm wide, glabrous with a few long trichomes at base; sheaths glabrous or hispid with papilla-based trichomes; ligule membranous with ragged (erose) margin, 1.0–4.7 mm long. INFLORESCENCES raceme with 4–7 flexuous branches, 1.5–4.0 cm long, rachis dark purple; spikelets obovate, 2.0–2.5 mm long, glabrous or softly pubescent on margins, subacute tip; first glume absent, second glume as long as spikelet, 3-nerved; first lemma equaling spikelet, 3-nerved, second lemma crustaceous, shiny; palea same as second lemma.

FRUITS caryopsis, obovate, pubescent, glume and lemma attached.

Special Identifying Features

Perennial from creeping stolons; leaf and sheath glabrous; ligule membranous with ragged margin; dark purple rachis on raceme inflorescence with 4–7 flexuous branches.

Toxic Properties

See comments under *Paspalum dilatatum.*

Inflorescence

Bahiagrass

Paspalum notatum Flüggé · Poaceae · Grass Family

Synonyms

None

Habit, Habitat, and Origin

Tufted or spreading perennial; to 8.0 dm tall; fields, pastures, lawns, turf, roadsides, railroad beds, and waste sites; native of South America.

Seedling Characteristics

Leaf blades glabrous, flat or folded; ligule membranous, pubescent.

Mature Plant Characteristics

ROOTS fibrous, with thick, coarsely scaly rhizomes. STEMS erect, to 5.0–8.0 dm tall, glabrous. LEAVES flat or folded, 2.5 dm long, glabrous or pilose on margins at base, shorter than flowering stem; sheaths glabrous; ligule membranous, pubescent from behind, 0.4 mm long. INFLORESCENCES 2 or rarely 3 racemes, 10.0–12.0 cm long, ascending, forming a V; spikelets 2.0–4.0 mm long, 2.0–2.8 mm wide, solitary, alternating on either side of rachis, 2 rows, overlapping, pointing outward; first glume absent, second glume equaling spikelet, 5-nerved, glabrous; lemma as second glume; palea coriaceous, shiny. FRUITS caryopsis, 3.0–4.0 mm long, with glume, lemma, and palea attached.

Special Identifying Features

Tufted or spreading perennial; terminal pair of racemes forming a V; rhizomes thick, scaly.

Toxic Properties

See comments under *Paspalum dilatatum*.

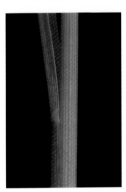

TOP Seeds
BOTTOM Collar

LEFT Inflorescence
RIGHT Seedling

1mm

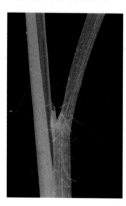

TOP Seeds
BOTTOM Collar

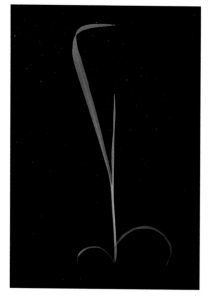

Seedling

Inflorescence

Vaseygrass

Paspalum urvillei Steud. • Poaceae • Grass Family

Synonyms
None

Habit, Habitat, and Origin
Erect, coarse, tufted perennial; to 2.0 m tall; fields, pastures, lawns, turf, roadsides, railroad beds, and waste sites; native of South America.

Seedling Characteristics
Leaf sheaths pubescent, ligule membranous and pointed.

Mature Plant Characteristics
ROOTS fibrous with very short rhizomes. STEMS 0.7–2.0 m tall, tufted, erect, nodes pubescent or smooth. LEAVES 2.0–57.0 cm long, 3.0–12.0 mm wide, smooth except for long, stiff trichomes at base; sheaths pubescent basally, smooth apically; ligule a long, pointed membrane, 2.0–7.7 mm long. INFLORESCENCES 4–30 erect branches (racemes), 2.0–12.0 cm long; spikelets with 2 florets, 2.0–2.7 mm long, 1.1–1.5 mm wide, tip pointed; first glume absent, second glume as long as spikelet, margins pubescent; sterile lemma as long as spikelet, pubescent, with pointed tip; fertile lemma 1.6–2.1 mm long, surface with bumps, tip rounded; palea as fertile lemma. FRUITS caryopsis, glumes and sterile lemma attached.

Special Identifying Features
Ligule membranous, long, pointed; spikelets pubescent; long, stiff trichomes at base of leaf.

Toxic Properties
See comments under *Paspalum dilatatum*.

Reed Canarygrass

Phalaris arundinacea L. • Poaceae • Grass Family

Synonyms

None

Habit, Habitat, and Origin

Erect perennial from creeping rhizomes; to 1.6 m tall;
wetlands, marshes, and roadside ditches; native of North
America.

Seedling Characteristics

Leaf blades bluish green, thin;
ligule membranous, sandpapery
or smooth.

Mature Plant Characteristics

ROOTS fibrous from creeping rhizomes. STEMS 0.4–1.6 m tall, erect,
smooth. LEAVES 3.0–41.0 cm long,
0.6–2.0 cm wide, flat, sandpa-

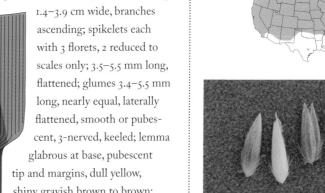

pery or smooth on both sides; sheaths
smooth; ligule membranous, 2.4–7.0 mm
long. INFLORESCENCES 5.0–18.0 cm long,
1.4–3.9 cm wide, branches
ascending; spikelets each
with 3 florets, 2 reduced to
scales only; 3.5–5.5 mm long,
flattened; glumes 3.4–5.5 mm
long, nearly equal, laterally
flattened, smooth or pubescent, 3-nerved, keeled; lemma
glabrous at base, pubescent
tip and margins, dull yellow,
shiny grayish brown to brown;
palea 3.0–4.1 mm long, 2-nerved,
occasionally pubescent. FRUITS
caryopsis enclosed in entire upper
floret.

Special Identifying Features

Creeping rhizome; flat leaves, membranous ligule; inflorescence elongate,
spikelike, with erect branches; spikelets
with 2 nearly equal keeled glumes.

Toxic Properties

None reported.

Inflorescence

TOP Seeds
MIDDLE Seedling
BOTTOM Collar

Common Reed

Phragmites australis (Cav.) Trin. ex Steud. • Poaceae • Grass Family

Synonyms

Cane grass, phragmites, reed grass; *Phragmites communis* Trin.

Habit, Habitat, and Origin

Erect perennial from rhizomes or stolons; to 4.3 m tall; streams, lake borders, marshes, and other wetlands; native of North America.

Seedling Characteristics

Leaves rolled in bud, glabrous, with rough edges; sheath glabrous, open; ligule membranous, fringed with trichomes.

Mature Plant Characteristics

Inflorescence

ROOTS fibrous from rhizomes or stolons. STEMS erect, sometimes bent at base, round, glabrous, hollow, 1.3–4.3 m tall. LEAVES linear, rolled inside bud, flat at maturity, gradually tapering to sharp tip, glabrous with rough margins, thick midvein below, 10.0–60.0 cm long; sheaths round, open, glabrous or sometimes pubescent on margins; ligule membranous, fringed with trichomes, 0.3–1.2 mm long. INFLORESCENCES dense panicle, plumose, branched, oblong to obovoid, 15.0–50.0 cm tall, subtended by silky trichomes, ascending and nodding at maturity; spikelets obovate to obtriangular, somewhat flattened, 10.0–16.0 mm long, with 3–7 florets, rachilla covered with long trichomes between florets; glumes glabrous, lanceolate, first glume 2.9–7.0 mm, second glume 5.0–10.0 mm with acute tip; lemma lanceolate, glabrous, 8.0–12.0 mm long, sharp tip; palea much shorter than lemmas. FRUITS caryopsis, oblong, round to slightly flattened, short beak, 2.0–3.0 mm long, seldom produced.

Special Identifying Features

Tall, stout, warm-season perennial in dense colonies; stems hollow; inflorescence a dense, branched, plumose panicle subtended by a tuft of silky trichomes.

Toxic Properties

Plants can be infected with the fungus that causes ergot.

TOP **Seeds**
MIDDLE **Seedling**
BOTTOM **Collar**

Annual Bluegrass

Poa annua L. • Poaceae • Grass Family

Synonyms

Poanna

Habit, Habitat, and Origin

Erect or bending annual, often
rooting at lower nodes, to 35.0
cm tall; cultivated areas, fields,
pastures, lawns, turf, roadsides,
railroad beds, and waste sites;
native of Europe.

Seedling Characteristics

Sheath and blade glabrous, ligule
membranous.

Mature Plant Characteristics

ROOTS fibrous. STEMS erect or bending, 1.0–35.0 cm tall, often rooting at
lower nodes. LEAVES 1.0–14.0 cm long,
1.0–5.0 mm wide, glabrous on both
surfaces, flat, soft; sheaths glabrous;
ligule membranous, to 1.8 mm
long. INFLORESCENCES panicle
2.0–7.0 cm long, 1.0–4.0 cm
wide, open, pyramidal; spikelets
3–6-flowered, 3.8–5.6 mm long;
glumes 1.5–2.9 mm long, first
glume shorter than second;
lemma 2.2–3.3 mm long, nerves
pubescent, glabrous at base;
palea 1.9–2.8 mm long, with 2 long-
pubescent keels. FRUITS caryopsis,
lemma and palea attached.

Special Identifying Features

Glabrous, loose sheath; bright
green, open, pyramidal panicle;
lemma pubescent on nerves.

Toxic Properties

Plants can be infected with the fungus
that causes ergot.

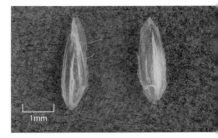

1mm

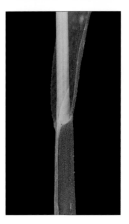

TOP **Seeds**
BOTTOM **Collar**

ABOVE **Seedling**
LEFT **Inflorescence**

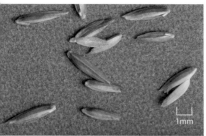

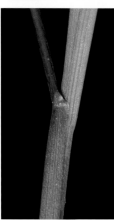

Kentucky Bluegrass

Poa pratensis L. • Poaceae • Grass Family

Synonyms

None

Habit, Habitat, and Origin

Erect perennial from creeping rhizomes, tufts or loose colonies; to 1.0 m tall; cultivated areas, fields, pastures, lawns, turf, roadsides, railroad beds, and waste sites; native of Europe.

Seedling Characteristics

Leaves flat to folded, smooth or slightly rough, boat-shaped tip; ligule a short, blunt membrane with uneven margin.

Mature Plant Characteristics

ROOTS fibrous from creeping rhizomes. **STEMS** erect to curving upward from base, 0.1–1.0 m tall, round to slightly flattened, glabrous. **LEAVES** linear, flat or sometimes folded at base, 1.0–25.0 cm long, glabrous or rough near base, margins rough, grooved above on both sides of midvein, boat-shaped tip; sheaths closed one-half of their length, round, glabrous or sometimes rough, auricle absent or inconspicuous; ligule membranous, 1.0–2.0 mm long, blunt, truncate to rounded, margin somewhat uneven. **INFLORESCENCES** open panicle, 3.0–15.0 cm long, spreading or ascending branches; spikelets compressed, 3–5-flowered, 3.0–6.0 mm long, keel pubescent; glumes acute, first glume 1-veined, 1.7–3.0 mm long, second glume 3-veined, 2.2–3.5 mm long; lemma elliptic, acute tip, 2.5–4.0 mm long, 5-veined, with tuft of long, cobwebby trichomes at base; palea elliptic, shorter than lem-

mas. **FRUITS** caryopsis, elliptic, slightly flattened on one side, reddish brown, shiny, 1.5–2.2 mm long.

Special Identifying Features

Erect, cool-season perennial from rhizomes; ligule membrane short; leaves with boat-shaped tip; panicle inflorescence.

Toxic Properties

Reported to be cyanogenic but not generally dangerous; plant can be infected with the fungus that causes ergot.

Inflorescence

Itchgrass

Rottboellia cochinchinensis (Lour.) W. D. Clayton · Poaceae · Grass Family

Synonyms

Rottboellia exaltata (L.) L. f.

Habit, Habitat, and Origin

Erect or ascending annual from prop-rooted base; to 3.0 m tall; cultivated areas, fields, pastures, roadsides, railroad beds, and waste sites; native of India; listed as a Federal Noxious Weed.

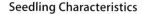

Seedling Characteristics

Sheath and blade pubescent with stiff, long trichomes; ligule membranous, sometimes with a minute fringe of trichomes.

Mature Plant Characteristics

ROOTS fibrous and prop. STEMS erect or ascending to 3.0 m tall from prop-rooted base. LEAVES blade 11.0–50.0 cm long, 4.0–19.0 (to 30.0) mm wide, usually stiffly pubescent on both surfaces; sheaths pubescent with long, stiff trichomes; ligule membranous, to 3.1 mm long, entire or with minute fringe of scattered trichomes. INFLORESCENCES terminal raceme, nearly cylindrical, stout, 6.0–14.0 cm long, 2.0–4.0 mm wide; spikelets 2-flowered, 4.2–5.4 (to 7.0) mm long, in pairs, one sessile and fertile, the other pediceled and neuter; glumes nearly equal, hardened, as long as spikelet, first glume 2-keeled upward, second glume keeled; lemma nearly as long as glumes, papery; palea slightly shorter than lemma, papery. FRUITS caryopsis; entire rachis joint falling with both pediceled and sessile spikelets attached; sessile spikelet eventually breaking between glumes and lemma.

Special Identifying Features

Sheath and blade with long, stiff trichomes; prop roots present; inflorescences pencil-like and jointed.

Toxic Properties

None reported.

Inflorescence

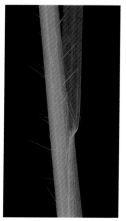

TOP Seeds
MIDDLE Seedling
BOTTOM Collar

TOP **Seeds**
BOTTOM **Collar**

Giant Foxtail

Setaria faberi Herrm. • Poaceae • Grass Family

Synonyms

Chinese millet

Habit, Habitat, and Origin

Erect, tufted summer annual; to 2.0 m tall; cultivated areas, fields, pastures, lawns, turf, roadsides, railroad beds, and waste sites; native of southeastern Asia.

Seedling Characteristics

Leaf sheath margins pubescent; upper side of leaves pubescent; ligule fringed, membranous.

Mature Plant Characteristics

ROOTS fibrous, tan. STEMS bent at nodes, rarely branching, glabrous, to 1.2 m tall. LEAVES linear, 10.0–30.0 cm long, 3.0–20.0 mm wide, pubescent on upper side only; sheaths glabrous except on margins; ligule with membrane fringed with trichomes, 1.5–2.0 mm long. INFLORESCENCES spikelike panicle, 5.0–20.0 cm long, 1.0–2.5 cm wide, cylindrical, nodding; spikelets 2.5–3.0 mm long, 1.2–1.5 mm wide, subsessile, 3–6 bristles below each spikelet, bristles up to 1.0 cm long; first glume to 1.0 mm long, second glume to 2.2 mm long; lemma equaling spikelet in length; palea equaling spikelet, strongly wrinkled. FRUITS caryopsis, shed with glumes attached.

Special Identifying Features

Conspicuous nodding panicle; long trichomes scattered on upper leaf surface.

Toxic Properties

The long awns can injure grazing animals.

Seedling

Inflorescence

Foxtail Millet

Setaria italica (L.) Beauv. • Poaceae • Grass Family

Inflorescence

Synonyms

Common millet, German or Hungarian millet, Hungarian grass, Italian millet

Habit, Habitat, and Origin

Erect, tufted, stout summer annual; to 1.5 m tall; cultivated areas, fields, pastures, lawns, turf, roadsides, railroad beds, and waste sites; possibly native of temperate regions of Asia.

Seedling Characteristics

Leaf glabrous to scabrous, pubescent margins; ligule a fringe of trichomes from short, membranous base.

Mature Plant Characteristics

ROOTS fibrous. **STEMS** erect, hollow, glabrous, bent and often pubescent at nodes, sometimes branched at base, 0.4–1.5 m tall. **LEAVES** to 30.0 cm long, 5.0–21.0 mm wide, glabrous or scabrous; sheaths closed, pubescent or ciliate margin near mouth, otherwise glabrous; ligule a fringe of trichomes from short, membranous base, to 2.5 mm long. **INFLORESCENCES** panicle, often nodding, lobed, cylindrical or tapering to tip, 2.0–20.0 cm long, 1.0–4.0 cm diameter, bristles scabrous and pointing upward; spikelets 2 florets, pedicillate, 1.8–2.5 mm long, 1.0–1.2 mm wide, elliptic, with 1–3 bristles, yellow, green, or purple, 5.0–10.0 mm long; first glume one-third as long as spikelet, second glume two-thirds to almost as long as spikelet; lemma equaling spikelet; palea shorter than lemma, finely transversely wrinkled. **FRUITS** caryopsis; smooth, shiny; tawny to red, brown, or black; enclosed within entire spikelet or with only glumes and bristles attached.

Special Identifying Features

Stout, hollow stem; ligule a fringe of trichomes from short, membranous base; 3 or fewer long bristles per spikelet.

Toxic Properties

See comments under *Setaria faberi*.

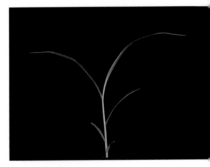

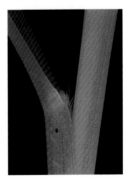

TOP Seeds
MIDDLE Seedling
BOTTOM Collar

Knotroot Foxtail

Setaria parviflora (Poir.) Kerguélen • Poaceae • Grass Family

Synonyms

Knotroot bristlegrass; *Setaria geniculata* auct. non (Lam.) Beauv.

Habit, Habitat, and Origin

Erect or spreading perennial from short, knotty rhizomes; to 1.2 m tall; cultivated areas, fields, pastures, lawns, turf, roadsides, railroad beds, and waste sites; native of the Americas.

Seedling Characteristics

Leaf blade scabrous above, sheath glabrous, ligule a fringed membrane.

Mature Plant Characteristics

ROOTS fibrous but with a short, knotty rhizome. **STEMS** geniculate to erect, sometimes branching and rooting at nodes, smooth, to 1.2 m tall. **LEAVES** blades 6.0–25.0 cm long, 1.9 mm wide, sometimes pubescent at mouth only; sheaths, at least lower ones, usually keeled, glabrous; ligule a fringed membrane 0.4–13.0 mm. **INFLORESCENCES** densely flowered panicle, 1.0–10.0 cm long, 0.5–2.9 cm wide, cylindrical, yellow or purple; spikelets in groups of 1 fertile and 1–2 sterile, 2.0–3.1 mm long, pediceled, with 4–12 bristles 1.0–15.0 mm long; first glume one-third as long as spikelet, 3-nerved, second glume up to twice as long as spikelet, 5-nerved; lemma equaling spikelet, 5–7-nerved; palea equaling spikelet, distinctly transversely rugose. **FRUITS** cary-

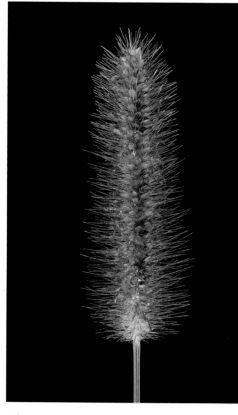

Inflorescence

opsis; enclosed within entire spikelet, or glumes may be absent.

Special Identifying Features

Seedheads yellow or purple; 4–12 bristles per spikelet; with a short, knotty rhizome; glabrous sheath margins.

Toxic Properties

See comments under *Setaria faberi*.

TOP **Seeds**
MIDDLE **Rhizome**
BOTTOM **Collar**

Yellow Foxtail

Setaria pumila (Poir.) Roem. & Schult. • Poaceae • Grass Family

Synonyms

Setaria glauca (L.) Beauv.

Habit, Habitat, and Origin

Erect, tufted summer annual, branching at base; to 1.3 m tall; cultivated areas, fields, pastures, lawns, turf, roadsides, railroad beds, and waste sites; native of tropical and warm regions of Eurasia.

Seedling Characteristics

Leaf blade and sheath glabrous, ligule membranous.

Mature Plant Characteristics

ROOTS fibrous. STEMS 20.0–130.0 cm tall, erect, branching at base, glabrous. LEAVES to 30.0 cm long, 4.0–10.0 mm wide, glabrous; sheaths closed,

glabrous except for trichomes at mouth and along margin; ligule a membrane fringed with trichomes, to 2.0 mm long. INFLORESCENCES panicle, 3.0–15.0 cm long, yellowish when mature, cylindrical, approximately 1.5–2.5 cm wide; spikelets 2 florets, 3.0–3.5 mm long, 1.6–2.2 mm wide, with 5 or more bristles; first glume up to twice as long as spikelet, second glume two-thirds as long as spikelet; lemma equaling spikelet; palea equaling spikelet, strongly transversely wrinkled. FRUITS caryopsis; enclosed within entire spikelet, or glumes may be absent.

Special Identifying Features

Yellowish inflorescence, 5 or more bristles per spikelet; no rhizome; base of leaf blade pubescent.

Toxic Properties

See comments under *Setaria faberi.*

Inflorescence

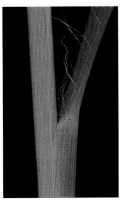

TOP **Seeds**
MIDDLE **Seedling**
BOTTOM **Collar**

TOP Seeds
BOTTOM Collar

Bristly Foxtail

Setaria verticillata (L.) Beauv. • Poaceae • Grass Family

Synonyms

Bur bristlegrass

Habit, Habitat, and Origin

Erect summer annual, branched at base; to 1.0 m tall; cultivated areas, fields, pastures, lawns, turf, roadsides, railroad beds, and waste sites; native of Europe and Africa.

Seedling Characteristics

Leaf blades smooth with rough margins, ligule a fringed membrane.

Mature Plant Characteristics

ROOTS fibrous. STEMS erect, glabrous, much branched at base, to 1.0 m tall. LEAVES erect, glabrous to scabrous, rough margins, tendency to spiral, 10.0–40.0 cm long; sheaths glabrous to scabrous, keeled, sometimes ciliate on upper margins; ligule a fringed membrane, 1.0–2.2 mm long. INFLORESCENCES erect, spike-shaped panicle, 2.0–11.0 cm long, 0.6–1.5 cm wide; spikelets 2 florets, 1.8–2.3 mm long, 1 or 2 bristles at base of each spikelet, barbs pointed downward; first glume 0.8–1.2 mm long, second glume 1.7–2.2 mm long; lemma minutely papillate, 1.8–2.2 mm long on lower floret, sterile; upper floret perfect, 1.7–2.1 mm long; palea about half as long as spikelet. FRUITS caryopsis, enclosed in floret with barbed bristle.

Special Identifying Features

No trichomes on leaf; erect, spike-shaped panicle; downward-barbed bristles that adhere to objects.

Toxic Properties

See comments under *Setaria faberi*.

ABOVE Seedling
RIGHT Inflorescences

LEFT **Inflorescence**
BELOW **Seedling**

Green Foxtail

Setaria viridis (L.) Beauv. • Poaceae • Grass Family

Synonyms

Green bristlegrass

Habit, Habitat, and Origin

Erect, tufted summer annual, branching at base; to 1.0 m tall; cultivated areas, fields, pastures, lawns, turf, roadsides, railroad beds, and waste sites; native of Eurasia.

Seedling Characteristics

Leaf sheaths glabrous with pubescent margin, blades glabrous, ligule membranous.

Mature Plant Characteristics

ROOTS fibrous. **STEMS** erect, bent at nodes, may be branched at base, to 1.0 m tall, glabrous. **LEAVES** to 30.0 cm long, 5.0–15.0 mm wide, glabrous or scabrous; sheaths closed, pubescent margin near mouth, otherwise glabrous; ligule a short membrane fringed with trichomes, to 2.0 mm long. **INFLORESCENCES** panicle, cylindrical or tapering to tip, 2.0–15.0 cm long, 1.0–1.5 cm diameter; spikelets 2 florets, pediceled, 1.8–2.5 mm long, 1.0–1.2 mm wide, with 1–3 bristles, green or purple, 5.0–10.0 mm long; first glume one-third as long as spikelet, second glume two-thirds to almost as long as spikelet; lemma equaling spikelet length; palea shorter than lemma, finely transversely wrinkled. **FRUITS** caryopsis, usually enclosed in whole spikelet or without glumes and bristles.

Special Identifying Features

Inflorescence with 3 or fewer bristles per spikelet; no trichomes on leaf.

Toxic Properties

See comments under *Setaria faberi*.

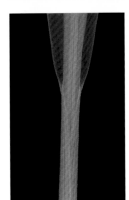

TOP **Seed**
BOTTOM **Collar**

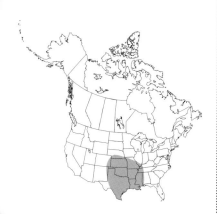

Culm

Inflorescence

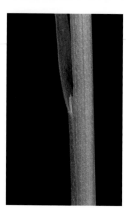

TOP **Seeds**
BOTTOM **Collar**

Sorghum-almum

Sorghum almum Parodi • Poaceae • Grass Family

Synonyms

Columbus grass

Habit, Habitat, and Origin

Weak perennial from thick underground rhizomes; to 4.5 m tall; cultivated areas, fields, pastures, roadsides, railroad beds, and waste sites; native of South America, introduced from Argentina.

Seedling Characteristics

Leaf sheaths and leaves glabrous; ligule fringed, membranous.

Mature Plant Characteristics

ROOTS fibrous from rhizomes with short, curled tips. **STEMS** stout, erect, to 4.5 m tall. **LEAVES** glabrous, 20.0–60.0 cm long, wider than Johnsongrass; sheaths open, glabrous except for a few trichomes; ligule a prominent membrane fringed with trichomes, 2.0–5.0 mm long.

INFLORESCENCES open panicle, 15.0–40.0 cm long, composed of numerous whorled branches; spikelets 2-flowered, one fertile and sessile, 4.0–5.5 mm long, the other pediceled and male or neuter, 4.0–6.5 mm long, dropping with pedicel attached; glumes equal to floret, pubescent; lemma thin; palea thin, with awn 5.0–13.0 mm long or absent. **FRUITS** caryopsis, usually hulled, dark reddish brown.

Special Identifying Features

Similar to Johnsongrass but differentiated by reduced perennating habit, male and neuter spikelets dropping with pedicel attached, and stout branched stems; also shatters more readily than Johnsongrass.

Toxic Properties

Foliage of *Sorghum* species is reported to be cyanogenic.

Shattercane

Sorghum bicolor (L.) Moench ssp. *arundinaceum* (Desv.) de Wet & Harlan • Poaceae • Grass Family

Synonyms

None

Habit, Habitat, and Origin

Erect, single to tufted, slender to robust annual, with prop roots; to more than 5.0 m tall; cultivated areas, fields, roadsides, and waste sites; native of Africa.

Seedling Characteristics

Leaf sheath pubescent or glabrous, blade often pubescent at base on both surfaces; ligule a fringed membrane.

Mature Plant Characteristics

ROOTS fibrous with adventitious prop roots. **STEMS** erect, 3.0–5.0 m or more tall, slender or robust. **LEAVES** flat, 13.0–75.0 cm long, 0.5–7.0 cm wide, glabrous or pubescent; sheaths glabrous to pubescent; ligule membranous, to 2.6 mm long, fringed with trichomes to 1.3 mm long. **INFLORESCENCES** open panicle, 4.0–6.0 dm long, 1.5–4.0 dm wide, branches slightly ascending to somewhat pendulous; spikelets 2-flowered, in pairs: one sessile and fertile, 4.9–7.7 mm long and 1.8–2.7.0 mm wide, the other pediceled and sterile or staminate; glumes thin or firm, about as long as spikelet, pubescent; lemma similar to sessile spikelet; lower lemma 4.5–6.4 mm long, usually with awn 4.0–7.0 mm long; upper lemma slightly shorter, awnless; palea thin, as long as lemma. **FRUITS** caryopsis, shed with sessile and pediceled spikelets attached.

Special Identifying Features

Large annual, height exceeding 5.0 m; prop roots; ligule membranous, fringed with trichomes; inflorescence large, open; spikelets pubescent, awned, falling easily and early.

Toxic Properties

See comments under *Sorghum almum*.

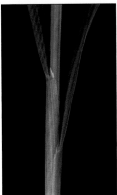

TOP Seed
MIDDLE Seedling
BOTTOM Collar
LEFT Inflorescences

Johnsongrass

Sorghum halepense (L.) Pers. • Poaceae • Grass Family

Synonyms

None

Habit, Habitat, and Origin

Coarse perennial from thick, scaly underground rhizomes; to 3.5 m tall; cultivated areas, fields, pastures, roadsides, railroad beds, open woodlands, and waste sites; native of southern Eurasia east to India.

Seedling Characteristics

Leaf sheaths and leaves glabrous; ligule fringed, membranous.

Mature Plant Characteristics

ROOTS fibrous, from thick rhizomes. STEMS erect, to 3.5 m tall. LEAVES 20.0–60.0 cm long, 10.0–30.0 mm wide, glabrous; sheaths open, glabrous except for a few trichomes at mouth on some plants; ligule 2.0–5.0 mm long, prominent membrane with fringe of trichomes. INFLORESCENCES open panicle, 15.0–50.0 cm long, with numerous whorled branches; spikelets 2-flowered, one fertile, 4.0–5.5 long, sessile, the other pediceled and male or neuter, 4.0–6.5 mm long; glumes equal to floret, pubescent; lemma thin; palea thin, with awn 5.0–13.0 mm long or absent. FRUITS caryopsis, dark reddish brown, usually shed hulled.

Special Identifying Features

Thick, scalelike rhizomes.

Toxic Properties

See comments under *Sorghum almum*.

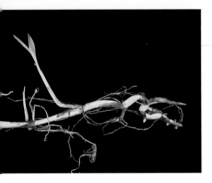

TOP **Seeds**
BOTTOM **Rhizome**

Seedling

Inflorescence

Smutted heads

Inflorescence

1mm

TOP Seeds
BOTTOM Seedling

Smutgrass

Sporobolus indicus (L.) R. Br. • Poaceae • Grass Family

Synonyms

Rattail smutgrass; *Sporobolus poiretii* (Roem. & Schult.) A. S. Hitchc.

Habit, Habitat, and Origin

Erect, tufted perennial; to 1.1 tall; fields, pastures, roadsides, railroad beds, and waste sites; native of Eurasia.

Seedling Characteristics

Leaf blades folded, smooth; leaf sheath smooth; ligule membranous, tiny.

Mature Plant Characteristics

ROOTS fibrous. **STEMS** 0.3–1.1 m tall, erect. **LEAVES** 15.0–48.0 cm long, 1.0–5.0 mm wide, usually folded but can be flat or rolled, smooth; sheaths smooth; ligule a tiny membranous scale, to 0.1

mm long. **INFLORESCENCES** 9.0–41.0 cm long, spikelike, with appressed branches or sometimes with ascending branches; spikelets 1-flowered, 1.6–2.1 mm long, solitary; first glume 0.4–1.0 mm long, second glume 0.6–1.3 mm long; lemma ovate, membranous, glabrous, acute or obtuse, 1.8–2.7 mm long; palea 1.4–1.9 mm long, 2-nerved. **FRUITS** caryopsis, shed separately.

Special Identifying Features

Dark green, tufted, erect plant; leaves smooth, usually folded; seedhead usually elongate and spikelike; often infected with black smut fungus.

Toxic Properties

None reported.

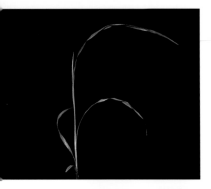

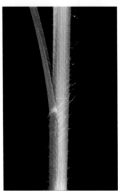

Inflorescence

Eastern Gammagrass

Tripsacum dactyloides (L.) L. • Poaceae • Grass Family

Synonyms

Bullgrass

Habit, Habitat, and Origin

Perennial from thick, knotty rhizomes, growing in large colonies or clumps; to 3.0 m tall; fields, pastures, roadsides, railroad beds, wet ditches, and waste sites; native of North America.

Seedling Characteristics

Leaf glabrous with scabrous margins, sheath glabrous, ligule a ciliate ring of trichomes from inconspicuous membrane.

Mature Plant Characteristics

ROOTS fibrous from short, thick, knotty rhizomes. STEMS erect, 1.5–3.0 m tall, decumbent below, solid, slightly flattened, glabrous. LEAVES flat, 30.0–75.0 cm long, 1.0–3.0 cm wide, glabrous with scabrous margins; sheaths glabrous, oval to slightly flattened; ligule a ciliate ring of trichomes from an incon- spicuous or lacerate membrane, trun- cate. INFLORESCENCES single, spicate, racemose, 12.0–25.0 cm long, or 2–3 erect spikelike racemose branches; spikelets unisexual: staminate spikelet above, 2-flowered, in pairs on one side of continuous rachis, 6.0–10.0 mm long; pistillate spikelet below, subsessile, solitary, hard, bony, awnless, 6.0–8.0 mm long; glumes 1- or 2-flowered; stami- nate spikelet glumes equal, firm, keeled, pistillate; spikelet glumes equal, indurate, shiny, fused with rachis; lemma delicate, thin, hyaline, membranous, reduced; palea as lemma. FRUITS caryopsis, broadly conical, red, 4.0 mm long.

Special Identifying Features

Perennial from thick rhizomes; growing in clumps; leaf and sheath glabrous; ligule a short ciliate ring of trichomes; monoecious with staminate and pistillate spikelets in same inflores- cence; flowering April–November.

Toxic Properties

None reported.

Browntop Signalgrass

Urochloa fusca (Sw.) B. F. Hansen & Wunderlin • Poaceae • Grass Family

Synonyms

Browntop brachiaria; browntop panic grass, browntop panicum; *Urochloa fasciculata* (Sw.) R. Webster

Habit, Habitat, and Origin

Erect or much-branched, tufted annual, spreading from a reclining base; to 1.2 m tall; cultivated areas, fields, pastures, roadsides, open woodlands, and waste sites; native of Central and South America.

Seedling Characteristics

Leaf blades glabrous, ligule a ring of trichomes.

Mature Plant Characteristics

ROOTS densely fibrous root system. STEMS pubescent and often curving upward from a reclining base, with profuse rooting at nodes. LEAVES flat and glabrous, 4.0–20.0 cm long, 6.0–20.0 mm wide, narrowing slightly at base and tapering to tip; sheaths glabrous to papillose-long; ligule a ring of trichomes less than 1.0 mm long. INFLORESCENCES panicle, 5.0–15.0 cm long, with closely pressed or spreading, mostly simple branches, 1.0–8.0 cm long; spikelets rounded, obovate, 2.0–3.0 mm long, yellow to brown, glabrous; first glume thin, one-fourth to one-third as long as spikelet, second glume and lower lemma with fine cross-grains nearing middle; lemma of upper floret rough, nearly as long as spikelet, blunt-tipped; palea as lower lemma with fine cross-grains. FRUITS caryopsis, tan to light brown, late May to September.

Special Identifying Features

Glabrous, gently tapering leaves from ascending stem arising from a reclining base; rooted at nodes.

Toxic Properties

None reported.

Inflorescence

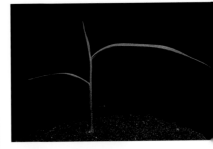

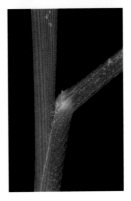

TOP Seeds
MIDDLE Seedling
BOTTOM Collar

TOP **Seeds**
BOTTOM **Collar**

Seedling

Inflorescence

Guineagrass

Urochloa maxima (Jacq.) R. Webster · Poaceae · Grass Family

Synonyms

Green panicgrass; *Panicum maximum* Jacq.

Habit, Habitat, and Origin

Extremely variable, annual or rarely perennial from short stolons; 2.0–2.5 m tall; cultivated areas, fields, pastures, roadsides, railroad beds, and waste sites; native of Africa.

Seedling Characteristics

Glabrous or pubescent; sheaths glabrous with ciliate margin, blades glabrous to scabrous; ligule a ciliate fringe of trichomes.

Mature Plant Characteristics

ROOTS fibrous from short stolons. STEMS erect or tufted, from fibrous roots or short stolons, nodes often with dense, coarse hair, 2.0–2.5 m tall. LEAVES surface glabrous or pubescent, margins scabrous, 30.0–75.0 cm long, to 3.5 cm wide, sometimes with short, stiff trichomes on upper surface near base; sheaths covered with short, stiff trichomes, papillose or glabrous, pubescent at collar; ligule membranous with a ciliate fringe, 4.0–6.0 mm long. INFLORESCENCES panicle with ascending, whorled, stiff branches, 20.0–50.0 cm long; spikelets clustered, short pedicels, ellipsoid, 3.0–3.9 mm long; first glume one-third as long as spikelet, second glume and lemma equal to spikelet; palea transversely rugose, 2.3–2.5 mm long. FRUITS caryopsis, transversely rugose.

Special Identifying Features

Lowest set of panicle branches whorled, fruits with transverse wrinkles.

Toxic Properties

Reported to be cyanogenic, but risk of poisoning is considered low.

Broadleaf Signalgrass

Urochloa platyphylla (Nash) R. D. Webster • Poaceae • Grass Family

Synonyms

Bracharia platyphylla (Griseb.) Nash

Habit, Habitat, and Origin

Decumbent, spreading, branched summer annual, bent and rooting at nodes; to 90.0 cm; cultivated areas, fields, pastures, lawns, turf, roadsides, railroad beds, moist ditches, and waste sites; native of southeastern and south-central United States.

Seedling Characteristics

Leaf sheaths pubescent; blades glabrous; ligule fringed, membranous; leaf margins with distinct trichomes.

Mature Plant Characteristics

ROOTS fibrous. STEMS decumbent, bent at nodes, often branched and rooting at lower nodes, to 90.0 cm tall. LEAVES 4.0–15.0 cm long, 6.0–15.0 mm wide, glabrous; sheaths with pubescent margin, lower leaves often pubescent throughout; ligule a narrow membrane fringed with trichomes, to 0.8 mm long. INFLORESCENCES raceme, 30.0 cm long, 10.0 cm wide, with 2–6 branches, each 3.0–9.0 cm long; spikelets 1-flowered, ovate, flattened toward summit, 3.5–5.4 mm long, 1.8–2.3 mm wide; first glume about one-third as long as spikelet, second glume equal to spikelet; lemma similar to second glume; palea 2.5–3.2 mm long, firm, finely wrinkled, with inrolled margins. FRUITS caryopsis; entire spikelet shed, including glumes.

Special Identifying Features

Crease near tip of leaves; decumbent habit; distinctive spikelets.

Toxic Properties

None reported.

1mm

TOP Seeds
MIDDLE Seedling
BOTTOM Collar

Inflorescence

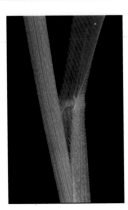

TOP Seeds
BOTTOM Collar

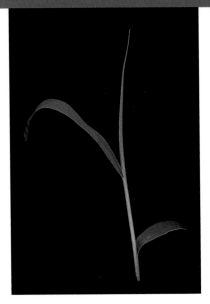

Seedling

Inflorescence

Browntop Millet

Urochloa ramosa (L.) Nguyen · Poaceae · Grass Family

Synonyms

Browntop panicum; *Brachiaria ramosa* (L.) Stapf., *Panicum ramosum* L.

Habit, Habitat, and Origin

Erect to decumbent, loosely tufted annual; to 0.7 m tall; cultivated areas, fields, pastures, lawns, turf, roadsides, railroad beds, and waste sites; native of southeastern Asia.

Seedling Characteristics

Leaf blades broad, flat-glabrous; leaf sheath glabrous to pubescent; ligule ciliate.

Mature Plant Characteristics

ROOTS fibrous. STEMS erect to decumbent, to 0.7 m tall, nodes pubescent. LEAVES 2.0–33.0 cm long, 2.0–20.0 mm wide, flat-glabrous; sheaths glabrous to pubescent; ligule a membrane fringed with trichomes to 1.5 mm long. INFLORESCENCES erect, 3.0–10.0 cm long, with straight and spreading branches; spikelets 2-flowered, 2.5–3.5 mm long, glabrous or pubescent; first glume 1.2–2.0 mm long, second glume as long as spikelet; sterile lemma as long as spikelet, fertile lemma slightly shorter than spikelet, surface roughened; palea slightly shorter than fertile lemma, surface roughened. FRUITS caryopsis, entire spikelet shed.

Special Identifying Features

First glume is on branch side of spikelet; broad, flat-glabrous leaf blades; inflorescence branches stick out like signal flags, spikelets less than 3.5 mm long.

Toxic Properties

None reported.

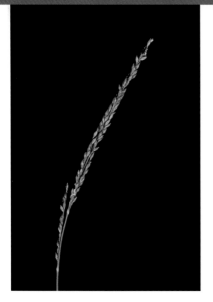

Inflorescence

Seedling

Texas Millet

Urochloa texana (Buckl.) R. Webster · Poaceae · Grass Family

Synonyms

Buffalograss, Coloradograss, Texas panicum; *Panicum texanum* Buckl.

Habit, Habitat, and Origin

Decumbent or creeping, coarse, tufted summer annual; to 80.0 cm tall; cultivated areas, fields, pastures, roadsides, railroad beds, and waste sites; native of southern United States, perhaps Texas.

Seedling Characteristics

Pubescent leaf sheath, blades with soft pubescence on both surfaces, ligule membranous.

Mature Plant Characteristics

ROOTS fibrous. **STEMS** 40.0–80.0 cm tall, tufted, erect from base or creeping, decumbent, and rooting at lower nodes; to 1.5 m long, nodes with soft pubescence. **LEAVES** 8.0–27.0 cm long, 7.0–20.0 mm wide, soft pubescence on both surfaces, with short fine trichomes; sheaths pubescent with short or long trichomes to nearly glabrous; ligule short, membrane fringed with dense trichomes to 1.0–1.8 mm long. **INFLORESCENCES** simple panicle, 7.0–25.0 cm long, with short, erect, appressed or slightly spreading branches; spikelets with single fertile floret, 5.0–6.0 mm long; first glume two-thirds as long as spikelet, second glume and lemma as long as spikelet; palea of fertile floret transversely wrinkled, 3.5 mm. **FRUITS** caryopsis; entire spikelet shed, or without glumes.

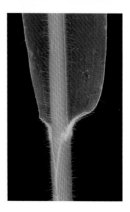

TOP **Seeds**
BOTTOM **Collar**

Special Identifying Features

Soft, velvety, pubescent leaves.

Toxic Properties

None reported.

Inflorescences

Waterhyacinth

Eichhornia crassipes (Mart.) Solms • Pontederiaceae • Pickerel-weed Family

Synonyms

Aquape, falkumbhi

Habit, Habitat, and Origin

Free-floating perennial; to 5.0 dm tall; in rafts or colonies on ponds, lakes, streams, and rivers; native of the Americas.

Seedling Characteristics

Young leaves oval, petioled with basal swelling.

Mature Plant Characteristics

ROOTS long, pendant roots of free-floating plant producing numerous stolons, sometimes rooted in mud, reproducing vegetatively. STEMS stout, erect, often connected by stolons. LEAVES blades orbicular to reniform, 4.0–12.0 cm wide, petioles of basal leaves usually inflated. INFLORESCENCES produced from summer until frost, spicate, 2–8-flowered, blue, zygomorphic, arising from spathe of a contracted panicle, 4.0–15.0 cm wide; perianth 5.0–7.0 cm wide, lilac or rarely white, upper lobe bearing a violet blotch with yellow center; stamens 6, inserted on perianth tube, the 3 stamens opposite lower lip included; ovary trilocular, ripening into 3-valved; nectaries septal, 6 or 7. FRUITS ellipsoid capsule with many seeds, self-pollinated but usually reproduces vegetatively. SEEDS ovate, 0.5–0.9 mm long, 0.3–0.5 mm wide, 8–14-winged, longitudinally ribbed.

Special Identifying Features

Free-floating plants, often covering water surface; basal leaves with inflated petioles; showy blue-lilac flowers with violet blotch and yellow center.

Toxic Properties

None reported.

Ducksalad

Heteranthera limosa (Sw.) Willd. • Pontederiaceae • Pickerel-weed Family

Synonyms

Longleaf mudplantain

Habit, Habitat, and Origin

Aquatic, freshwater annual or perennial, tufted but spreading from rhizomes rooted in mud; to 15.0 cm tall; wet cultivated areas, rice fields, ditches, and shallow open water; native of the Americas.

Seedling Characteristics

Triad of 3 fleshy, entire or acuminate leaves.

Mature Plant Characteristics

ROOTS fibrous, with slender rhizome. STEMS erect, fleshy, green, to 15.0 cm tall, tending to root at nodes. LEAVES blades to 10.0 cm long including petiole 3.0–6.0 cm long, 0.4–3.3 cm wide, narrowed to acute, obtuse, or slightly heart-shaped base. INFLORESCENCES each inflorescence with 1 flower, white or blue, tube 15.0–44.0 mm long, spathes 0.9–4.5 cm long, smooth. FRUITS cylindrical capsule 10.0–14.0 mm long. SEEDS ovoid, 0.5–0.8 mm long, 0.2–0.6 mm wide, black to gray, reticulate, 9–14-ribbed.

1mm

Special Identifying Features

Aquatic plant, may be submerged; leaves longer than wide; blue or white solitary flowers.

Toxic Properties

None reported.

TOP **Seeds**
BOTTOM **Infestation**

TOP **Flower**
BOTTOM **Seedling**

Roundleaf Mudplantain

Heteranthera reniformis Ruiz & Pavon • Pontederiaceae • Pickerel-weed Family

Synonyms

Kidneyleaf mudplantain

Habit, Habitat, and Origin

Aquatic, creeping, floating, submerged, or rooted annual or perennial herb; to 0.8 m long; wet cultivated areas, rice fields, ditches, and shallow water in open areas; native of North America.

Seedling Characteristics

Emersed or submerged, leaves initially linear, becoming rounded.

Mature Plant Characteristics

ROOTS fibrous with slender rhizome. **STEMS** submerged, creeping, branching and rooting at nodes, to 0.8 m tall. **LEAVES** in rosettes, blades cordate-reniform, twice as wide as long, petioles 1.5–3.0 cm long. **INFLORESCENCES** 3–10-flowered raceme; perianth white, pale blue, or mauve blue, yellow spot within central dark blotch, inserted in short-peduncled spathe, tube 1.0–3.0 cm long, stamens 3. **FRUITS** capsule about 13.0 mm long, filled with numerous seeds. **SEEDS** ovoid, 1.1–2.1 mm long, 0.6–0.9 mm wide, 11–14-ribbed, with reticulate pattern.

Special Identifying Features

Round or reniform leaves; 2–10 white or pale blue flowers on raceme.

Toxic Properties

None reported.

TOP **Seeds**
BOTTOM **Seedling**

Flowers

Inflorescence

Pickerelweed

Pontederia cordata L. · Pontederiaceae · Pickerel-weed Family

Synonyms

Pontederia lancifolia (Muhl.) Torr.

Habit, Habitat, and Origin

Erect to ascending, stout, aquatic, herbaceous perennial; to 12.0 dm tall; shallow water at edges of ponds, lakes, and slow-moving streams; native of North America.

Seedling Characteristics

Young leaves ovate to elliptical; petiole two to three times longer than blade.

Mature Plant Characteristics

ROOTS thick, enlarged, fibrous, from nodes along short, thick, submerged rhizome. STEMS erect, 3.0–12.0 dm tall, stout, short, somewhat succulent, and at least partly submerged. LEAVES basal, 7.0–20.0 cm long, ovate to lanceolate, cordate-sagittate, narrowed at base, variable, emergent, arising from sheaths; petiole 5.0–30.0 cm long. INFLORESCENCES asymmetrical, above water, in dense elongated spike 5.0–15.0 cm long, with 6 blue to violet petaloid tepals, funnelform below to form a nonfused corolla tube; upper corolla lobe with reniform yellow spot; covered by glandular trichomes, subtended by pair of bracts. FRUITS ellipsoid utricle, 1-seeded, beaked, 4.0–8.0 mm long, ridged, toothed crest, covering base of perianth. SEEDS reddish brown, glutinous, ovoid, 3.0–4.0 mm long, 2.0–2.5 mm wide.

Special Identifying Features

Aquatic perennial; leaves glandular-pubescent; flowers emergent, blue to lavender; uppermost corolla lobe with bean-shaped yellow patch.

Toxic Properties

None reported.

TOP Seed
MIDDLE Seedling
BOTTOM Young plant

Inflorescences

Mature plants

TOP **Seed dispersal**
BOTTOM **Young plants**

Common Cattail

Typha latifolia L. • Typhaceae • Cattail Family

Synonyms

Brown-headed cattail, bullrush, great reedmace, marsh beetle, nailrod

Habit, Habitat, and Origin

Tall perennial with large, creeping rhizomes, usually an emergent aquatic, often densely colonial; to 3.0 m tall; native of North America.

Seedling Characteristics

Inconspicuous, grasslike.

Mature Plant Characteristics

ROOTS fibrous with large, creeping rhizomes. **STEMS** 1.0–3.0 m tall, cylindrical, pithy. **LEAVES** linear, thick, flattened, 5.0–24.0 mm wide, taller than inflorescence, firm but spongy; sheath open at throat, sheathing base of stem. **INFLORESCENCES** staminate flowers above pistillate flowers on spikes, contiguous or less than 2.5 cm apart; staminate spike 4.0–16.0 cm long, with reduced stamens; pistillate spike dark brown, 5.0–20.0 cm long, 1.2–3.5 cm diameter, thicker toward base, stipitate, unilocular superior ovaries, with stipes bearing long, slender bristles. **FRUITS** 1-seeded, 1.0 cm long, with many white trichomes arising at base. **SEEDS** achene, 1.0 mm long, thin-walled, splitting in water.

Special Identifying Features

Dense, brown to gray, spadix-like inflorescence; linear, thick leaves taller than seedhead; staminate and pistillate parts of spike normally contiguous at top of long, unbranched stem.

Toxic Properties

May disturb digestion in livestock; edible by humans.

GLOSSARY

Acaulescent Without an obvious aboveground stem, usually because the stem is belowground or emerges only slightly above the soil surface; opposite of *caulescent*.

Achene A simple, one-seeded fruit whose seed and pericarp are attached at only one point. The seed is the product of a one-loculed ovary. This common type of fruit is found in several families including Asteraceae and Polygonaceae.

Actinomorphic Referring to a flower with more than two planes of symmetry. A line drawn from the tip of any petal through the flower's center will result in mirror images; also called *regular* or *radial*.

Acuminate Having a sharp point with tapering, concave sides.

Acute Having a sharp point with tapering, straight, or slightly convex sides.

Adnate Fused; usually referring to unlike organs.

Adventitious roots Roots that arise from stems or leaves instead of from the interior of other roots. Both grasses and broadleaf plants may have adventitious roots.

Aerenchyma Tissue containing air spaces between the cells facilitating gas exchange and maintaining buoyancy, especially in the roots and stems of aquatic plants.

Aggregate fruit Fruits that have several mature ovaries attached to the receptacle of a single flower. The fruit of blackberry, for example, is actually an aggregate of small drupes. This fruit type is common in the rose family (Rosaceae).

Alternate leaves Growth pattern in which only one leaf is attached to each node. The leaves may spiral up the stem or may occur on opposite sides of the stem as they ascend. This is the most common leaf arrangement.

Angular Having sharp edges or corners. Common angular seed shapes include cylindrical, rectangular, and wedge (resembling an orange slice, as in the morningglory family, Convolvulaceae).

Annual A plant that germinates, grows, flowers, produces seed, and dies within 12 months or less.

Anther The portion of the stamen that contains pollen. The anther usually comprises one to four pollen sacs joined by a connection that is an extension of the filament.

Anthesis Period during which flowers are fully expanded and pollination occurs.

Antrorse Pointing toward the apex of the organ to which it is attached.

Apex (pl. apices) The tip of a structure farthest from the point of attachment.

Apical Located at the apex, or tip.

Appressed Pressed flat or very close against a surface.

Aquatic Adapted to living in water, as opposed to terrestrial.

Armed Bearing sharp barbs, prickles, spines, or thorns.

Aromatic Having a fragrant, sweet-smelling, or pungent odor.

Ascending Growth pattern in which the stem and branches gradually slope or curve upward.

Auricle An earlobe-shaped projection; usually refers to structures growing from the base of the sheath at the collar region of some grasses (Poaceae).

Awn A stiff, narrow, bristlelike structure, such as those found on the glumes and lemmas of grasses (Poaceae).

Axillary Arising from a leaf axil. Buds borne in a leaf axil, for example, are called axillary or lateral buds.

Barb A structure with a tip that bends backward like a fishhook.

Barbed Referring to bristles, awns, or prickles with terminal, lateral rigid hooks that are bent sharply backward.

Bark The outer layers of a woody trunk or stem. Bark can be flaky, furrowed, plated, scaly, or smooth.

Basal whorl Leaves arranged around the base of the stem. These are most common in biennial plants. See *rosette*.

Beak A long, narrow, firm, prominent tip.

Bearded Bearing a tuft of long or stiff hairs.

Berry A simple, fleshy, indehiscent fruit with two or more seeds inside a pericarp that is soft at maturity. A very thin exocarp covers the fruit.

Berries develop from superior or inferior ovaries.

Biennial A plant whose life cycle usually takes two years to complete. Biennials germinate and grow vegetatively during the first year (often in a rosette stage) and flower, make seed, and die at the end of the second year.

Bilabiate With two lips, as the corolla of many irregular flowers.

Bilateral symmetry An arrangement in which similar parts fall on either side of a central axis. A line drawn vertically through the center of a bilaterally symmetric flower is the only way to divide it into mirror images; also *irregular* or *zygomorphic*.

Bipinnatifid Referring to pinnate lobes that are themselves cleft into pinnate lobes.

Bole The trunk of a tree.

Bract A modified leaf structure attached to the stem beneath a flower or an inflorescence. Bracts are usually smaller than a plant's true leaves or differ in form. Also the small, dry, leaf-like structure found at the base of a cone scale in conifers (Pinaceae and other conifer families).

Branch A division from an axis.

Bristle A short, stiff hair or hairlike structure such as the reduced perianth segments found in members of the sedge family (Cyperaceae); also *seta*.

Bud An undeveloped shoot or flower.

Bulb A short, thick, vertical stem usually consisting of nongreen fleshy leaf bases surrounded by scalelike leaves. Bulbs usually grow underground and are perennial.

Bulblet A small bulb. Bulblets occur aboveground, usually in leaf axils.

Bur A rough, hooked, or barbed structure. Burs are often associated with fruit and facilitate seed dispersal.

Calyx The outermost whorl of a flower. The calyx is usually composed of green sepals and covers the flower when it is in the bud stage.

Campanulate Having a bell-shaped corolla with a

tube that is as long as or longer than it is wide and has a flaring lip or lobes.

Capitate Pin-headed or in a headlike cluster. Flowers of plants in the sunflower family (Asteraceae) are typically capitate.

Capsule A simple, dry, dehiscent fruit that is composed of more than one carpel, each containing from one to many seeds. Capsules release their seeds, or dehisce, in various ways after they become mature and dry.

Canescent Having a gray or white color caused by a covering of fine pubescence.

Carpel The basic unit of the female part of the flower (also gynoecium or pistil). Flowers may have a single carpel or two or more carpels, which may be fused into a single unit. Carpels are usually enclosed with an inner chamber called a locule that contains the ovules. A bean pod is a single carpel.

Cartilaginous Tough and hard, but flexible; having the texture of cartilage.

Caryopsis A simple, indehiscent, one-seeded, dry fruit with a completely fused seed and pericarp; also called a *grain*. Only grasses (Poaceae) have caryopses.

Caudex The persistent stem base of a perennial herbaceous plant. The caudex is often woody, occurs near or just below the soil surface, and is the structure that bears new shoots or leaves each year.

Cauline Belonging or attached to the stem. Cauline leaves are attached at intervals along an aerial stem; opposite of *basal*.

Cespitose Very short-stemmed, much-branched, and growing in dense tufts, mats, or mounds; also *caespitose*.

Chaff Dry, small, thin membranous bracts or scales; also the bracts on the receptacle of the head inflorescence of plants of the sunflower family (Asteraceae).

Ciliate Referring to a margin fringed with fine hairs.

Circumscissile Dehiscing along a circular line around the top of a capsule creating a ring or lid that comes off. See *pyxis*.

Cladode A stem section that functions like a leaf; referring in particular to the stem joints of plants in the family Cactaceae. Also *cladophyll*.

Climbing-twining Growing upward in a spiral pattern. The stems of vines climb by wrapping around other plant stems, fence posts, or other supporting structure.

Closed A sheath of a grass or sedge's leaf collar with margins that are fused together down the stem to the next node; common in the genus *Bromus* (oats, family Poaceae).

Collar The region where the leaf sheath and blade meet on a grass plant.

Colonial Growing in groups (colonies) that are usually connected by underground parts.

Columella The central axis that supports one of the mericarps of a schizocarp fruit in plants of the carrot (Apiaceae) and spurge (Euphorbiaceae) families.

Coma A tuft of long, silky hairs, usually on the tip of a seed, that facilitates dispersal, as in the milkweeds (Asclepiadaceae).

Compound Consisting of two or more parts; opposite of *simple*.

Compound dichasium A type of cyme with multiple pairs of flowers arising from the axils of opposite bracts at the base of the pedicels of older flowers. This is the most common type of cyme.

Compressed Flat or flattened.

Connate Having two or more like parts fused into a united structure. Unfused parts of the same kind are said to be *distinct*.

Cool-season Referring to plants with a fall-winter-spring life cycle. Typically these are annuals that germinate in the fall or early spring and flower by late spring or early summer, but the term also refers to biennials or perennials that grow most vigorously in the fall, winter, or early spring months.

Cordate Heart-shaped.

Coriaceous Having a leathery texture.

Corm A short, vertical, enlarged, fleshy underground stem that is usually wider than high. Corms are covered by dry, scalelike leaves.

Corolla The inner whorl of the perianth that is composed of petals. The petals of the corolla are usually a color other than green and are the showy part of a flower.

Corona A petal- or crownlike appendage or series of united appendages; often referring to the crownlike structure between the petals and stamens of some flowers.

Corymb An indeterminate flat or convex-topped inflorescence with flowers borne on pedicels of different lengths that arise from different points on the rachis.

Cotyledon The primary leaves (or leaf) in the embryo; also called *seed leaves*. Monocots have one cotyledon; dicots have two.

Creeping Growing along the ground either above or just below the surface. Creeping stems usually produce roots at the nodes; also used to describe tight-clinging or twining vines.

Crenate Having shallow, rounded teeth, as a leaf margin.

Crown The basal portion of the stem of an herbaceous perennial; also the top part of a branched tree above the trunk.

Crustaceous Having a brittle, dry, and hard texture.

Culm The aerial stem of a grass, sedge, or rush. Culms are often hollow or jointed.

Cuneate Triangular or wedge-shaped with straight sides that taper to a point.

Cyathium An inflorescence with a cuplike involucre that contains a single pistillate flower surrounded by small staminate flowers that have one stamen each. Cyathia are characteristic of the genus *Euphorbia* (Euphorbiaceae).

Cyme Flat-topped to convex flower clusters, with the oldest flowers in the center. A simple cyme (dichasium) is a cluster of three flowers with the oldest flower in the center. See also *compound dichasium*.

Cymose Referring to flowers in a cyme.

Deciduous Describing a non-evergreen plant that drops its leaves at the end of the growing season.

Decompound Compound more than once, as when leaflets are further divided.

Decumbent With the stem and branches lying flat on the ground but the tips pointing upward; also called *spreading ascending*.

Decurrent Extending down an axis from the point of insertion.

Dehiscent Having a pericarp that dries as it matures and splits open at maturity to release the contents. Dehiscent fruits split open to release seeds; dehiscent anthers open to release pollen.

Deltoid Triangular or delta-shaped, with two or three sides that are equal in length.

Dentate Having coarse, sharp teeth that point outward at right angles to the margin.

Diffuse Widely or loosely branched.

Digitate With lobes, veins, narrow leaflets, or other structures arising from one point, like the fingers of a hand.

Dioecious Having male (staminate) flowers and female (pistillate) flowers on different individual plants.

Diploid A condition in which each cell contains two sets of homologous chromosomes. Haploid cells have only one set of homologous chromosomes; tetraploid cells have four sets.

Disarticulating Breaking apart into segments, usually at a joint.

Discoid In the shape of a disk; also a head composed only of disk flowers found in members of the sunflower family (Asteraceae).

Disk The enlarged or fleshy receptacle at the base of the ovary within the perianth. In the sunflower family (Asteraceae), the disk is the central portion of the head bearing the flowers; also the basal plate of a bulb.

Disk flower The perfect, regular flower in the central disk portion of the head in plants belonging to the sunflower family (Asteraceae).

Dissected Deeply divided into narrow segments.

Distal Near the tip, away from the point of attachment or base.

Distichous Arranged in two ranks on opposite sides of the stem and in the same plane.

Distinct Not fused; the opposite of *connate*.

Dorsal Referring to the back side, outer, or lower adaxial surface of a structure or organ; the opposite of *ventral*.

Drupe An indehiscent, simple, fleshy fruit with three distinct layers in the pericarp. The inner layer (endocarp) is a hard "stone" that covers the seed; the middle layer (mesocarp) is soft and fleshy at maturity; and the outer layer (exocarp) is a usually thin protective layer. Drupes are also called *stone fruit*; for example, cherries and peaches.

Ellipsoid A three-dimensional shape with an elliptic outline that is broadest in the middle and tapers to opposite ends.

Elliptic In the form of an ellipse; that is, having curved sides that are widest at the middle and taper equally to both ends.

Elongate Stretched or drawn out; much longer than wide.

Embryo The rudimentary or nascent immature plant within a seed.

Emergent Referring to an aquatic plant with a portion of the shoot above the water's surface and the plant base and roots submerged, for example, members of the genus *Pontederia*; also called *emersed*.

Emersed See *emergent*.

Endocarp The inner layer of the pericarp next to the seed.

Entire Having a smooth, continuous margin.

Erect Growth form in which the main stem is more or less perpendicular to the ground. Erect plants have upright, usually vertical growth.

Erose With an irregular or jagged margin that appears eroded or gnawed.

Excurved Curved out and away from an axis.

Farinose Covered with a mealy, granular powder.

Fascicle A compact bundle or cluster of plant parts.

Fibrous Resembling or consisting of fibers. See also *fibrous roots*.

Fibrous roots A root system of many slender roots of similar diameter. Fibrous roots are common in grasses and other monocots (Liliopsida).

Filament The stalk of a stamen that supports the anther; also any thin, threadlike structure.

Filiform Long and slender like a filament or thread.

Flaccid Limp, floppy, lacking turgor.

Flexuous Sequentially curved or bent in opposite directions, wavy or zigzagged.

Floating aquatic plant One of two types of water plants. Floating attached plants, such as *Bacopa*, are rooted to the soil at the bottom of the body of water, but the leaves usually float on the surface. Free-floating unattached plants, such as *Eichhornia*, float on the surface; the roots hang free in the water and are not attached to the soil.

Floret A type of small flower found in the grass (Poaceae) and sunflower (Asteraceae) families. The floret of grasses consists of the lemma, the palea, and the small flower they cover. The florets of sunflowers are the small individual flowers that make up the flower head.

Flower The organ of sexual reproduction of angiosperms (Magnoliophyta). A flower has a highly modified axis bearing one or more carpels (pistils), one or more stamens, or both.

Foliate Referring to or having leaves.

Foliolate Referring to or having leaflets.

Follicle A simple, dry, dehiscent fruit with one carpel that opens (dehisces) along only one side at maturity. A follicle has one locule that contains the seeds.

Fringed Having hairs along the margin.

Frond The divided leaf of a fern or palm.

Fruit A mature ovary plus any accessory floral or vegetative parts that are attached to and enlarge and ripen with it. Fruits normally contain seeds, and their main function is seed dispersal.

Funiculus The stalk that connects the ovule to the placenta or ovary wall; the stalk in the seed that meets the ovule at the hilum.

Fusiform Spindle-shaped, often widest at the middle and tapering toward both ends.

Geniculate Bent abruptly or sharply, like a knee joint.

Genus The first part of the Latin binomial (two-part) name that is unique for every plant. Plants within the same genus are closely related and share morphological characteristics. Genus names are Latin or Latinized nouns and are always capitalized.

Germination The development of seed into a seedling, or a spore into a plantlet.

Glabrous Having a smooth surface without trichomes, scales, glands, or other surface structures. Glabrous surfaces are often glossy or shiny.

Gland A structure that contains or secretes a mucilaginous or sugary fluid; often associated with certain types of trichomes or flowers.

Glandular Pubescent with gland-bearing hairs. The glands may also be sessile (attached directly to the surface without plant hairs).

Glaucous Having a whitish or bluish coating or bloom that often can be easily rubbed off.

Globose Having a spherical structure, like a globe; also *globular*.

Globular See *globose*.

Glochidiate Having or bearing apically barbed trichomes or bristles.

Glomerule A small, compact, dense, often rounded inflorescence of small flowers; a glomerate cluster.

Glume Either of the two sterile, dry, chaffy paired bracts at the base of a spikelet.

Granular Referring to a rough surface that is or appears to be covered with small granules or grainlike particles; also called *granulate* or *granulose*.

Granulate See *granular*.

Granule A very small grain.

Granulose See *granular*.

Hastate Triangular or arrowhead-shaped with basal lobes that point outward or curve away from the petiole.

Haustoria (sing. *haustorium*) The rootlike organs through which a parasitic plant attaches to a host plant to obtain nourishment and water.

Head An inflorescence consisting of a dense cluster of small, sessile flowers such as those found in members of the sunflower family (Asteraceae); also called a *capitulum*.

Hemispherical Shaped like one-half of a sphere.

Herb A plant with few or no persistent woody parts aboveground. Herbaceous plants usually die completely or die back to the soil surface at the end of the growing season.

Herbaceous Nonwoody; with the characteristics of an herb.

Hilum The scar on the surface of a seed where the funiculus was attached to the ovule.

Hip The fleshy, berrylike fruit of roses. A hip comprises a receptacle and a hypanthium that enclose numerous achenes.

Hirsute Covered with long hairs that are coarse, rough, or stiff.

Hispid Having bristly, dense, stiff, erect hairs.

Hyaline Having a thin, membranous, transparent or translucent texture.

Hypanthium A floral structure formed by the fusion of the bases of the sepals, petals, stamens, and receptacle. The hypanthium arises from the receptacle and surrounds the ovary. A hypanthium can arise below the ovary (superior ovaries with perigynous insertion), or above the ovary (inferior ovaries with epigynous insertion).

Hypocotyl The part of a seedling stem below the cotyledons.

Hypogynous Having the parts of the flower attached directly to the receptacle below the ovary.

Imbricate Tiled or overlapping, like roof shingles.

Imperfect Referring to a unisexual flower with either functioning male stamens or functioning female pistils.

Incumbent Referring to the back side of the cotyledons lying parallel along the radicle.

Indehiscent Referring to a type of fruit that retains seeds within the pericarp or does not split open along a regular line or pattern.

Indurate Hard or hardened.

Inflorescence The arrangement of a group of flowers on a plant. The flowers arise along the floral axis, a branch system. An inflorescence, or "floral cluster," may be terminal or axillary; also, nontechnically, a *seed head*.

Internode The segment of stem between two nodes.

Introduced Not native to an area. In the context of this book, weeds that originated outside the continental United States or Canada are said to be introduced.

Involucre A whorl of bracts subtending a flower or inflorescence.

Involute More or less flattened with the margins rolled inward toward the upper surface.

Irregular Having a structure without an apparent plane of symmetry or one that can be divided in only one plane. Also called *zygomorphic* or *bilaterally symmetrical*. See *bilateral symmetry*.

Isthmus (pl. *isthmi*) A narrow strip of tissue that connects two broader parts of a structure.

Joint A node on the stem where branches or leaves originate.

Jointed With nodes or points where the outline is broken or bent. The term often refers to the nodes on grasses (Poaceae) where stems bend.

Keel A prominent or projecting longitudinal ridge; also the connate lower petals of a papillonaceous flower in the bean family (Fabaceae).

Labiate With a lip; usually referring to the liplike petals of members of the mint family (Lamiaceae).

Lacerate With a margin that is irregularly cut or cleft so that it appears torn.

Lamina The flat, expanded blade of a leaf or petal; also a blade or bladelike portion.

Lanceolate Having a long, narrow shape that is widest near the base and tapers to the apex; lance-shaped.

Latex Milky juice or sap such as that produced by milkweeds (Asclepiadaceae) and many spurges (Euphorbiaceae).

Leaf The photosynthetic organ of the typical plant, usually consisting of a flattened blade (lamina) and a slender, stalklike base (petiole).

Leaf axil The angle formed between the stem (or axis) and a leaf base.

Leaf blade The main part of the leaf. A leaf may be flattened and thin, thick and succulent, or needlelike. Leaves can be simple (one blade unit) or compound (with many blade units).

Leaflet The multiple apparent blade units of a compound leaf. Leaflets may be arranged on the leaf in a palmate or pinnate pattern.

Legume A simple, dry fruit with one carpel (locule) that dehisces longitudinally along two opposite sides to release its seeds; the fruit type characteristic of the bean family (Fabaceae).

Lemma The lower of the two bracts that enclose a grass flower (Poaceae). The lemma is below and outside the palea, the other bract enclosing the flower.

Lenticular Lens-shaped with both sides convex, the widest point at the center, and then tapering to the edge in cross section; a common seed shape.

Ligule A membrane or fringe of hair above the leaf collar on a grass plant.

Linear Long and narrow with almost parallel sides; most grasses (Poaceae) have linear leaves.

Lip A structure arranged or shaped like a lip, as one of the two parts of a bilabiate corolla of flowers of the mint family (Lamiaceae) or the labellum of an orchid (Orchidaceae.)

Lobe A rounded incision or segment of an organ.

Lobed Bearing one or more lobes that indent less than halfway to the midvein or base.

Lobulate Consisting of small lobes or lobules.

Locule The compartment within an ovary that usually contains the ovules; also called a *chamber* or *cell*.

Lodicule Small scalelike structures at the base of the ovary in a grass plant that swell to push the lemma and palea apart during flowering.

Loment An elongated fruit with a single carpel constricted between the seeds that splits into single-seeded segments; a type of modified legume.

Lyrate Cleft pinnately with an enlarged, rounded terminal lobe and smaller lateral lobes.

Membranous Having a thin, translucent, and flexible structure or texture.

Mericarp A segment of a dry schizocarp that splits away from the axis of the ovary.

-merous Suffix referring to the number of parts making up a structure, usually the number of parts in a corolla.

Midrib The central vein in a leaf blade. The midrib usually extends to the tip of the leaf; also *midvein*.

Midvein See *midrib*.

Monoecious Having separate male (staminate) and female (pistillate) flowers on the same individual plant.

Mouth The collar region or sheath summit of a grass leaf.

Mucro An abrupt, short, sharp tip; also see *mucronate*.

Mucronate Having a distinct short, sharp, abrupt point or projection (mucro) at the midvein.

Multiple fruit Compound fruits derived from closely adhering ovaries of several flowers in an inflorescence. Accessory tissues such as calyx parts and receptacles may become fleshy. Examples include mulberry, Osage orange, pineapple, and fig.

Muricate Roughened by short, hard, pointed projections.

Nascent Starting to develop but not yet completely formed.

Native A plant believed to have originated naturally in a geographic region. Herein it refers to plants believed to have been growing naturally within the continental United States or Canada before the first human inhabitats arrived.

Nectar A sweet, sugary, sticky fluid produced by flowers of some plants.

Nectary The gland that secretes nectar.

Nerve A prominent, simple vein or rib of a leaf, bract, or other structure.

Nodding Drooping or bending over and downward.

Node The joint where a leaf or bud is or was attached to a stem.

Nut A simple, dry, indehiscent one-seeded fruit.

Nutlet A small nut, or one of the deeply four-lobed sections of the mature ovary of plants in the borage (Boraginaceae), mint (Lamiaceae), and vervain (Verbenaceae) families.

Obconic Shaped like an inverted cone, with the attachment at the narrow end.

Obcordate Inverse heart-shaped, with the lobes and notch at the apex and the narrow point at the attached end.

Oblanceolate Long and narrow, widest near the apex and tapering at the base; inverted lanceolate.

Oblique Having slanting or unequal, asymmetrical sides that do not match; usually refers to a leaf base.

Oblong Rectangular, with nearly parallel sides and two to four times longer than wide.

Obovate Egg-shaped, with a narrow base and wide apex; also *obovoid*.

Obovoid See *obovate*.

Obtuse Having a rounded or blunt apex or base.

Ocrea A papery sheath around a swollen node formed by the fused stipules at the base of a leaf. Ocreae are common in the smartweed family (Polygonaceae); also *ochrea*.

Open Having margins that do not overlap. A sheath of a grass is open for a significant distance down the stem to the next node; opposite of *closed*. Inflorescences and flowers that have spreading, widely spaced parts are also said to be open.

Opposite Originating from the same node on opposite sides of a stem.

Orbicular Round, as a round leaf blade.

Ornamental A plant selected and cultivated for aesthetic or landscape use.

Ovary The basal part of a pistil that contains the ovules.

Ovate Rounded but longer than wide. An egg-shaped outline widest at the base is said to be ovate; also *ovoid*.

Ovoid Ovate or oval in cross section.

Ovule The egg-containing part of a pistil within the ovary; an immature seed that develops into the seed after fertilization.

Palea The upper of the two bracts that enclose a grass flower (Poaceae). The palea is above and on the inside of the lemma, the other bract enclosing the flower.

Palmate Describing a compound leaf with all leaflets originating from the same attachment point; also the venation pattern of leaves with veins that radiate from the same point at the base.

Pandurate Shaped like a violin.

Panicle An indeterminate, loose, multiple-branching inflorescence with individual flowers borne on pedicels; common in the grass family (Poaceae).

Papilla (pl. *papillae*) A short, almost microscopic, rounded bump or projection.

Papillose Having papillae.

Pappus A modified calyx made up of awns, hairs, or dry scales forming a crown at the apex of an achene on plants in the sunflower family (Asteraceae).

Parasite A plant that grows and obtains water and nutrients from another host plant, usually through haustorial roots.

Pedicel The stalklike stem that supports a single flower in an inflorescence. Flowers may also be *sessile*, or attached directly to the rachis without a stalk.

Peduncle The stalklike stem that supports an inflorescence or a solitary flower.

Pendulous Drooping or hanging down.

Pepo A simple, fleshy fruit that develops from an inferior ovary and has a thick rind (outer shell) at maturity. Pepos are most common in plants of the cucumber family (Cucurbitaceae).

Perennial A plant that lives for more than two years, reproducing both vegetatively by spreading roots or stems and by seed. Simple perennials survive and spread vegetatively but reproduce (i.e., make new plants) only by making seeds. Creeping perennials spread by rhizomes, runners, or other structures and reproduce both vegetatively and by seed.

Perfect A flower type with functioning male stamens and female pistils.

Perfoliate Referring to a sessile leaf or opposite leaves whose bases surround the stem, giving the appearance that the stem is passing through the leaf blade.

Perianth Collectively, the calyx and corolla of a flower.

Pericarp The wall of the ovary or fruit. The pericarp can vary in texture from soft and juicy to hard and dry, and may consist of one to three layers: the exocarp (the outer skin), the mesocarp (the middle layer that may become soft and juicy), and the endocarp (the inner layer next to the seed).

Perigynous Having the parts of the flower attached at the rim of a hypanthium that surrounds the ovary.

Persistent Remaining attached longer than required for the function of the organ.

Petal One unit of the corolla. The petals of a corolla may be polypetalous (distinct and separate) or sympetalous (connate or fused to some degree into a single ring or tube). A flower without petals is *apetalous*. Typical petals are brightly colored to attract pollenators to the flower.

Petiole The stalklike base of a leaf that attaches the leaf to the stem.

Petiolule The stalklike base of a leaflet in compound leaves.

Phyllodium An expanded petiole that functions as a leaf without having a true blade.

Pilose Covered with long, soft, straight hairs.

Pinnate Describing a compound leaf with leaflets arranged along a rachis (central continuation of the petiole) like the vanes of a feather. The venation pattern of leaves with lateral veins that arise at intervals along a central midvein is also said to be pinnate.

Pinnatifid Pinnately lobed or cleft, with a margin that extends more than halfway to the midrib.

Pinnule The ultimate leaflet or pinna when the leaf is two times or more compound pinnate.

Pistil The female reproductive organ (gynoecium) of a flower, comprising the ovary, style, and stigma. A simple pistil has one carpel; a compound pistil consists of two or more fused (connate) carpels.

Pistillate Having pistils but no stamens; a female flower.

Pith The region derived from ground meristem at the center of dicot stems and roots.

Plano-convex Flat on one side and convex on the opposite side.

Plumose Having fine hairs or bristles branched on both sides of an axis like a feather; plumelike.

Pod Any dehiscent, dry fruit.

Pollen The immature male gametophyte produced within the pollen sacs in the anthers.

Prickle Sharp, pointed protective structures arising from the outer surface of bark or epidermis.

Procumbent Referring to a stem lying or trailing flat on the ground but not rooting.

Prop root An adventitious root arising from the lower nodes that provides additional support for the stem; maize (*Zea mays*) is an example.

Prostrate With stem and branches lying flat on the ground.

Proximal Toward the point of attachment, or the base of a structure.

Pubescent General term for a surface with trichomes (plant hairs).

Punctate Covered with pits, shallow depressions, or colored dots; also *pitted*.

Pustulate Covered with pustules or blisters; also *pustulose*.

Pyriform Shaped like a pear.

Pyxis A type of capsule with circumscissile dehiscence in which the top of the fruit separates as a single piece or lid.

Raceme An unbranched inflorescence with flowers borne on pedicels.

Racemiform In the form of or resembling a raceme; also *racemose*.

Rachilla A secondary axis that bears the florets in inflorescences of grasses (Poaceae) and sedges (Cyperaceae).

Rachis The central axis of an elongated inflorescence.

Radial symmetry See *actinomorphic*.

Radicle The embryonic root of a germinating seed.

Ray The branch of an umbel inflorescence; also the straplike portion of a ray flower in sunflowers (Asteraceae).

Ray flower A peripheral zygomorphic flower in the head of many members of the sunflower family (Asteraceae).

Receptacle The enlarged end of the pedicel that bears the parts of a flower; in Asteraceae, the expanded, flattened top of the peduncle to which bracts and flowers are attached.

Recurved Curled or curved backward or downward.

Reflexed Abruptly bent or turned backward or downward.

Regular See *actinomorphic*.

Reniform Shaped like a kidney, as the seeds of beans (Fabaceae) and mallow (Malvaceae) and some leaves.

Repent Having a prostrate or creeping growth habit.

Reticulate With venation like a net or network.

Retrorse Pointed in a backward or downward direction.

Retuse Referring to a round or blunt leaf tip with a slight indentation or notch at the midvein.

Revolute Having margins that roll or curl under; opposite of *involute*.

Rhizome A perennial stem that spreads horizontally beneath the soil surface and gives rise to buds and adventitious roots.

Rhombic Shaped like a diamond with equilateral sides. A rhomboid is about as long as it is wide but is broadest at the middle.

Root A plant structure that does not have external nodes or buds. Branch roots initiate from the interior of other roots. Roots usually grow underground, but some plants produce roots that grow aboveground or in water.

Rootstock A rather general term usually applied to an underground perennial root or an underground perennial stem structure such as a rhizome.

Rosette A tight cluster of leaves that radiate from a central stem. Plants of the mustard family (Brassicaceae) and certain tribes of the sunflower family (Asteraceae) commonly have basal rosettes.

Rotate Having a monopetalous corolla with a flattish border, widely spreading lobes, and little or no tube, as in members of the figwort family (Scrophulariaceae).

Rugose Wrinkled.

Runner A slender, trailing stem or shoot that roots at the nodes.

Sagittate Triangular or arrowhead-shaped. A sagittate leaf has basal lobes that project downward or curve toward the petiole or stalk.

Sap The juice or fluid within plant tissues.

Scabrous Rough to the touch. Scabrous surfaces have a dull appearance and are common on plants with pointed or bulbous epidermal cells or short, stiff hairs.

Scale A small, thin, flat, dry, scarious bract associated with the small flowers of sedges (Cyperaceae); also a thin, flattened, microscopic feature of the epidermis.

Scandent With a climbing growth habit. Vines have a scandent habit.

Scape The leafless flowering stem that is the peduncle of a solitary flower.

Scarious Thin, dry, and membranous; lacking in green color and appearing shriveled.

Schizocarp A simple, usually dry fruit that splits into two longitudinally one-seeded mericarps.

Seed A mature, ripe ovule. The seed contains the embryo and stored food.

Seedhead A nontechnical term for an inflorescence.

Seedling A young plant that has just emerged from a seed and is still obtaining most of its nutrition from the seed reserves and cotyledons.

Segmented Deeply divided but not truly compound.

Sepal A part or division of the calyx. Sepals may be distinct, individual leaf, or bractlike, or may be fused to form a calyx tube. Sepals are usually green but may be showy like petals.

Septate Divided by one or more internal cross-partitions.

Septum A partition, especially the partitions separating the locules in a compound ovary.

Serrate With sharp teeth pointing toward the apex.

Sessile Attached directly to another structure without a stalk. A leaf without a petiole is an example of a sessile structure.

Sheath The lower part of the leaf that surrounds the stem on a grass plant; also any tubular structure that surrounds part or all of another plant structure.

Shoot A vegetative stem or branch.

Shrub A low, woody, perennial plant usually with multiple slender stems growing from the base and the potential to grow no taller than 6.0 m.

Silicle A simple, dry, dehiscent fruit with two short, broad carpels. Silicles split into three parts—two outer covers and a central partition—and are found in the mustard family (Brassicaceae).

Silique A simple, dry, dehiscent fruit with two long, narrow carpels. Siliques split into three parts—two outer covers and a thin central partition. Also found in the mustard family (Brassicaceae), siliques are longer and narrower than silicles.

Simple Unbranched or undivided; opposite of *compound*. Plant structures such as stems, roots, inflorescences, and leaves may be simple.

Sinuate Flat with a distinctly wavy margin moving in and out relative to the midrib.

Sinus The notch or cleft between two lobes or teeth.

Solitary Having a single flower at the end of a branch or in the axils of leaves. The peduncle of certain kinds of solitary flowers is called a scape.

Sori (sing. *sorus*) A cluster of sporangia on the surface of a fern frond.

Spathe A large, usually conspicuous bract resembling a leaf or petal that subtends an inflorescence.

Spatulate Having a wide apex abruptly tapering to a narrow base.

Species The second element of the Latin binomial unique to each plant. More commonly, the term *species name* refers to the entire binomial, both genus and species. The species name is often followed by an abbreviation that refers to the author or taxonomist (called an authority) who named the plant. For example, "L." is the abbreviation for Carolus Linnaeus, the "father" of plant taxonomy.

Spherical Shaped like a sphere, a round solid structure.

Spicate In a spikelike arrangement.

Spike An indeterminate, unbranched, elongate inflorescence with sessile flowers attached to a rachis.

Spikelet A small or secondary spike, composed of one to several flowers, that is subtended by glumes or scales. The spikelet is the primary inflorescence of grasses (Poaceae) and sedges (Cyperaceae).

Spine A modified leaf that is pointed and sharp. Stipular spines are sharp, pointed, modified stipules.

Sporangium (pl. *sporangia*) A case or sac bearing spores in ferns and other seedless plants.

Spore A reproductive cell in a sporangium on a fern frond.

Spreading Extending outward horizontally.

Stalk A narrow structure that supports a plant part.

Stamen The male reproductive organ (androecium) of a flower. The stamen is usually composed of a pollen-producing anther and a filament, although some anthers are sessile (without a filament).

Staminate Having stamens but no pistils; a male flower.

Stellate Star-shaped. Stellate hairs branch or radiate from a base.

Stem The main structural axis that bears the plant's leaves. Stems have distinct nodes and internodes and can be above or below the soil surface.

Stigma The part of the pistil at the tip of the style where pollen is deposited and germinates prior to fertilization.

Stipe The petiole of a fern leaf; also a stalk that supports an organ, such as the stalk that connects the receptacle and the ovary in some flowers.

Stipitate Borne on a stipe.

Stipule A pair of small appendages located at the base of certain leaves. Some plant species do not have stipules, or the stipules fall off shortly after the leaf develops.

Stolon A perennial stem that grows horizontally along the ground. Adventitious roots develop wherever the stolon touches the soil.

Stone A seed surrounded by a hard, stony endocarp at the center of a drupe; also *pyrene*.

Stone fruit See *drupe*.

Striate Covered by longitudinal lines or ridges.

Strigillose Minutely strigose.

Häfliger, E., U. Kühn, L. Hämet-Ahti, C. D. K. Cook, R. Faden, and F. Speta. 1982. Monocot Weeds 3: Monocot Weeds Excluding Grasses. Basel, Switzerland: CIBA-GEIGY. 132 pp.

Häfliger, E., and H. Scholz. 1980. Grass Weeds 1: Weeds of the Subfamily Panicoideae. Basel, Switzerland: CIBA-GEIGY. 142 pp.

———. 1981. Grass Weeds 2: Weeds of the Subfamilies Chloridoideae, Pooideae, and Oryzoideae. Basel, Switzerland: CIBA-GEIGY. 138 pp.

Häfliger, E., M. Wolf, C. D. K. Cook, T. J. Crovello, P. Hiepko, U. Kühn, N. K. B. Robson, H. Scholz, and E. U. Zajac. 1988. Dicot Weeds 1: Dicotyledonous Weeds of 13 Families. Basel, Switzerland: CIBA-GEIGY. 300 pp.

Haragan, P. D. 1991. Weeds of Kentucky and Adjacent States. Lexington: University Press of Kentucky. 278 pp.

Hardin, J. W. 1961. Poisonous Plants of North Carolina. Agricultural Experiment Station Bulletin 414. Raleigh: North Carolina State University. 128 pp.

Harper, R. M. 1944. Geological Survey of Alabama: Preliminary Report on the Weeds of Alabama. Bulletin 53. Wetumpka, Ala.: Wetumpka Printing. 275 pp.

Harrington, H. D. 1977. How to Identify Grasses and Grasslike Plants. Chicago: Swallow Press. 142 pp.

Harris, J. G., and M. W. Harris. 1994. Plant Identification Terminology: An Illustrated Glossary. Spring Lake, Utah: Spring Lake Publishing. 197 pp.

Hickey, M., and C. King. 2000. The Cambridge Illustrated Glossary of Botanical Terms. Cambridge: Cambridge University Press. 208 pp.

Hickman, J. C., ed. 1993. The Jepson Manual: Higher Plants of California. Berkeley: University of California Press. 1400 pp.

Hitchcock, A. S. 1950. Manual of the Grasses of the United States. 2nd ed. Revised by Agnes Chase. Reprint 1971. Vols. 1 and 2. New York: Dover 1051 pp.

Hitchcock, C. L., and A. Cronquist. 1973. Flora of the Pacific Northwest: An Illustrated Manual. Seattle: University of Washington Press. 730 pp.

Holm, L., J. Doll, E. Holm, J. Pancho, and J. Herberger. 1997. World Weeds: Natural Histories and Distribution. New York: John Wiley & Sons. 1129 pp.

Hultén, E. 1968. Flora of Alaska and Neighboring Territories: A Manual of the Vascular Plants. Stanford: Stanford University Press. 1008 pp.

Isley, Duane. 1990. Vascular Flora of the Southeastern United States. Vol. 3, pt. 2: Leguminosae (Fabaceae). Chapel Hill: University of North Carolina Press. 258 pp.

———. 1998. Native and Naturalized Leguminosae (Fabaceae) of the United States (Exclusive of Alaska and Hawaii). Provo, Utah: Monte L. Bean Life Science Museum and Brigham Young University. 1006 pp.

Kaufman, S. R., and Wallace Kaufman. 2007. Invasive Plants: Guide to Identification and the Impacts and Control of Common North American Species. Mechanicsburg, Pa.: Stackpole Books. 458 pp.

Kearney, T. H., R. H. Peebles, J. T. Howell, and E. McClintock. 1960. Arizona Flora. 2nd ed. Berkeley: University of California Press. 1085 pp.

Kingsbury, J. M. 1964. Poisonous Plants of the United States and Canada. Englewood Cliffs, N.J.: Prentice-Hall. 626 pp.

Langer, R. H. M. 1972. How Grasses Grow. London: Edward Arnold. 60 pp.

Lawrence, G. H. M. 1955. An Introduction to Plant Taxonomy. New York: Macmillan. 179 pp.

Mabberley, D. J. 1998. The Plant-Book: A Portable Dictionary of the Vascular Plants. 2nd ed. Cambridge: Cambridge University Press. 858 pp.

Martin, A. C., and W. D. Barkley. 1961. Seed Identification Manual. Berkeley: University of California Press. 221 pp.

McCarty, L. B., J. W. Everest, D. W. Hall, T. R. Murphy, and F. Yelverton. 2001. Color Atlas of Turfgrass Weeds. Chelsea, Mich.: Sleeping Bear Press. 269 pp.

McGregor, R. L., and T. M. Barkley, eds. 1986. Flora of the Great Plains. Lawrence: University Press of Kansas. 1392 pp.

Miller, J. H., and K. V. Miller. 1999. Forest Plants of the Southeast and Their Wildlife Uses. Athens: University of Georgia Press. 454 pp.

Morre, J. D., R. J. Hull, and J. L. Williams Jr. 1971. Plant Identification Using Family Characteristics. Lafayette, Ind.: Balt. 234 pp.

Muenscher, W. C. 1980. Weeds. 2nd ed. 1980. Ithaca, N.Y.: Cornell University Press. 586 pp.

Musil, A. F. 1978. Identification of Crop and Weed Seeds. Agricultural Handbook 219. Washington, D.C.: U.S. Government Printing Office. 215 pp.

Parker, K. F. 1972. An Illustrated Guide to Arizona Weeds. Tucson: University of Arizona Press. 327 pp.

Patterson, D. T., ed. 1989. Composite List of Weeds. Champaign, Ill.: Weed Science Society of America. 112 pp.

Radford, A. E., H. F. Ahles, and C. R. Bell. 1968. Manual of the Vascular Flora of the Carolinas. Chapel Hill: University of North Carolina Press. 1183 pp.

Reimer, D. N. 1984. Introduction to Freshwater Vegetation. Westport, Conn.: AVI Publishing. 207 pp.

Ross, M. A., and C. A. Lembi. 1985. Applied Weed Science. New York: Macmillan. 340 pp.

Small, J. K. 1933. Manual of the Southeastern Flora. New York: Published by the author. 1554 pp.

Spencer, E. R. 1968. All about Weeds. New York: Dover. 333 pp.

Stephens, H. A. 1973. Woody Plants of the North Central Plains. Lawrence: University Press of Kansas. 530 pp.

———. 1980. Poisonous Plants of the Central United States. Lawrence: University Press of Kansas. 165 pp.

Stubbendieck, J., M. J. Coffin, and L. M. Landholt. 2003. Weeds of the Great Plains. Lincoln: Nebraska Department of Agriculture. 605 pp.

Stubbendieck, J., G. Y. Friisoe, and M. R. Bolick. 1994. Weeds of Nebraska and the Great Plains. Lincoln: Nebraska Department of Agriculture. 588 pp.

Stubbendieck, J., S. L. Hatch, and C. H. Butterfield. 1994. North American Range Plants. Lincoln: University of Nebraska Press. 493 pp.

Stucky, J. M. 1980. Identifying Seedling and Mature Weeds Common in the Southeastern United States. Raleigh: North Carolina State University. 197 pp.

Uva, R. H., J. C. Neal, and J. M. DiTomaso. 1997. Weeds of the Northeast. Ithaca, N.Y.: Comstock Publishing Associates. 397 pp.

Voss, E. G. 1972. Michigan Flora. Part I. Gymnosperms and Monocots. Bloomfield Hills, Mich.: Cranbrook Institute of Science. 488 pp.

———. 1985. Michigan Flora. Part II. Dicots. Bloomfield Hills, Mich.: Cranbrook Institute of Science. 727 pp.

———. 1996. Michigan Flora. Part III. Dicots Continued. Bloomfield Hills, Mich.: Cranbrook Institute of Science. 622 pp.

Walters, D. R., and D. J. Keil. 1977. Vascular Plant Taxonomy. Dubuque, Iowa: Kendall/Hunt. 488 pp.

Weber, W. A., and R. C. Wittmann. 2001. Colorado Flora: Eastern Slope. 3rd ed. Boulder: University Press of Colorado. 521 pp.

———. 2001. Colorado Flora: Western Slope. 3rd ed. Boulder: University Press of Colorado. 488 pp.

Welsh, S. L., N. D. Atwood, S. Goodrich, and L. C. Higgins. 2003. A Utah Flora. 3rd ed. Revised. Provo, Utah: Brigham Young University. 912 pp.

Westbrooks, R. G., and J. W. Preacher. 1986. Poisonous Plants of Eastern North America. Columbia: University of South Carolina Press. 226 pp.

Wharton, M. E., and R. W. Barbour. 1973. Trees and Shrubs of Kentucky. Lexington: University Press of Kentucky. 582 pp.

Whitson, T. D., ed. 2001. Weeds of the West. Western Society of Weed Science. Jackson, Wyo.: Grand Teton Lithography. 628 pp.

Wunderlin, R. P. 1998. Guide to the Vascular Plants of Florida. Gainesville: University Press of Florida. 806 pp.

Yatskievych, G. 1999. Steyermark's Flora of Missouri. Vol. 1. St. Louis: Missouri Botanical Garden Press. 991 pp.

———. 2006. Steyermark's Flora of Missouri. Vol. 2. St. Louis: Missouri Botanical Garden Press. 1181 pp.

Zomlefer, W. B. 1994. Guide to Flowering Plant Families. Chapel Hill: University of North Carolina Press. 430 pp.

CONTRIBUTORS

S. D. Askew *Brassica nigra, Verbesina encelioides,* and *Viola sororia.*

R. Becker *Lythrum salicaria* and *Tanacetum vulgare.*

M. P. Blair *Heliopsis helianthoides, Plantago major,* and *Trifolium pretense.*

J. W. Boyd *Veronica persica* Poir.

D. C. Bridges *Andropogon virginicus, Galium aparine, Paspalum dilatatum, Saururus cernuus,* and *Urochloa fasciculata.*

C. T. Bryson *Abutilon theophrasti, Anoda cristata, Bacopa rotundifolia, Carex cherokeensis, Cerastium fontanum* ssp. *vulgare, Cirsium horridulum, Cynoglossum officinale, Cyperus compressus, Cyperus difformis, Cyperus entrerianus, Cyperus erythrorhizos, Cyperus esculentus, Cyperus iria, Cyperus odoratus, Cyperus rotundus, Cyperus strigosus, Datura stramonium, Equisetum hyemale, Eupatorium capillifolium, Fimbristylis annua, Fimbristylis miliacea, Geranium carolinianum, Geranium dissectum, Helianthus petiolaris, Hibiscus trionum, Hypericum perforatum, Kyllinga brevifolia, Lamium amplexicaule, Malva neglecta, Oenothera laciniata, Parthenocissus quinquefolia, Passiflora lutea, Phyllanthus urinaria, Physalis angulata, Physalis heterophylla, Phytolacca americana, Prunella vulgaris, Pteridium aquilinum, Schoenoplectus acutus, Sida rhombifolia, Sida spinosa, Solanum americanum, Solanum carolinense, Solanum elaeagnifolium, Solanum physalifolium, Solanum ptycanthum, Solanum rostratum, Solanum viarum, Stellaria media, Striga asiatica,* and *Vigna unguiculata.*

C. T. Bryson and N. C. Coile *Solanum capsicoides, Solanum dimidiatum, Solanum jamaicense, Solanum mammosum, Solanum sisymbriifolium, Solanum tampicense,* and *Solanum torvum.*

C. T. Bryson and M. L. Ketchersid *Fatoua villosa.*

J. Cardina *Acanthospermum hispidum, Allium vineale, Euphorbia cyathophora, Euphorbia heterophylla, Lonicera japonica, Matricaria discoidea, Mollugo verticillata, Soliva sessilis,* and *Toxicodendron radicans.*

J. Cardina and C. D. Elmore *Vicia sativa* ssp. *nigra.*

J. M. Chandler *Amphiachyris dracunculoides.*

T. R. Clason *Cirsium discolor, Rosa multiflora,* and *Verbena bracteata.*

M. S. DeFelice *Ageratina altissima, Agrostemma githago, Amaranthus tuberculatus, Ambrosia bidentata, Aristida oligantha, Arundo donax, Bidens bipinnata, Bromus inermis, Cichorium intybus, Dac-*

tylis glomerata, Distichlis spicata, Equisetum arvense, Gutierrezia sarothrae, Holcus lanatus, Leptochloa dubia, Lolium perenne, Lupinus perennis, Myosurus minimus, Ornithogalum umbellatum, Oxalis stricta, Panicum miliaceum, Panicum repens, Paspalum distichum, Paspalum lividum, Phragmites australis, Poa pratensis, Polygonum amphibium var. *emersum, Polygonum lapathifolium, Polygonum pensylvanicum, Polygonum persicaria, Ranunculus sardous, Rumex obtusifolius, Setaria italica, Setaria verticillata, Solidago canadensis, Teucrium canadense, Tripsacum dactyloides, Urochloa maxima,* and *Veronica peregrina.*

C. D. Elmore *Acalypha ostryifolia, Achillea millefolium, Ammannia coccinea, Apocynum cannabinum, Calystegia sepium, Caperonia palustris, Chamaecrista fasciculata, Chamaesyce humistrata, Chamaesyce hyssopifolia, Chamaesyce maculata, Chamaesyce nutans, Chenopodium ambrosioides, Coronopus didymus, Croton capitatus, Cucumis melo, Daucus carota, Helenium amarum, Helianthus ciliaris, Heteranthera reniformis, Jacquemontia tamnifolia, Lolium arundinaceum, Muhlenbergia frondosa, Muhlenbergia schreberi, Nicandra physalodes, Oenothera biennis, Paspalum notatum, Plantago lanceolata, Polygonum aviculare, Raphanus raphanistrum, Rumex crispus, Scoparia dulcis, Setaria parviflora, Silene latifolia, Silybum marianum, Solidago canadensis* var. *scabra, Sphenoclea zeylanica, Symphyotrichum divaricatum,* and *Vicia villosa.*

C. D. Elmore and C. T. Bryson *Abelmoschus esculentus, Digitaria ischaemum, Ipomoea coccinea, Ipomoea hederacea, Ipomoea hederacea* var. *integriuscula, Ipomoea lacunosa, Ipomoea purpurea, Ipomoea quamoclit, Ipomoea wrightii,* and *Urochloa platyphylla.*

C. D. Elmore, A. Evans, and R. Hayes *Echinodorus cordifolius, Sagittaria graminea, Sagittaria latifolia,* and *Sagittaria montevidensis.*

C. D. Elmore and D. W. Hall *Digitaria sanguinalis* and *Oryza sativa.*

C. D. Elmore, D. W. Hall, and C. T. Bryson *Digitaria ciliaris, Echinochloa colona, Echinochloa crus-galli,* and *Eleusine indica.*

C. D. Elmore, D. W. Hall, and L. R. Oliver *Cenchrus spinifex, Cynodon dactylon, Leptochloa panicea, Leptochloa panicoides, Panicum dichotomiflorum, Setaria faberi, Setaria pumila, Setaria viridis, Sorghum halepense,* and *Urochloa texana.*

C. D. Elmore, D. W. Hall, and E. J. Retzinger Jr. *Aegilops cylindrica, Bromus catharticus, Bromus secalinus, Bromus tectorum,* and *Lolium perenne* ssp. *multiflorum.*

C. D. Elmore, D. W. Hall, and L. M. Wax *Amaranthus albus, Amaranthus hybridus, Amaranthus palmeri, Amaranthus retroflexus, Amaranthus spinosus,* and *Amaranthus viridis.*

C. D. Elmore and P. W. Jordan *Acalypha virginica* and *Cannabis sativa.*

C. D. Elmore and W. A. Skroch *Convolvulus arvensis, Ipomoea cordatotriloba, Ipomoea cordatotriloba* var. *torreyana, Ipomoea pandurata,* and *Ipomoea turbinata.*

C. D. Elmore and J. F. Stritzke *Erodium cicutarium.*

C. D. Elmore and A. F. Wiese *Ambrosia grayi.*

S. A. Fitterer *Salvia lyrata* and *Salvia reflexa.*

J. P. Goodlett and C. T. Bryson *Aeschynomene rudis, Conyza ramosissima, Egeria densa, Glottidium vesicarium, Helianthus grosseserratus, Ipomoea aquatica, Lepidium densiflorum, Lindernia dubia, Oenothera speciosa, Senecio vulgaris, Solanum pseudocapsicum,* and *Trifolium repens.*

J. D. Green *Asclepias verticillata, Duchesnea indica, Eragrostis cilianensis, Euphorbia dentata, Perilla frutescens, Physalis longifolia* var. *subglabrata, Sonchus asper, Sonchus oleraceus,* and *Symphyotrichum pilosum.*

N. M. Hackett and D. S. Murray *Aeschynomene indica, Crotalaria spectabilis, Desmodium tortuosum, Hoffmannseggia glauca, Pueraria montana* var. *lobata, Senna obtusifolia, Senna occidentalis,* and *Sesbania herbacea.*

D. W. Hall *Alopecurus carolinianus, Avena fatua, Bromus commutatus, Bromus diandrus, Bromus japonicus, Cenchrus longispinus, Dactyloctenium aegyptium, Digitaria bicornis, Elymus repens, Eriochloa acuminata, Eriochloa contracta, Hordeum jubatum,*

Hordeum pusillum, Imperata cylindrica, Leptochloa fusca var. *fascicularis, Paspalum urvillei, Phalaris arundinacea, Poa annua, Rottboellia cochinchinensis, Sorghum bicolor* ssp. *arundinaceum, Sporobolus indicus,* and *Urochloa ramosa.*

P. Dalton Haragan *Barbarea vulgaris, Capsella bursa-pastoris, Cardamine parviflora, Chorispora tenella, Descurainia pinnata, Descurainia sophia, Erysimum repandum, Lepidium virginicum, Sinapis arvensis, Sisymbrium irio,* and *Thlaspi arvense.*

S. A. Harrison *Brassica rapa, Conium maculatum, Erigeron philadelphicus,* and *Packera glabella.*

D. H. Johnson *Cenchrus echinatus, Eragrostis pilosa, Rotala ramosior, Rubus trivialis,* and *Typha latifolia.*

D. L. Jordan *Dracopis amplexicaulis.*

P. W. Jordan *Bacopa rotundifolia, Hypericum perforatum,* and *Phyllanthus urinaria.*

J. W. Keeling *Argemone mexicana, Buchloë dactyloides, Lamium purpureum, Panicum capillare, Sisymbrium altissimum,* and *Taraxacum officinale.*

M. L. Ketchersid *Bromus hordeaceus, Conyza bonariensis, Euphorbia marginata, Grindelia squarrosa, Melilotus officinalis, Plantago aristata, Tragopogon dubius, Trifolium incarnatum,* and *Vernonia baldwinii.*

V. L. Maddox *Opuntia humifusa.*

C. D. Monks *Cleome gynandra, Leucanthemum vulgare,* and *Rudbeckia hirta.*

T. R. Murphy *Arctium minus, Coreopsis tinctoria, Diodia virginiana, Galinsoga quadriradiata,* and *Pyrrhopappus carolinianus.*

L. R. Oliver *Ambrosia artemisiifolia, Ambrosia trifida, Bidens frondosa, Carduus nutans, Centaurea biebersteinii, Centaurea cyanus, Centaurea solstitialis, Conyza canadensis, Eclipta prostrata, Helianthus annuus, Lactuca floridana, Lactuca serriola, Parthenium hysterophorus, Sonchus arvensis,* and *Xanthium strumarium.*

J. A. Pawlak and D. S. Murray *Amaranthus blitoides, Cirsium vulgare, Croton glandulosus* var. *septentrionalis,* and *Verbascum thapsus.*

E. J. Retzinger Jr. *Allium canadense, Alternanthera philoxeroides, Asclepias syriaca, Commelina communis, Commelina diffusa, Heteranthera limosa, Melochia corchorifolia, Portulaca oleracea, Ranunculus arvensis, Richardia scabra,* and *Trianthema portulacastrum.*

B. E. Serviss *Camelina microcarpa, Lathyrus latifolius, Pontederia cordata,* and *Valerianella radiata.*

W. A. Skroch *Ampelopsis arborea, Anthemis cotula, Brunnichia ovata, Campsis radicans, Cardiospermum halicacabum, Citrullus lanatus* var. *citroides, Cocculus carolinus, Cucumis anguria, Cynanchum laeve, Desmanthus illinoensis, Dianthus armeria, Nuttallanthus canadensis, Passiflora incarnata, Plantago rugelii,* and *Polygonum convolvulus.*

B. S. Smith, D. S. Murray, and R. J. Tyrl *Datura quercifolia, Verbascum blattaria,* and *Xanthium spinosum*

L. M. Sosnoskie *Lespedeza cuneata* and *Viola bicolor.*

J. F. Stritzke *Buglossoides arvense, Cirsium arvense, Dipsacus fullonum, Silene noctiflora,* and *Tribulus terrestris.*

W. K. Vencill *Chenopodium leptophyllum, Dichondra carolinensis, Eichhornia crassipes, Grindelia papposa, Helianthus tuberosus, Ludwigia decurrens, Ludwigia peploides, Rumex acetosella, Sorghum almum,* and *Vaccaria hispanica.*

T. M. Webster *Bowlesia incana, Commelina benghalensis, Diodia teres, Glechoma hederacea, Helenium autumnale, Helianthus divaricatus, Mirabilis nyctaginea, Saponaria officinalis,* and *Vernonia gigantea.*

A. F. Wiese *Chenopodium album, Chloris virgata, Cucurbita pepo* var. *texana, Flaveria trinervia, Kochia scoparia, Proboscidea louisianica, Salsola tragus,* and *Sicyos angulatus.*

A. F. Wiese and C. D. Elmore *Cucurbita foetidissima.*

INDEX